ADVANCED LOCATION-BASED TECHNOLOGIES AND SERVICES

ADVANCED LOCATION-BASED TECHNOLOGIES AND SERVICES

Edited by
Hassan A. Karimi

CRC Press
Taylor & Francis Group
Boca Raton   London   New York

CRC Press is an imprint of the
Taylor & Francis Group, an **informa** business

CRC Press
Taylor & Francis Group
6000 Broken Sound Parkway NW, Suite 300
Boca Raton, FL 33487-2742

© 2013 by Taylor & Francis Group, LLC
CRC Press is an imprint of Taylor & Francis Group, an Informa business

No claim to original U.S. Government works

Printed on acid-free paper
Version Date: 20130426

International Standard Book Number-13: 978-1-1380-7286-2 (Paperback)
International Standard Book Number-13: 978-1-4665-1818-6 (Hardback)

Visit the Taylor & Francis Web site at
http://www.taylorandfrancis.com

and the CRC Press Web site at
http://www.crcpress.com

Contents

SECTION 1   Techniques and Technologies

SECTION 2   New Trends

SECTION 3   Services

Preface

The rapid increase in the number of mobile device users worldwide has given rise to many new services and applications. Of these services and applications, location-based services (LBS) have become pervasive, and the demand for them is growing exponentially. Among other factors, ubiquitous computing, which is commonplace on cell phones, and location-awareness capabilities of cell phones (e.g., smartphones) can be attributed to this high demand for LBS. Although early cell phones had to be retrofitted to become location-aware (by equipping them with positioning sensors) in order to provide LBS, the new generation of cell phones (e.g., smartphones) is LBS-enabled and users need only to choose from a pool of mobile services and applications. For example, a location-based mobile application that has become a common feature on smartphones and that is in high demand by users is navigation services. Another example is advertising where relevant applications are continually gaining the interest of businesses and consumers. Key to all location-based services and applications are *location*, *mobility*, and *connectivity*, all central to computations and decision-making activities.

There have been several books addressing topics on or related to LBS during the past decade. Given the nature of LBS, which are primarily technologically centered, new technologies and developments are continually paving the way for new LBS. These new systems, while still operating on the original principles, provide new applications and opportunities. For this reason, there is a need for new books every few years, such as this one, to discuss the latest developments with respect to location-based technologies, services, and applications. This book is divided into three parts, each containing four chapters. Part I is focused on techniques and technologies, Part II on new trends, and Part III on services. Following is a summary of the content featured in each chapter of the book.

Chapter 1 provides a detailed survey of positioning approaches and technologies since localization is an essential component of any LBS. In this chapter, the latest developments in global navigation satellite systems such as Global Positioning System (GPS), GLObal'naya Navigatsionnay Sputnikovaya Sistema (GLONASS), Galileo, and COMPASS; WiFi positioning systems; ultra-wideband positioning; radio frequency identification (RFID); dead reckoning systems (DRS) such as inertial navigation systems (INS); and optical tracking systems such as charged-coupled device (CCD) cameras, video cameras, and infrared cameras are discussed.

Chapter 2 discusses and analyzes WiFi location fingerprinting, an inexpensive positioning technology for indoor navigation. The chapter begins with the basics of location fingerprinting and then goes on to provide details of topics such as types of location fingerprinting, location fingerprint databases for indoor areas, positioning algorithms using location fingerprinting, performance of location fingerprinting, and the amount of effort that goes into building fingerprint databases.

One common primary function in LBS, as with all geospatial applications, is geocoding, which takes addresses of businesses or customers and converts them to x–y coordinates so that they can be used in computations and decision-making activities. Chapter 3 discusses and analyzes geocoding techniques and technologies for LBS. The chapter adresses the relationships between geocoding and LBS, describing input (addresses), reference databases required for geocoding, and interpolation algorithms. The chapter also discusses different geocoding systems, geocoding in different location-based applications, the new trend of collecting LBS data through crowdsourcing services, 3D and indoor geocoding, and spatiotemporal geocoding.

Chapter 4 discusses and analyzes multimodal route planning. The chapter begins with a discussion of the need for multimodal route planning in LBS. It then describes the theoretical foundations for multimodal transportation networks including two typical modeling approaches, that is, one-planar graphs with multilabeled links/edges and multiple graphs connected with linking edges, plug-and-play multimodal graph sets and multimodal routing algorithms, and real-world applications of multimodal route planning.

During the past decade, location-based social networks (LBSNs) have gained the attention of researchers and practitioners. Starting with the basics of LBSNs, Chapter 5 discusses location information in LBSNs, the concept of check-ins, and examples of LBSN services. The chapter then discusses research activities dealing with LBSNs including usage of LBSN services, social aspects of LBSNs, spatial and temporal activities, privacy concerns in LBSNs, and emerging uses of LBSNs.

Chapter 6 discusses geo-crowdsourcing, which is a new trend in collecting and sharing location data and information through volunteered geographic information (VGI). The chapter provides definitions of geo-crowdsourcing and example applications. After an overview of several existing geo-crowdsourcing services, the chapter focuses on describing the OpenStreetMap project, which is one of the most widely used geo-crowdsourcing services to date. The chapter provides a historical view of the OpenStreetMap project and a view of current OpenStreetMap activities, including the types of data being collected and the size of data collected to date. The chapter ends with a discussion of challenges and issues relevant to geo-crowdsourcing.

One major activity made possible by geo-crowdsourcing services is collaborative mapping where volunteers can participate in collecting map data of specific types and in specific geographic areas. Today, it is reasonable to assume that general users have GPS-equipped mobile devices (such as smartphones) at their disposal to collect map data. Given this trend, Chapter 7 discusses automatic generation of pedestrian paths, needed in many LBS and applications such as navigation and physical activity studies. The chapter defines a pedestrian path as well as its different types and presents an algorithm that uses geo-crowdsourced GPS traces on pedestrian paths to generate them automatically.

CyberGIS, which is GIS based on cyberinfrastructure, facilitates effective collection, access, and analysis of geospatial data. In Chapter 8, a CyberGIS environment is discussed for the collection, management, access, analysis, and visualization of location-based social media data. Of all types of location-based social media data, Twitter feeds are emphasized in this chapter. The chapter provides details of a CyberGIS architecture and design, spatiotemporal data analysis possible in the CyberGIS environment (focused on cluster detection of disease risk), flow mapping, and case studies (including detection of flulike diseases and cluster detection of moving trajectories).

The increased interest in pedestrian navigation is led by researchers to develop new techniques and technologies, by practitioners to provide new services, and by users to receive pedestrian navigation assistance. Chapter 9 is focused on pedestrian navigation services. The chapter is focused on map matching algorithms suitable for navigation assistance for pedestrian and wheelchair users. The chapter explains the general foundation for map matching, which is the core module in any navigation system/service, and presents new map matching algorithms that integrate accelerometer, gyroscope, GPS, and visual odometry (images and videos) technologies, and discusses a prototype of the multisensor map matching approach on Android phones.

Since the debut of GPS and the availability of GPS-based systems and products (e.g., navigation applications) in the market, location privacy has been a significant concern to users. Location privacy has particularly become one of the topics of interest in location-based services and applications. Chapter 10 is focused on security and privacy in LBS. The chapter discusses information security and privacy, the ecosystem of location information in LBS, LBS classification based on protection needs, security and privacy approaches for LBS (e.g., location obfuscation, pseudonymization, anonymization), location-based access control (e.g., GEO-RBAC), and location authentication.

Chapter 11 is focused on understanding and analyzing the effects that LBS have on the environment. The chapter discusses the issue of users' dependency on LBS, in particular navigation services that are very popular LBS. The chapter starts by analyzing the notion of automation, outlining the tasks that are usually automated, and explaining how automation is perceived by users of systems and services. It then presents the case study of navigation services to analyze LBS and spatial learning. The chapter also discusses LBS used to enhance or assist with educational activities.

Developing LBS and applications is only possible through the integration of several heterogeneous data sources, technologies, and software tools. Such integration, as with integration in other systems and services, requires adherence to standards agreed upon by the geospatial community, the computing community, and the Internet community, among others. Chapter 12 discusses LBS standards set by the Open Geospatial Consortium (OGC). The chapter examines key OGC standards relevant to LBS, OGC KML (a background to OpenLS), OGC collaboration with other standards organizations and location services, World Wide Web Consortium (W3C) points of interest (POIs) collaboration, recent OGC location services standards work, and current OGC activities related to LBS.

Although people have many different technologies at their disposal today, only a few profoundly impact people's activities by assisting them in addressing common and challenging problems. Of the currently available technologies, those that enable LBS and applications are becoming integral to important daily life activities. The continuous advancements in LBS, the tremendous demand for LBS, and the potential for innovation in new LBS and applications are the main reasons for providing periodic updates (e.g., through books) on the latest developments in LBS. This book will be of value to researchers (e.g., faculty, students) and practitioners (e.g., system developers, engineers), among others, who are interested in exploring the potential of the latest developments in LBS.

Editor

Dr. Hassan Karimi is a professor and director of the Geoinformatics Laboratory in the School of Information Sciences at the University of Pittsburgh. Karimi's research is focused on navigation, location-based services, location-aware social networking, geospatial information systems, mobile computing, computational geometry, grid/distributed/parallel computing, and spatial databases. Karimi's research has resulted in more than 150 publications in peer-reviewed journals and conference proceedings, as well as in many workshops and presentations at national and international forums. Karimi has published the following books: *Universal Navigation on Smartphones* (sole author), published by Springer in 2011; *CAD and GIS Integration* (lead editor), published by Taylor & Francis in 2010; *Handbook of Research on Geoinformatics* (sole editor), published by IGI in 2009; and *Telegeoinformatics: Location-Based Computing and Services* (lead editor), published by Taylor & Francis in 2004.

Section 1

Techniques and Technologies

1 Positioning and Tracking Approaches and Technologies

Dorota Grejner-Brzezinska and Allison Kealy

CONTENTS

1.1   INTRODUCTION

Telegeoinformatics is a discipline that has emerged from the convergence of geo-informatics, telecommunications, and mobile computing technologies. Geoinformatics, as a distinctive component of telegeoinformatics, is based around an increasingly diverse range of technologies that enable the acquisition, analysis, and dissemination of spatial information. The traditional technologies used in geoinformatics such as remotely sensed airborne and satellite imagery, Global Navigation Satellite Systems (GNSS), and geographic information systems (GISs) have now been augmented with new technologies and techniques including the availability of new signals of opportunity for positioning, for example, pseudolites and wireless local area networks (WiFi); the ability to rapidly capture finely grained spatial datasets using terrestrial LIDAR and wireless sensor networks; more efficient and relevant spatial computing approaches; and decentralized processing algorithms and enhanced image processing, modeling, and geospatial visual analytics.

These developments in geoinformatics combined with the increasing ubiquity of mobile computing devices have driven the development of modern location-based services (LBS) such as mobile games, navigation assistance, personal security, emergency response, vehicle and asset tracking, environmental monitoring, and mobile commerce. Enabling Internet access on the go using wireless data connections (such as cellular [GPRS] or 4G networks), mobile computing platforms have increased in their levels of sophistication with regard to their processing capacity as well as user interactivity. They have developed to the stage where much of the utility afforded by current generation geoinformatics technologies and techniques can be readily integrated, accessed, and exploited by mobile users. Key to the successful development and uptake of LBS is the availability of an accurate and reliable position solution that can be readily embedded in these mobile computing devices. For example, if the user is requesting directions to the nearest post office, the mobile device should be able to locate the user, forward his/her coordinates to the LBS provider where the navigation instructions are generated, and sent back to the user. To adequately respond to the mobile user's request, not only must his or her location be known, but the position uncertainty, updates to the position, knowledge of the mobile terminal's operating environment, and the user's personal navigation preferences must be known. To accomplish this requires an efficient integration of the position solution with robust spatial computing, analysis, and communications tools. This integration establishes the relationship between the user's position and other information databases, and is key to the success of any LBS. Web-based GISs and spatial data infrastructures (SDIs) can therefore serve not only as a resource repository but also as an instruction and decision clearinghouse, enabling mobile network-based decision making.

The major focus of this chapter is on positioning and tracking techniques supporting LBS (e.g., GPS). Tracking is commonly defined as a combination of positioning and telemetry providing a full tracking solution for mobile vehicles, where position is computed and then transmitted via a communications network to a control center. However, a positioning system maintaining a continuous log of the object's trajectory is also referred to as a tracking system; this latter definition will be used in this chapter. Several positioning techniques known as *radiolocation* are presented, with a special emphasis on GNSS. Although GPS is the primary choice for providing a positioning capability for current LBS, other emerging satellite positioning systems such as the Russian GLONASS and European GALILEO are expected to make significant contributions to LBS in the near future and are therefore presented in this chapter. The fundamental operational characteristics of all current and proposed GNSS including their signal structures and positioning techniques will be presented. In addition, the range of LBS requiring a position solution in GNSS-challenged environments such as inside buildings or within urban canyons has established a trend toward mobile computing devices integrating multiple positioning technologies to augment the core GNSS solution. In fact, for some LBS, solutions based solely on an alternative GNSS positioning technology are now commercially available; these include WiFi and ultra-wideband (UWB). A review of these alternative positioning technologies is presented with a view to deploying them as autonomous positioning solutions as well as complementary technologies that can augment GNSS.

1.2 GLOBAL NAVIGATION SATELLITE SYSTEMS (GNSS)

GNSS is the generic term for satellite navigation systems that provide three-dimensional positioning (and timing) information at a global scale, in any weather condition, and at any time of the day using radio signal ranging techniques from multiple satellites. Currently, only two GNSSs have reached their full operational capability (FOC). The most well-known is GPS, which was developed and is operated by the United States Department of Defense and has been fully operational since 1993. The other GNSS in operation since 1996 is GLONASS, developed by the former Soviet Union and now operated and maintained by Russia. Two other major GNSSs are currently under development: (1) Galileo, developed by the European Union, and (2) Compass/Beidou2 by China. In addition, regional navigation satellite systems (RNSS) such as the Quasi-Zenith Satellite System (QZSS) in Japan and satellite-based augmentation systems (SBAS) such as India's GAGAN (GPS Aided Geo Augmented Navigation) are being designed to be interoperable or compatible with GNSS and therefore need to be considered as part of a future GNSS positioning capability. In this section, GPS will be used to illustrate the terminology, operational characteristics, and measurement principles that underpin current and future GNSS, RNSS, and SBAS.

1.2.1 GLOBAL POSITIONING SYSTEM (GPS)

The NAVSTAR (NAVigation System Timing And Ranging) GPS is a satellite-based, all-weather, continuous, global radionavigation and time-transfer system, designed,

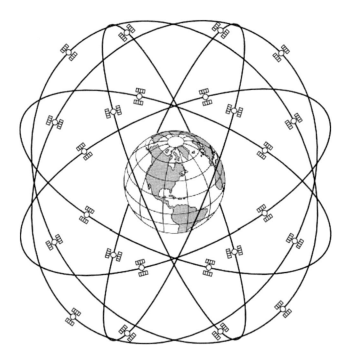

FIGURE 1.1   The GPS constellation. (From Rizos, C., 2002, Introducing the Global Positioning System, in: *Manual of Geospatial Science and Technology*, J. Bossler, J. Jensen, R. McMaster, and C. Rizos (eds.), Taylor & Francis, London. With permission.)

financed, deployed, and operated by the U.S. Department of Defense (DOD). The concept of NAVSTAR was initiated in 1973 through the joint efforts of the U.S. Army, the Navy, and the Air Force. The first GPS satellite was launched in 1978, and in 1993 the system was declared fully operational. GPS technology was designed with the following primary objectives:

- Suitability for all classes of platforms (aircraft, ship, land based, and space), and a wide variety of dynamics
- Real-time positioning, velocity, and time determination capability
- Availability of the positioning results on a single global geodetic datum
- Restricting the highest accuracy to a certain class of users (military)
- Redundancy provisions to ensure the survivability of the system
- Providing the service to an unlimited number of users worldwide
- Low cost and low power users' unit

The nominal GPS Operational Constellation consists of 24 satellites (21 plus 3 spares) that orbit the earth at the altitude of ~20,000 km in 12 hours (Figure 1.1). The satellites approximately repeat the same track and configuration once a day, advancing by roughly 4 minutes each day. They are placed in six nearly circular orbital planes, inclined at about 55 degrees with respect to the equatorial plane, with nominally four satellites in each plane. This configuration assures the simultaneous visibility of five to eight satellites at any point on earth. The current constellation (July 2012) consists of 32 satellites comprising 11 Block IIA, 12 Block IIR, 2 Block IIF, and 7 Block IIR-M satellites. Since February 22, 1978—the launch of the first GPS Block I satellite—the system evolved through several spacecraft designs, focused primarily on the increased design life, extended operation time without a contact from the control system (autonomous operation), better frequency standards (clocks), and the provision for the accuracy manipulation, controlled by DOD.

Block I satellites are referred to as the original concept validation satellites; Block II satellites are the first full-scale operational satellites (first launch February 1989); Block IIA satellites are the second series of operational satellites (first launch November 1990) and Block IIR satellites, the operational replenishment satellites (first launch January 1997). The block IIR-M satellites (first launch September 2005) transmit a second civil signal (L2C) and a military signal (M) code on the fundamental L1 and L2 frequencies. With increased power, the L2C civil signal offers better penetration in obscured environments as well as mitigation of ionospheric effects when combined with the existing civil signal. The M code has been designed with greater security and resistance to jamming and will eventually replace the current P(Y) code on L1 and L2. A third signal (L5) was officially delivered with the launch of the GPS IIF satellites (first launch May 2010). This signal offers increased signal redundancy and opportunities for enhanced positioning accuracy, integrity, and reliability. Another signal (L1C), designed to assure compatibility with international GNSSs such as the European Galileo is designed to be launched as part of the GPS III satellites. It is expected that the first GPS III satellite will be available for launch in 2015 (Divis, 2012). Information about the current status of the constellation can be found, for example, at the U.S. Coast Guard Navigation Center (2012).

GPS consists of three segments: (1) *satellite segment*, the satellite constellation itself; (2) the *user segment*, including all GPS receivers used in a variety of civilian and military applications; and (3) the *control segment*, responsible for maintaining proper operation of the system (Figure 1.2). Since the satellites have a tendency to drift from their assigned orbital positions, primarily due to so-called orbit perturbations caused by the earth, moon, and planets gravitational pull, solar radiation pressure, and so forth, they have to be constantly monitored by the Control Segment to determine the satellites' exact location in space. The control segment consists of a master control station, an alternate master control station, 12 command and control antennas, and 16 monitoring sites (6 from the Air Force and 10 from the National Geospatial-Intelligence Agency [NGA]), each checking the exact altitude, position, speed, and overall health of the orbiting satellites 24 hours a day. Based on these observations, the position coordinates and clock bias, drift, and drift rate can be predicted for each satellite, and then transmitted to the satellite for the retransmission back to the users. The satellite position is parameterized in terms of predicted ephemeris, expressed in earth-centered, earth-fixed (ECEF) reference frame, known as World Geodetic System 1984 (WGS84). The clock parameters are provided in a form of polynomial coefficients, and together with the predicted ephemeris are broadcast to the users in

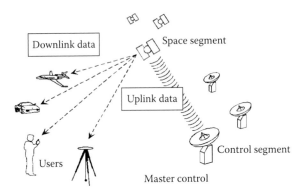

FIGURE 1.2 GPS segments. (From Rizos, C., 2002, Introducing the Global Positioning System, in: *Manual of Geospatial Science and Technology*, J. Bossler, J. Jensen, R. McMaster, and C. Rizos (eds.), Taylor & Francis, London. With permission.)

the GPS navigation message. The accuracy of the predicted orbit is typically at a few-meter level.

According to the National Coordination Office for Space-Based Positioning, Navigation, and Timing (2012), the next generation GPS operational control system (OCX) will be developed by the Raytheon Company and "will add many new capabilities to the GPS control segment, including the ability to fully control the modernized civil signals (L2C, L5, and L1C)." The modernized OCX will be delivered in increments, with Block 1 replacing the existing command and control segment and supporting the mission operations of the initial GPS III satellites. "This version will introduce the full capabilities of the L2C navigation signal. OCX Block 1 is scheduled to enter service in 2016. OCX Block 2 will support, monitor, and control additional navigation signals, including L1C and L5. OCX Block 3 will support new capabilities added to future versions of GPS III."

1.2.1.1 GPS Signal Structure

The signal generated by a GPS satellite oscillator contains three primary components: (1) pure sinusoidal waves or carriers (L1 and L2 with frequencies of 54×10.23 MHz and 120×10.23 MHz, respectively),[*] (2) pseudorandom noise (PRN) codes, and (3) the navigation message. There are two legacy PRN codes: the precise P(Y)-code, superimposed on L1 and L2 carriers; and coarse-acquisition C/A-code, superimposed on L1 carrier. The L2C and M codes are additional PRN codes transmitted on the modernized satellites (see Figure 1.3). All signals transmitted by GPS satellites are coherently derived from a fundamental frequency of 10.23 MHz, as shown in Table 1.1. The frequency separation between L1 and L2 is 347.82 MHz or 28.3%, and it is sufficient to permit accurate dual-frequency estimation of the ionospheric group delay affecting the GPS observables (see Section 1.2.1.2).

---

[*] $f_o = 10.23$ MHz is the fundamental frequency of GPS. The GPS satellites also transmit an L3 signal at 1381.05 MHz (135×10.23 MHz), associated with their dual role as a nuclear burst detection satellite as well as S-band telemetry signal.

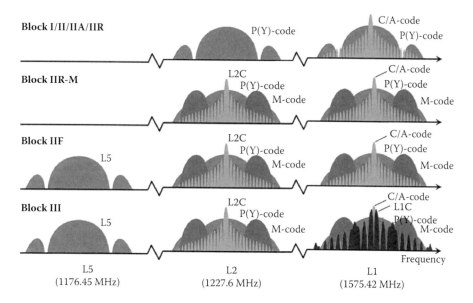

Block I/II/IIA/IIR

P(Y)-code

C/A-code
P(Y)-code

Block IIR-M

L2C
P(Y)-code
M-code

C/A-code
P(Y)-code
M-code

Block IIF

L5

L2C
P(Y)-code
M-code

C/A-code
P(Y)-code
M-code

Block III

L5

L2C
P(Y)-code
M-code

C/A-code
L1C
P(Y)-code
M-code

Frequency

L5
(1176.45 MHz)

L2
(1227.6 MHz)

L1
(1575.42 MHz)

FIGURE 1.3 GNSS signal characteristics. (From Heggarty C., 2008, Evolution of the Global Navigation Satellite System (GNSS), http://ieeexplore.ieee.org/ieee_pilot/articles/96jproc12/jproc-CHegarty-2006090/article.html, accessed July 20, 2012. With permission.)

TABLE 1.1

Basic Components of the GPS Satellite Signal

Component	Frequency (MHz)	Ratio of Fundamental Frequency (f_o)	Wavelength (cm)
Fundamental frequency (f_o)	10.23	1	2932.6
L1 Carrier	1,575.42	154	19.04
L2 Carrier	1,227.60	120	24.45
L5 Carrier	1,176.45	115	25.48
P-code	10.23	1	2932.6
C/A code	1.023	1/10	29326
W-code	0.5115	1/20	58651
L2C-code	1.023	1/10	29326
Navigation message	50×10^{-6}	1/204,600	N/A

GPS satellites transmit a C/A-code and a unique one-week long segment of P-code, which is the satellite's designated ID, ranging from 1 to 32. PRN is a very complicated digital code, that is, a sequence of "on" and "off" pulses that looks almost like random electrical noise. This carrier modulation enables the measurement of the signal travel time between the satellite and the receiver (user), which is a fundamental GPS observable (see Section 1.2.1.2). Access to the C/A code is provided to all users and is designated as the standard positioning service (SPS). Under the *Anti-Spoofing* (AS) policy imposed by the DOD, the additional W-code is implemented to encrypt the P-code into the Y-code, available exclusively to the military users, and designated as the precise positioning service (PPS). PPS guarantees positioning accuracy of at least 22 m (95% of the time) horizontally, and 27.7 m vertically, while the guaranteed positioning accuracy of SPS is 17 m (95% of the time) horizontally, and 37 m (95% of the time) vertically. However, most of the time the practical accuracy of SPS is much higher (see Section 1.2.4.2) with well-designed receivers observing approximately 3 to 5 m horizontal positioning accuracy 95% of the time. Under AS, civilian receivers have had to use special signal tracking techniques to recover observables on L2, since no civilian code was available on L2 (see, for example, Hofman-Wellenhof et al., 2001). The L2C signal transmitted by modernized GPS satellites will enable easier access to the L2 carrier and consequently more robust satellite tracking.

1.2.1.2 GPS Observables and the Error Sources

There are two fundamental types of GPS observables: *pseudorange* and *carrier phase,* both subject to measurement errors of systematic and random nature. For example, systematic errors due to the ionosphere or troposphere can delay or advance the GPS signal, and cause the measured range to be different from the true range by some systematic amount. Other errors, such as the receiver noise, are considered random. The following sections provide an overview of the GPS observables and the primary error sources.

1.2.1.2.1 Systematic Errors

Satellite orbital errors (errors in predicted ephemeris), reference station position errors (in differential/relative positioning), satellite and receiver clock errors, and effects of the propagation media represent bias and errors that have to be properly accommodated before positioning solutions can be obtained. The biases can be either mathematically modeled and accounted for in the measurement model (such as tropospheric correction), or special observation and data reduction techniques must be applied to remove their effects. For example, dual-frequency observables can be used to mitigate or remove the effects of the ionospheric signal delay, and differential GPS or relative positioning can be used to mitigate the effects of imperfectness of the satellite and receiver clock or broadcast ephemeris errors.

1.2.1.2.2 Errors Due to Propagation Media

The presence of free electrons in the *ionosphere* causes a nonlinear dispersion of electromagnetic waves traveling through the ionized medium, affecting their speed, direction, and frequency. The largest effect is on the signal speed, and as a result the GPS pseudorange is delayed (and thus, is measured too long), while the phase advances (and thus, is measured too short). The total effect can reach up to 150 m, and is a function of the total electron content (TEC) along the signal's path, the frequency of the signal itself, the geographic location and the time of observation, time of the year, and the period within the 11-year sun spot cycle (the last peak in ionospheric activity was 2001, with the next to occur in 2013). The effects of the ionosphere can be removed from the GPS observable by combining dual-frequency data in so-called iono-free linear combination and by relative processing as will be explained later (see also Hofman-Wellenhof et al., 2001).

The *troposphere* is a nondispersive medium for all frequencies below 15 GHz, thus phase and pseudorange observables on both L1 and L2 frequencies are delayed by the same amount. Since the amount of delay is not frequency dependent, elimination of the tropospheric effect by dual-frequency observable is not possible but is accomplished using mathematical models of the tropospheric delay. The total effect in zenith direction reaches ~2.5 m and increases with the cosecant of the elevation angle up to ~25 m at 5° elevation angle. The tropospheric delay consists of the *dry* component (about 90% of the total tropospheric refraction), proportional to the density of the gas molecules in the atmosphere and the *wet refractivity* due to the polar nature of the water molecules. In general, empirical models, which are functions of temperature, pressure, and relative humidity, can eliminate the major part (90%–95%) of the tropospheric effect from the GPS observables.

In addition, GPS signals can experience *multipath*, which is a result of an interaction of the incoming signal with the objects in the antenna surrounding. It causes multiple reflections and diffractions, and as a result, the signal arrives at the antenna via direct and indirect paths. These signals interfere with each other, resulting in an error in the measured pseudorange or carrier phase, degrading the positioning accuracy. The magnitude of multipath effect tends to be random and unpredictable, varying with satellite geometry, location, and type of reflective surfaces in the antenna surrounding, and can reach 1 to 5 cm for the carrier phases and 10 to 20 m for the code pseudoranges (Hofman-Wellenhof et al., 2001). Properly designed choke ring antennas can almost entirely eliminate this problem for the surface waves and the signals reflected from the ground.

1.2.1.3 Mathematical Models of Pseudorange and Carrier Phase

Pseudorange is a geometric range between the transmitter and the receiver, distorted by the propagation media and the lack of synchronization between the satellite and the receiver clocks. It is recovered from the measured *time difference* between the epoch of the signal transmission and the epoch of its reception by the receiver. The actual time measurement is performed with the use of the PRN code. In principle, the receiver and the satellite generate the same PRN sequence. The arriving signal is delayed with respect to the replica generated by the receiver, as it travels ~20,000 km. In order to find how much the satellite's signal is delayed, the receiver-replicated signal is delayed until it falls into synchronization with the incoming signal. The amount by which the receiver's version of the signal is delayed is equal to the travel time of the satellite's version (Figure 1.4). The travel time, Δt (~0.06 s), is converted to a range measurement by multiplying it by the speed of light, c.

There are two types of pseudoranges: C/A-code pseudorange and P-code pseudorange. The precision of the pseudorange measurement is partly determined by the wavelength of the chip in the PRN code. Thus, the shorter the wavelength, the more precise the range measurement would be. Consequently, the P-code range measurement precision (noise) of 10 to 30 cm is about 10 times higher than that of the C/A code. The pseudorange observation can be expressed as a function of the unknown

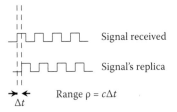

FIGURE 1.4 Principles of pseudorange measurement based on time observation.

receiver coordinates, satellite and receiver clock errors and the signal propagation
errors (Equation 1.1).

$$P_{r,1}^s = \rho_r^s + \frac{I_r^s}{f_1^2} + T_r^s + c\left(dt_r - dt^s\right) + M_{r,1}^s + e_{r,1}^s$$

$$P_{r,2}^s = \rho_r^s + \frac{I_r^s}{f_2^2} + T_r^s + c\left(dt_r - dt^s\right) + M_{r,2}^s + e_{r,2}^s \qquad (1.1)$$

$$\rho_r^s = sqrt\left[\left(X^s - X_r\right)^2 + \left(Y^s - Y_r\right)^2 + \left(Z^s - Z_r\right)^2\right] \qquad (1.2)$$

where

$P_{r,1}^s, P_{r,2}^s$	pseudoranges measured between receiver r and satellite s on L1 and L2
ρ_r^s	geometric distance between satellite s and receiver r
$\dfrac{I_r^s}{f_1^2}, \dfrac{I_r^s}{f_2^2}$	range error caused by ionospheric signal delay on L1 and L2
dt_r:	r-th receiver clock error (unknown)
dts:	s-th satellite clock error (known from the navigation message)
c:	vacuum speed of light
T_r^s	multipath on pseudorange observables on L1 and L2
$M_{r,1}^s, M_{r,2}^s$	range error caused by tropospheric delay between satellite s and receiver r (estimated from a model)
$e_{r,1}^s, e_{r,2}^s$	measurement noise for pseudorange on L1 and L2
X^s, Y^s, Z^s	coordinates of satellite s (known from the navigation message)
X_r, Y_r, Z_r	coordinates of receiver r (unknown)
f_1, f_2:	carrier frequencies of L1 and L2

Carrier phase is defined as a difference between the phase of the incoming carrier
signal and the phase of the reference signal generated by the receiver. Since at the initial
epoch of the signal acquisition, the receiver can measure only a fractional phase, the
carrier phase observable contains the initial unknown *integer ambiguity, N.* Integer
ambiguity is a number of full-phase cycles between the receiver and the satellite at the
starting epoch, which remains constant as long as the signal tracking is continuous.

After the initial epoch, the receiver can count the number of integer cycles that are being tracked. Thus, the carrier phase observable (ϕ) can be expressed as a sum of the fractional part measured with millimeter-level precision and the integer number of cycles counted since the starting epoch. The integer ambiguity can be determined using special techniques referred to as *ambiguity resolution algorithms*. Once the ambiguity is resolved, the carrier phase observable can be used to determine the user's location. The carrier phase can be converted to a phase-range observable, Φ (Equation 1.3), by multiplying the measured phase, ϕ, with the corresponding wavelength, λ. This observable is used in applications where the highest accuracy is required.

$$\Phi_{r,1}^s = \rho_r^s - \frac{I_r^s}{f_1^2} + T_r^s + \lambda_1 N_{r,1}^s + c\left(dt_r - dt^s\right) + m_{r,1}^s + \varepsilon_{r,1}^s$$

$$\Phi_{r,2}^s = \rho_r^s - \frac{I_r^s}{f_2^2} + T_r^s + \lambda_2 N_{r,2}^s + c\left(dt_r - dt^s\right) + m_{r,2}^s + \varepsilon_{r,2}^s \tag{1.3}$$

where

$\Phi_{r,1}^s, \Phi_{r,2}^s$ phase ranges (in meters) measured between station r and satellite s on L1 and L2

$N_{r,1}^s, N_{r,2}^s$ initial integer ambiguities on L1 and L2, corresponding to receiver r and satellite s $\lambda_1 \approx 19$ cm and $\lambda_2 \approx 24$ cm are wavelengths of L1 and L2

$m_{r,1}^s, m_{r,2}^s$ multipath error on carrier phase observables on L1 and L2

$\varepsilon_{r,1}^s, \varepsilon_{r,2}^s$ measurement noise for carrier phase observables on L1 and L2

Another observation that is sometimes provided by GPS receivers, and is primarily used in kinematic applications for velocity estimation, is *instantaneous Doppler frequency*. It is defined as a time change of the phase range, and thus, if available, it is measured on the code phase (Lachapelle, 1990).

Equation (1.2), which is a nonlinear part of Equations (1.1) and (1.3), requires Taylor series expansion to enable the estimation of the unknown user coordinates. In addition, Equations (1.1) and (1.3) can be solved for other parameters, such as user clock error.

Secondary (nuisance) parameters in the above equations are satellite and clock errors, tropospheric and ionospheric errors, multipath, and integer ambiguities. These are usually removed by differential (relative) GPS processing (see Section 1.2.1.5.2), by empirical modeling (troposphere), or by processing of dual-frequency signals (ionosphere). As already mentioned, ambiguities must be resolved prior to users' position estimation.

1.2.1.4 Positioning with GPS

The main principle behind positioning with GPS is trilateration in space, based on the measurement of a range (pseudorange or phase range) between the receiver and the satellites (Figure 1.5). Essentially, the problem can be specified as follows: given the position vectors of GPS satellites (such as ρ^s of satellite s in Figure 1.5)

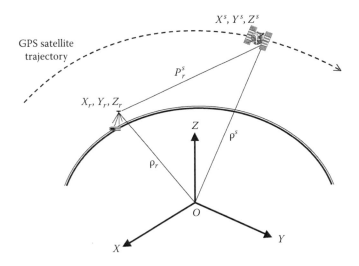

FIGURE 1.5   Range from satellite s to ground receiver r.

tracked by a receiver r, and given a set of range measurements (such as P_r^s) to these satellites, determine a position vector of the user, ρ_r. A single range measurement to a satellite places the user somewhere on a sphere with a radius equal to the measured range. Three simultaneously measured ranges to three different satellites place the user on the intersection of three spheres, which correspond to two points in space. One of them is usually an impossible solution that can be discarded by the receiver. Even though there are three fundamental unknowns (coordinates of the user's receiver), a minimum of four satellites must be simultaneously observed to provide a unique solution in space (Figure 1.6).

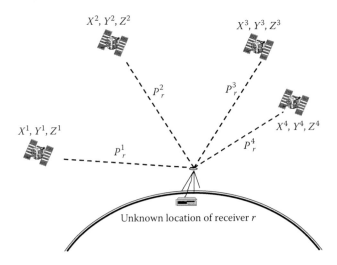

FIGURE 1.6   **(See color insert.)** Determination of position in space by ranging to multiple satellites.

As already mentioned, the fundamental GPS observable is the signal travel time between the satellite and the receiver. However, the receiver clock that measures the time is not perfect and may introduce an error to the measured pseudorange (even though we limit our discussion here to pseudoranges, the same applies to the carrier phase measurement that is indirectly related to the signal transit time, as the phase of the received signal can be related to the phase at the epoch of transmission in terms of the signal transit time). Thus, in order to determine the most accurate range, the receiver clock correction must be estimated and its effect removed from the observed range. Hence, a fourth pseudorange measurement is needed, since the total number of unknowns, including the receiver clock, is now four. If more than four satellites are observed, a least squares solution is employed to derive the optimal solution.

1.2.1.5 Point versus Relative Positioning

There are two primary GPS positioning modes: *point positioning* (or *absolute positioning*) and *relative positioning*. However, there are several different strategies for GPS data collection and processing, relevant to both positioning modes. In general, GPS can be used in static and kinematic modes, using both pseudorange and carrier phase data. GPS data can be collected and then postprocessed at a later time or processed in real time, depending on the application and the accuracy requirements. In general, postprocessing in relative mode provides the best accuracy.

1.2.1.5.1 Point (Absolute) Positioning

In point, or absolute positioning, a single receiver observes pseudoranges to multiple satellites to determine the user's location. For the positioning of the moving receiver, the number of unknowns per epoch equates to three receiver coordinates plus a receiver clock correction term. In the static mode with multiple epochs of observations there are three receiver coordinates and n receiver clock error terms, each corresponding to a separate epoch of observation *1* to *n*. The satellite geometry and any unmodeled errors will directly affect the accuracy of the absolute positioning.

Precise point positioning (PPP) is a technique that is gaining popularity across the high-precision GPS user community. In PPP, a single GPS receiver can achieve centimeter to decimeter level accuracy through processing of the undifferenced carrier phase measurements on L1 and L2 with precise GPS orbit and clock data products. The increasing use of PPP has led to the development of online PPP tools that enable users to obtain position information in national or global reference frames, for example, the Natural Resources Canada (NRCan) Canadian Spatial Reference System-PPP (CSRS-PPP) available since November 2003 (Langley, 2008).

1.2.1.5.2 Relative Positioning

The relative positioning technique (also referred to as differential GPS [DGPS] or real-time kinematic [RTK]) employs at least two receivers, a reference (base) receiver, whose coordinates must be known; and the user's receiver, whose coordinates can be determined relative to the reference receiver. Thus, the major objective of relative positioning is to estimate the 3D baseline vector between the reference receiver and the unknown location. Using the known coordinates of the reference

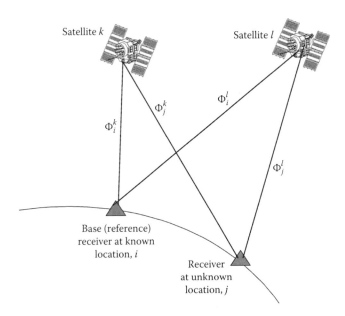

FIGURE 1.7   Between-receiver phase-range double differencing.

receiver and the estimated ΔX, ΔY, and ΔZ baseline components, the user's receiver coordinates in WGS84 can be readily computed. Naturally, the user's WGS84 coordinates can be further transformed to any selected reference system.

A relative (differenced) observable is obtained by differencing the simultaneous measurements to the same satellites observed by the reference and the user receivers. The most important advantage of relative positioning is the removal of the systematic error sources (common to the base station and the user) from the observable, leading to the increased positioning accuracy. Since for short to medium baselines (up to ~60–70 km), the systematic errors in GPS observables due to troposphere, satellite clock, and broadcast ephemeris errors are of similar magnitude (i.e., they are spatially and temporally correlated), the relative positioning allows for a removal or at least a significant mitigation of these error sources, when the observables are differenced. In addition, for baselines longer than 10 km, the so-called ionosphere-free linear combination must be used (if dual-frequency data are available) to mitigate the effects of the ionosphere (see Hofman-Wellenhof et al., 2001).

The primary differential modes are (1) single differencing mode, (2) double differencing mode, and (3) triple differencing mode. The differencing can be performed between receivers, between satellites, and between epochs of observations. The single-differenced (between-receiver) measurement, $\Phi_{i,j}^{k}$, is obtained by differencing two observables to the satellite k, tracked simultaneously by two receivers i (reference) and j (user): $\Phi_{i,j}^{k} = \Phi_{i}^{k} - \Phi_{j}^{k}$ (see Figure 1.7).

By differencing observables from two receivers, i and j, observing two satellites, k and l, or simply by differencing two single differences to satellites k and l, one arrives at the double-differenced (between-receiver/between-satellite differencing)

TABLE 1.2

Differencing Modes and Their Error Characteristics

Error source	Single Difference	Double Difference
Ionosphere	Reduced, depending on the baseline length	Reduced, depending on the baseline length
Troposphere	Reduced, depending on the baseline length	Reduced, depending on the baseline length
Satellite clock	Eliminated	Eliminated
Receiver clock	Present	Eliminated
Broadcast ephemeris	Reduced, depending on the baseline length	Reduced, depending on the baseline length
Ambiguity term	Present	Present
Noise level w.r.t. one-way observable	Increased by $\sqrt{2}$	Increased by 2

measurement: $\Phi_{i,j}^{k,l} = \Phi_i^k - \Phi_j^k - \Phi_i^l + \Phi_j^l = \Phi_{i,j}^k - \Phi_{i,j}^l$. The double difference is the most commonly used differential observable.

Furthermore, differencing two double differences, separated by the time interval $dt = t_2\text{-}t_1$, renders the triple-differenced measurement, $\Phi_{i,j}^{k,l}(dt) = \Phi_{i,j}^{k,l}(t_2) - \Phi_{i,j}^{k,l}(t_1)$, which in the case of the carrier phase observables effectively cancels the initial ambiguity term. Differencing can be applied to both pseudorange and carrier phase. However, for the best positioning accuracy with carrier phase double differences, the initial ambiguity term should be first resolved and fixed to the integer value. Relative positioning may be performed in static and kinematic modes, in real time (see the next section) or, for the highest accuracy, in postprocessing.

Table 1.2 shows the error characteristics for between-receiver single and between-receiver/between-satellite double differenced data.

1.2.1.6　GPS Correction Services

As explained earlier, the GPS error sources are spatially and temporally correlated for short to medium base–user separation. Thus, if the base and satellite locations are known (satellite location is known from broadcast ephemeris), the errors in the measurements can be estimated at specified time intervals and made available to nearby users through a wireless communication as *differential corrections*. These corrections can be used to remove the errors from the observables collected at the user's (unknown) location. This mode of positioning uses DGPS services to mitigate the effects of the measurement errors, leading to the increased positioning accuracy in real time. DGPS services are commonly provided by the government, industry, and professional organizations, and enable the user to use only one GPS receiver collecting pseudorange data, while still achieving superior accuracy as compared to the point-positioning mode. Naturally, to use a DGPS service, in real-time, the user must be equipped with additional hardware capable of receiving and processing the differential corrections. In an alternative implementation of DGPS, RTK or high precision DGPS the reference station broadcasts the observables (carrier phase and pseudorange) instead of the differential corrections. In that case, the user unit has to perform the relative positioning, as described in Section 1.2.1.5.2.

DGPS services normally involve some type of wireless communications systems. Today, mobile phone networks are widely used for DGPS data transmissions; however VHF or UHF systems and geostationary satellites are used for wide area DGPS (WADGPS). WADGPS involves multiple GPS base stations that track all GPS satellites in view and, based on their precisely known locations and satellite broadcast ephemeris, estimates the errors in the GPS pseudoranges. This information is used to generate pseudorange corrections that are subsequently sent to the master control station, which uploads checked and weighted corrections to the communication geostationary satellite, which in turn transmits the corrections to the users. The positioning accuracy of WADGPS, such as OmniSTAR VBS L1 code pseudorange correction service, is better than 1 m more than 95% of the time, and the L1/L2 solutions (OmniSTAR HP) used with dual-frequency receivers provide horizontal accuracies better than 6 cm 95% of the time (OmniSTAR, 2012). Example DGPS services include Federal Aviation Administration (FAA)-supported satellite-based augmentation system (SBAS), the Wide Area Augmentation System (WAAS), ground-based DGPS services, referred to as Local Area DGPS (LADGPS), such as U.S. Coast Guard and Canadian Coast Guard services, or FAA-supported Local Area Augmentation System (LAAS). LADGPS supports real-time positioning typically over distances of up to a few hundred kilometers, using corrections generated by a single base station (Rizos, 2002a). WAAS and LAAS are currently under implementation, with a major objective of supporting aviation navigation and precision approach (Federal Aviation Administration, 2012; U.S. Coast Guard, 2012). The accuracy of WAAS was expected at ~7.6 m at 95% of the time, but it is already significantly better than these specifications (~2 m and less horizontal RMS) (Lachapelle et al., 2002). WAAS is a public service and any user equipped with the appropriate receiver may have access to it.

More recently, a large number of countries have established regional networks of continuously operating reference stations (CORS) that support a range of applications, especially those requiring the highest accuracy in postprocessing or in real time. U.S. government agencies, such as National Geodetic Survey (NGS) and the Department of Transportation (DOT), or international organizations, such as the International GPS Service (IGS), deploy and operate these networks. Normally, all users have a free access to the archived data that can be used as a reference (base data) in carrier phase or range data processing in relative mode. Alternatively, network-based positioning using carrier-phase observations with a single user receiver in real time can be accomplished with local specialized networks, which can estimate and transmit carrier phase corrections (see, e.g., Raquet and Lachapelle, 2001). These network, real-time kinematic NRTK techniques enable a user with a single receiver to receive corrections from within a network comprising CORS sites with separations of up to 70 km. The techniques underpinning the generation, interpolation, and communication techniques of NRTK can be found in Al-Shaery (2011) and Dai (2003).

1.2.1.7 How Accurate is GPS?

The positioning accuracy of GPS depends on several factors, such as the number and the geometry of the observations collected, the mode of observation (point versus relative positioning), type of observation used (pseudorange or carrier phase), the measurement model used, and the level of biases and errors affecting the observables. Depending on the design of the GPS receiver and the aforementioned factors, the positioning accuracy varies from 17 m to better than 1 cm when carrier phases are used in relative positioning mode. In order to obtain better than 17 m accuracy with pseudoranges, differential positioning or DGPS services must be employed.

The geometric factor, geometric dilution of precision (GDOP), reflects the instantaneous geometry related to a single point. Normally, more satellites yield smaller DOP value, and GDOP of 6 and less indicates good geometry (a value around 2 indicates a very strong geometry). Other DOP factors, such as position DOP (PDOP), vertical DOP (VDOP), and relative DOP (RDOP), the last one related to the satellite geometry with respect to a baseline, can be also used as the quality indicators. These DOPs are normally computed by GPS receivers and provided in real time as the quality assessment. Formulas for computing various geometric factors are provided in Hofman-Wellenhof et al. (2001).

Other factors affecting the GPS positioning accuracy depend on (1) whether the user is stationary or moving (static versus kinematic mode), (2) whether the positioning is performed in real time or in postprocessing, (3) the data reduction algorithm, (4) the degree of redundancy in the solution, and (5) the measurement noise level. The typical accuracies achievable by commercial GPS receiver systems today (the Trimble R7™ is shown here) are summarized in Table 1.3. As the accuracy of GPS depends on several hardware and environmental factors, as well as the survey

TABLE 1.3
Currently Achievable GPS Accuracy

Point Positioning		Relative Positioning	
		Positioning Mode	
SPS	17 m, 95%	Static GNSS	Horizontal: 3 mm + 0.1 ppm RMS
			Vertical: 3.5 mm + 0.4 ppm RMS
PPP	Depends on observation	Real-time kinematic	
	period duration	Single baseline <30 km	Horizontal: 8 mm + 1 ppm RMS
	4 cm with 2 hours and 1 cm		Vertical: 15 mm + 1 ppm RMS
	after 12 hours	Network RTK	Horizontal: 8 mm + 0.5 ppm RMS
			Vertical: 15 mm + 0.5 ppm RMS
		Code differential GNSS	Horizontal: 0.25 m + 1 ppm RMS
		positioning	Vertical: 0.50 m + 1 ppm RMS
			SBAS: typically <5 m 3DRMS

Source: Trimble Corporation, 2012, Trimble R7: Trimble R-Track Technology for GPS Modernization, http://www.trimble.com/trimbler7.shtml (accessed July 23, 2012).

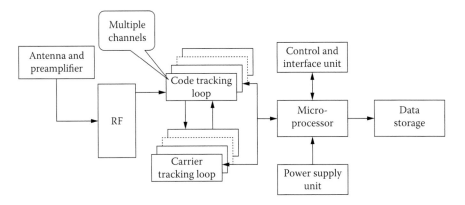

FIGURE 1.8 Basic components of a GPS receiver. (From Grejner-Brzezinska, D. A., 2002, GPS Instrumentation Issues, in: *Manual of Geospatial Science and Technology*, J. Bossler, J. Jensen, R. McMaster, and C. Rizos (eds.), Chapter 10, Taylor & Francis, New York. With permission.)

geometry among others, the accuracy levels listed in Table 1.3 should be understood as the best achievable accuracy.

1.2.1.8   GPS Instrumentation

Over the past two decades, civilian as well as military GPS instrumentation has evolved through several stages of design and implementation focused primarily on achieving an enhanced reliability of positioning and timing, modularization, and miniaturization. By far, the majority of the receivers manufactured today are of the C/A-code single-frequency type. However, for the high-precision geodetic applications the dual-frequency solution is a standard. Even though the civilian and military receivers as well as application-oriented instruments have evolved in different directions, one might pose the following question: *Are all GPS receivers essentially the same, apart from functionality and user software?* The general answer is, yes, all GPS receivers support essentially the same functionality blocks, even if their implementation differs for different types of receivers.

The following are the primary components of a generic GPS receiver (Figure 1.8): antenna and preamplifier, radio-frequency (RF) front-end section, a signal tracker block, microprocessor, control/interface unit, data storage device, and power supply (Langley, 1991; Parkinson and Spilker, 1996; Grejner-Brzezinska, 2002). Any GPS receiver must carry out the following tasks:

- Select the satellites to be tracked based on GDOP and the almanac[*]
- Search and acquire each of the GPS satellite signals selected
- Recover navigation data for every satellite
- Track the satellites, measure pseudorange and/or carrier phase
- Provide position/velocity information

---

[*] The almanac is a set of parameters included in the GPS satellite navigation message that is used by a receiver to predict the approximate location of a satellite. The almanac contains information on all of the satellites in the constellation.

- Accept user commands and display results via control unit or a PC
- Record the data for postprocessing (optional)
- Transmit the data to another receiver via radio modem for real-time solutions (optional)

An important characteristic of the RF section is the number of channels and hence the number of satellites that can be tracked simultaneously. Older receivers had a limited number of channels (even as little as one), which required sequencing through satellites to acquire enough information for 3D positioning. Modern GPS receivers are based on dedicated channel architecture, where every channel tracks one satellite on the L1 or L2 frequency.

1.2.2 GLONASS

The Russian satellite navigation system GLONASS (GLObal'naya Navigatsionnay Sputnikovaya Sistema) replicates many of the signal transmission and positioning principles of GPS. GLONASS attained full operating capabilities in 1996 with a full constellation of 24 satellites. With a subsequent decline to 8 satellites in 2001, the revitalized GLONASS constellation currently comprises 31 satellites. Developed to provide worldwide positioning capabilities, GLONASS is designed like GPS with space, control, and user segments. The GLONASS satellites comprising the space segment is currently made up of 24 operational satellites, 4 spares, 2 in maintenance, and 1 in test mode. The GLONASS satellites are arranged in three orbital planes each containing eight satellites (see Table 1.4 for additional information comparing the GPS and GLONASS orbits). Like GPS, each GLONASS satellite transmits two frequencies (L1 and L2) within the L band of the radio frequency spectrum. However, although all GPS satellites transmit on the same two

TABLE 1.4
Comparison of GPS and GLONASS Signals

	GPS	GLONASS
Carrier signals	L1: 1,575.42 MHz	L1: $(1,602 + k \times 9/16)$ MHz
	L2: 1,227.60 MHz	L2: $(1,246 + k \times 7/16)$ MHz
		k = Channel number
Codes	Different for each satellite	Same for all satellites
	C/A-code on L1	C/A-code on L1
	P-code on L1 and L2	P-code on L1 and L2
Code frequency	C/A-code: 1,023 MHz	C/A-code: 0.511 MHz
	P-code: 10.23 MHz	P-code: 5.11 MHz
Clock data	Clock offset, frequency offset, frequency rate	Clock frequency offset
Orbital data	Modified Keplerian orbital elements every hour	Satellite position, velocity, and acceleration every half hour

Source: Kleusnerg, A., 2009, GPS and GLONASS, http://gauss.gge.unb.ca/gpsworld/EarlyInnovation-Columns/Innov.1990.11-12.pdf (accessed July 20, 2012).

frequencies, each GLONASS satellite transmits on a different L band frequency. Therefore, although GPS receivers use a code division multiple access technique with a unique code modulated on the carrier to identify individual satellites, each GLONASS satellite has an individual frequency (determined as a function of the satellite's frequency channel number) assigned to it, a frequency division multiple access technique is used, and the same code can be modulated onto the signal.

Similar to GPS, GLONASS is also part of a modernization program where the GLONASS-M satellites first launched in 2003 now broadcast civilian codes on both the L1 and L2 frequencies. The first GLONASS K satellite launched in 2011 is undergoing testing of the transmission of a third signal L3 that will offer the same benefits as the GPS L5 signal. GLONASS K satellites will also use CDMA (as well as continue to use FDMA) signals to simplify combined receiver design and to facilitate compatibility with other GNSS. The GLONASS control segment comprises the System Control Centre located in Moscow and the telemetry and tracking stations located across the Russian territory. The GLONASS control segment performs tasks comparable to those of the GPS control segment.

The GLONASS user segment, like GPS, consists of the receiver hardware that provide positioning, velocity, and precise timing to any user. Hardware designed to use the GLONASS signals have to be able to handle the different satellite transmission frequencies; consequently a GPS receiver cannot be used to track and process the GLONASS signals. Other complications arising from combining GPS and GLONASS measurements come from differences in their coordinate and timing reference systems. Whereas GPS computes coordinates in WGS84 and is time referenced to UTC as maintained by the United States Naval Observatory (USNO), coordinates computed using GLONASS are referenced to PZ-90 (Parametry Zemli 1990 Goda [Parameters of the Earth Year]) and time referenced to UTC as maintained by Russia UTC(SU). Commercial receivers are available today that can readily track both the GPS and GLONASS signals and make the relevant computations that account for timing offsets and differences in determining the satellite coordinates. These receivers are typically used for high-accuracy applications that benefit from the enhanced performance in terms of measurement redundancy and signal availability.

1.2.3   GALILEO

Galileo is the European civilian Global Navigation Satellite System and is an ongoing initiative coordinated by the European Union and the European Space Agency (ESA). Intended to complement GPS and GLONASS, the expected Galileo constellation will comprise 30 satellites (24 operational and 3 spares). The fully deployed Galileo system will consist of 30 satellites (27 operational plus 3 active spares), positioned in three circular, inclined medium earth orbit (MEO) planes 23,222 km above the earth, and an inclination of 56 degrees to the equator.

The Galileo constellation was initiated with the launches of GIOVE-A and -B, in 2005 and 2008, respectively. These were followed by the launch of two Galileo in orbit validation satellites in October 2011. For civilian users, the interoperability and compatibility of Galileo and GPS is realized by having two common center frequencies in E5a/L5 and L1 (other Galileo frequencies are E5b and E6) as well as adequate geodetic coordinate and time reference frames. It is expected that Galileo will provide three early services in 2014 or 2015 based on an initial constellation of 18 satellites: an initial open service, an initial public regulated service, and an initial search-and-rescue service (European Space Agency, 2012).

1.2.4   COMPASS

In April 2012, China launched the 12th and 13th satellites in its Compass Navigation Satellite System. Based on the first generation BeiDou-1 system established in 2003 for the area of the greater China, Compass formally declared initial operation service in December 2011. China aims to deliver a fully operational Compass constellation by 2020 that will consist of 35 satellites, including 5 geostationary orbit satellites, 27 MEOs, and 3 in inclined geostationary orbit (IGSO). The Compass constellation currently consists of three MEOs, four GEOs, and five IGSOs. These satellites will broadcast signals on four carrier frequency bands and are intended to be compatible with GPS, GLONASS, and Galileo receivers. Compass will offer four service levels: an open service with positioning accuracy of within 10 m, an authorized service, a wide area differential positioning service with positioning accuracy of 1 m, and a short message service (SMS) of up to 120 Chinese characters.

1.2.5   SATELLITE AUGMENTATION SYSTEMS

The Quasi-Zenith Satellite System (QZSS) is Japan's RNSS that successfully launched its first satellite on September 11, 2010. In 2011, the Japanese government made the decision to accelerate the QZSS deployment in order to reach a four-satellite constellation by the end of the decade, while aiming at a final seven-satellite constellation in the future, Inside GNSS (2012). The QZSS satellite orbits have been designed so that at least one satellite is always near zenith over Japan. The orbits are periodic highly elliptical orbit (HEO) and therefore each satellite appears almost overhead most of the time (i.e., more than 12 hours a day with an elevation above 70°). This gives rise to the term *quasi-zenith* for which the system is named. The six signals planned for QZSS (L1-C/A, L1C, L2C, L5, L1-SAIF [interoperable with GPS-SBAS], LEX [QZSS Experimental Signal for high precision (3 cm level) service and compatible with Galileo E6 signal) are designed to have compatibility and interoperability with existing and future modernized GNSS signals.

India is also developing its own space-based augmentation system (SBAS), known as the GAGAN (GPS Aided Geo Augmented Navigation) system. The project, which would enhance the accuracy and integrity of GNSS signals to meet precision approach requirements in the civil aviation industry (over Indian airspace), had its first satellite launch in May 2011. With a second successful launch in September 2012. The full GAGAN constellation will consist of three satellites transmitting SBAS navigation data on the L1 and L5 frequencies and it is expected that the full constellation will be completed by June 2013. India also has plans to establish a seven-satellite constellation RNSS called the Indian Regional Navigation Satellite System (IRNSS).

Other operational SBAS include the U.S.'s Wide Area Augmentation System (WAAS), the European Geostationary Navigation Overlay System (EGNOS), and

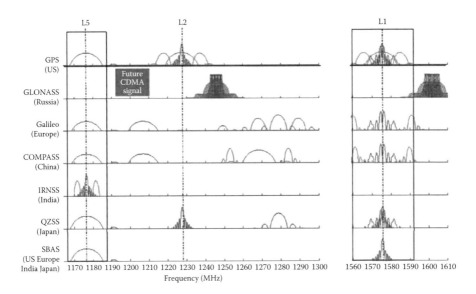

FIGURE 1.9   (See color insert.) Proposed Signal Overlay for all available GNSS, RNSS and SBAS (Turner, 2010). With Permission.

Japan's MTSAT Satellite Augmentation System (MSAS). Figure 1.9 shows the proposed signal overlay for all available GNSS, SBAS, and RNSS (Turner, 2010).

1.3   ALTERNATIVES TO GNSS

The traditional drivers underpinning the development of alternative or augmentation systems for positioning have arisen from the need to provide GNSS-like or enhanced GNSS performance in environments that completely or partially obscure GNSS signals. For example, in the civil aviation industry, GNSS receivers are augmented with the inertial navigation sensors to support aviation integrity requirements. Over the past decade, unprecedented growth in consumer LBS has generated new requirements for positioning in indoor and urban environments. In these environments GNSS signals are either unavailable or too degraded to be useful and alternative systems are required. In this section we review alternative positioning solutions to GNSS.

Mautz (2012) presents a comprehensive overview of indoor positioning technologies in Table 1.5 and Figure 1.10.

1.3.1   Pseudolites

Pseudo-satellite or pseudolite (PL) can be regarded as a minisatellite that can be used for autonomous navigation and positioning in the indoor or outdoor environments. The principle of pseudolite-based positioning systems is directly derived from the global positioning technology for outdoor environments. The system is able to triangulate the position of an object by accurately measuring the distances from

TABLE 1.5
Indoor Positioning Technologies

Technology	Typical Accuracy	Typical Coverage (m)	Typical Measuring Principle	Application
Cameras	0.1 mm–dm	1–10	Angle measurements from images	Metrology, robot navigation
Infrared	cm–m	1–5	Thermal imaging, active beacons	People detection, tracking
Tactile and polar	μm–mm	3–2000	Mechanical, interferometry	Automotive, metrology
Sound	cm	2–10	Distances from time of arrival	Hospitals, tracking
WLAN/WiFi	m	20–50	Fingerprinting	Pedestrian navigation, LBS
RFID	dm–m	1–50	Proximity detection, fingerprinting	Pedestrian navigation
Ultra-wideband	c	1–50	Body reflection, time of arrival	Pedestrian navigation
High sensitive GNSS	10 m	Global	Parallel correlation, assistant GPS	LBS
Pseudolites	cm–dm	10–1000	Carrier phase ranging	GNSS challenged pit mines
Inertial Navigation	1%	10–100	Dead reckoning	Pedestrian navigation
Magnetics Systems	mm–cm	1–20	Fingerprinting and ranging	Hospitals, mines
Infrastructure Systems	cm–m	Building	Fingerprinting, capacitance	Ambient assisted living

Source: Mautz, R., 2012, Indoor Positioning Technologies, Habilitation thesis, ETH Zurich.

the object to the array of pseudolites, whose location coordinates are known in a selected reference frame. Other applications of pseudolites include precision landing systems, such as LAAS, discussed in Section 1.2.1.6, outdoor navigation, and other system augmentation (such as GPS augmentation). In outdoor applications, the most commonly used type of pseudolites is a GPS pseudolite. It is a ground-based transmitter, which sends a GPS-like signal to support positioning and navigation in situations where the satellite constellation may be insufficient. PLs are usually located on building rooftops, high poles, or any high location in the vicinity of the survey area, resulting in a relatively low elevation angle, as compared to GPS satellites. The majority of GPS pseudolites transmit signals on L1 carrier (1575.42 MHz), and the more advanced systems can also transmit on L2 carrier (1227.6MHz). PLs can be designed to both receive and transmit ranging signals (transceivers) and thus can be used to self-determine their own location. With some firmware modification, standard GPS receivers can be used to track PL signals.

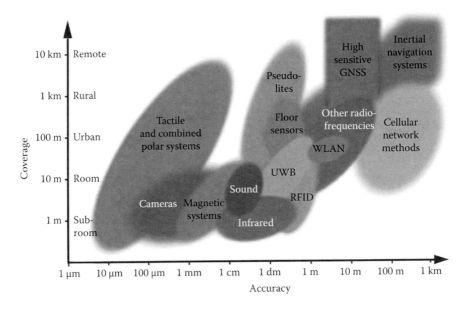

FIGURE 1.10 (**See color insert.**) Relationship between coverage and accuracy for a range of indoor positioning technologies. (From Mautz, R., 2012, Indoor Positioning Technologies, Habilitation thesis, ETH Zurich. With permission.)

The ionospheric and tropospheric errors do not apply to most of the pseudolite applications (tropospheric errors apply only to the outdoor situation). However, the most important error sources are multipath and the near-far problem, where the PL transmitter can be very close to the receiving antenna, as compared to, for example, GPS satellites. The methods most commonly used to mitigate these problems are the proper transmitter and receiver design and the appropriate signal structure. For example, one possible method for eliminating the near-far problem is the technique of pulsing the pseudolite's signal, while a higher chipping rate (CR) can mitigate the undesired effect of multipath (Progri and Michalson, 2001). In addition, to mitigating multipath errors, helical antennas are usually employed for the transmission of the pseudolite signals. Other possible problems related to the use of pseudolites are (1) any errors in PL location will have a significant impact on the receiving antenna coordinates due to the short distance between the receiver and the PL; (2) since PL is stationary, its location bias is constant, and its effect on position coordinates of the receiver depends on the geometry between the PL and the receiver; and (3) a differential technique may eliminate fewer error sources, as opposed to the differential GPS, especially if significant range differences exist between the receiver and the PLs in the array. For more information on pseudolites and their applications, the reader is referred to Barltrop et al. (1996), Elrod and Van Direndonck (1996), Wang et al. (2001), Progri and Michalson (2001), and Grejner-Brzezinska et al. (2002).

Another kind of pseudolite refers to those that utilize a signal structure that is similar to that of GNSS but uses a different part of the frequency spectrum. The

Australian Locata system is an example of such a system. Operating in the 2.4 GHz ISM band, two frequencies are transmitted by ground-based transmitters that are time synchronized. A Locata receiver can use these signals to deliver under certain configurations and operating conditions centimeter-level positioning accuracy in indoor environments (Locata Corp., 2012).

1.3.2 POSITIONING METHODS BASED ON CELLULAR NETWORKS

The majority of alternative positioning technologies used in LBS operate on the principle of radiolocation, in which the position of an object is determined from processing known characteristics of received radio waves. The signal parameters most commonly used in radionavigation are angle of arrival, time of arrival, signal strength, and signal multipath signature matching. The time of arrival and the signal strength can be directly converted to the range measurements. These techniques are presented here in the context of cellular network positioning systems, and their broad applicability to other positioning technologies is discussed.

In cellular networks, the most popular technique of finding the user's location is triangulation, based either on angular or distance observations (or some combination of both), between the mobile terminal and the base stations. The base stations in the radio-location techniques are either cellular service towers or GPS satellites. Thus, in general, the technologies for finding the user's location in LBS can be divided into *network-based* or *satellite-based* (currently primarily GPS-based) systems. Another classification is based on the actual device that performs the positioning solution, that is, mobile user or the base station (control center), leading to *mobile terminal (user)-centric, network-centric*, or *hybrid solutions*. In the network-centric systems, the user's position is determined by the base station and sent back to the user's set, while in the terminal-centric solution, the position computation is performed by the user's set.

In this section, we present an overview of the three main location techniques used in alternative wireless positioning technologies: (1) mobile terminal (user)-centric, (2) network-centric, and (3) hybrid solutions (Caffery and Stuber, 1998; Hellebrandt and Mathar, 1999; Djuknic and Richton, 2001; Hein et al., 2001; Abnizova et al., 2002; Andersson, 2002; Francica, 2002; SnapTrack, 2002). The summary characteristics of these methods are presented in Tables 1.6 and 1.7.

1.3.2.1 Terminal-Centric Positioning Methods

The terminal-centric methods rely on the positioning software installed in the mobile terminal (see Table 1.6). They are further divided into:

- GPS method
- Network-assisted GPS (A-GPS)
- Enhanced observed time difference (E-OTD); this method can also be used in the network-centric mode, according to Andersson (2001)

The GPS method uses ranging signals directly from a number of GPS satellites and provides instantaneous point-positioning information, with the accuracy of 5 to

50 m, depending on the availability of GPS signals. A-GPS uses an assisting net-work of GPS receivers that can provide information enabling a significant reduction of the time-to-first-fix (TTFF) from 20–45 s to 1–8 s. For example, the timing and navigation data for GPS satellites may be provided by the network, which means that the receiver does not need to wait until the broadcast navigation message is read. It only needs to acquire the signal to compute its position almost instantly. For the timing information to be available through the network, the network and GPS would have to be synchronized to the same time reference. In essence, the assistance data make it possible for the receiver to make the time measurements (equivalent to ranges) to GPS satellites without having to decode the actual GPS message, which significantly speeds up the positioning process. According to Andersson (2001) the assistance data is normally broadcast every hour, and thus it has very little impact on the network's operability. More details on positioning with GPS can be found in Section 1.2.1.4 of this chapter.

Another terminal-centric solution is the E-OTD, which measures the time of the signal arrival from multiple base stations (within the wireless network) at the mobile device. The time differences between the signal arrivals from different base stations are used to determine the user's location with respect to the base stations, provided that the base stations' coordinates are known and the base stations send time-synchronized signals. For the positioning and timing purposes, the base sta-tions might be equipped with stationary GPS receivers. Thus, the base stations in E-OTD serve as reference points, similar to GPS satellites. However, this method is not subject to limitations in signal availability affecting GPS. The positioning accu-racy of E-OTD is about 100 to 125 m. Since E-OTD requires monitoring equipment at virtually every base station, it adds to the cost of LBS.

1.3.2.2 Network-Centric and Hybrid Positioning Methods

The main network-centric methods are:

- Cell global identity with timing advance (CGI-TA)
- Time of arrival (TOA)
- Uplink time difference of arrival (TDOA)
- Angle of Arrival (AOA)
- Location (multipath) pattern matching
- Received signal strength (RSS)

CGI uses the *cell ID* to locate the user within the cell, where the cell is defined as a coverage area of a base station (the tower nearest to the user). It is an inexpen-sive method, compatible with the existing devices, with the accuracy limited to the size of the cell, which may range from 10–500 m (indoor micro cell) to an outdoor macrocell reaching several kilometers (Andersson, 2002). CGI is often supplemented by the timing advance (TA) information that provides the time between the start of a radio frame and the data burst (Figure 1.11). This enables the adjustment of a mobile set's transmit time to correctly align the time at which its signal arrives at the base.

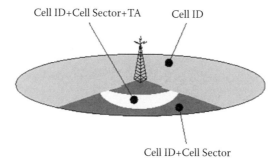

Cell ID+Cell Sector+TA   Cell ID

Cell ID+Cell Sector

FIGURE 1.11   Enhanced Cell ID Uses Basic Cell ID and other Information to estimate a position. (SnapTrack, 2002).

These measurements can be used to determine the distance from the user to the base, further reducing the position error (SnapTrack, 2002).

TOA is based on the travel time information (equivalent to a range) between the base station and the mobile terminal. In essence, the user's location can be found by triangulation, at the intersection of three (or more) arcs centered at the tower locations, with radii equal to the measured distances (Figure 1.12). The actual observation is the signal travel time, $t_U^r - t_B^t$, which is converted to a distance by multiplying it by the speed of light, where t_U^r and t_B^t denote the time of signal arrival at the user (*U*) and the time of signal transmission at the base (*B*), respectively. For a higher accuracy, the signal delay corrections, such as the tropospheric correction, might be applied. The basic observation equation (Equation 1.4), which is used to determine

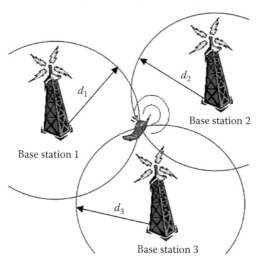

d_1 d_2

Base station 2

Base station 1

d_3

Base station 3

FIGURE 1.12   **(See color insert.)** Triangulation of the user's position based on the distance measurements to three base stations (cell towers). The latitude and longitude of the user are obtained as the intersection of three circles centered at the towers, with radii of d_1, d_2, and d_3.

the user's location, represents the measured distance from the user to a base station as a function of the coordinates of U (x_U, y_U) and B (x_B, y_B). Since the coordinates of the base station are known, linearized equations (similar to Equation 1.4) written for multiple base stations allow for the estimation of the user's coordinates by the least squares adjustment. It should be mentioned here that the base station time synchronization required by this method might need additional hardware and software support to achieve the required timing accuracy. It is achieved with the use of so-called location measurement units (LMUs) placed at known locations. LMUs, similarly to the mobile user, receive the signal from the surrounding towers. This information, combined with LMUs known position coordinates, enables the estimation of the clock offsets between pairs of base stations.

$$d_{UB} = \left(t_U^r - t_B^t\right)c = \sqrt{\left(x_U - x_B\right)^2 + \left(y_U - y_B\right)^2} \tag{1.4}$$

The concept of *TDOA* is similar to E-OTD; however, in TDOA the time of the user's signal arrival is measured by the network of base stations that observe the apparent arrival time differences (equivalent to distance differences) between pairs of sites. Since each base station is usually at a different distance from the caller, the signal arrives at the stations at slightly different times. The receivers, synchronized by an atomic clock (provided, for example, by GPS), send the user's voice call and timing data to the mobile switch, where the times are compared and computed to generate the coordinates (latitude and longitude) of the caller. To calculate the distance difference between the two base stations, a hyperbola is defined, with each base station located at one of its foci. The intersection of the hyperbolas defined by different pairs of base stations determines the 2D location of the mobile terminal (Balbach, 2000; Hein et al., 2000, 2001). A minimum of three stations must receive the signal to enable the user's location estimation as an intersection of two hyperbolas. The basic observation equation (Equation 1.5), which is used to determine the user's location, measures the distance from the user to two base stations (A, B).

$$d_{UB} - d_{UA} = \left(t_U^r - t_B^t\right)c - \left(t_U^r - t_A^t\right)c$$

$$= \sqrt{\left(x_U - x_B\right)^2 + \left(y_U - y_B\right)^2} - \sqrt{\left(x_U - x_A\right)^2 + \left(y_U - y_A\right)^2} \tag{1.5}$$

The *AOA* method is based on the observation of the angle of signal arrival by at least two cell towers. The towers that receive the signals measure the direction of the signal (azimuth) and send this information to the AOA equipment, which determines the user's location by triangulation using basic trigonometric formulas. The accuracy of AOA is rather high but may be limited by the signal interference and multipath, especially in urban areas. Much better and more reliable results are obtained by combining AOA with TOA (Deitel et al., 2002).

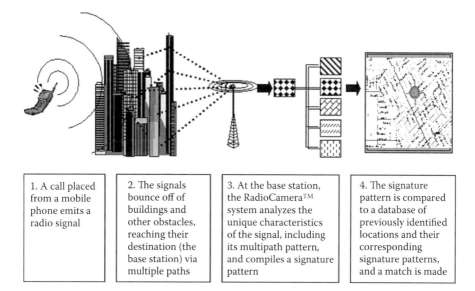

1. A call placed from a mobile phone emits a radio signal	2. The signals bounce off of buildings and other obstacles, reaching their destination (the base station) via multiple paths	3. At the base station, the RadioCamera™ system analyzes the unique characteristics of the signal, including its multipath pattern, and compiles a signature pattern	4. The signature pattern is compared to a database of previously identified locations and their corresponding signature patterns, and a match is made

FIGURE 1.13 **(See color insert.)** Location pattern matching.

The *location (multipath) pattern matching* method uses multipath signature in the vicinity of the mobile user to find its location. The user's terminal sends a signal that gets scattered by bouncing off the objects on its way to the cell tower. Thus, the cell tower receives a multipath signal and compares its signature with the multipath location database, which defines locations by their unique multipath characteristics (Figure 1.13). An example implementation, developed by the U.S. Wireless Corp. (2002), uses the location pattern matching technology (RadioCamera™) by measuring the radio signal's distinct radio frequency patterns and multipath characteristics to determine the user's location. With this method, the subscribers do not need any special updates to the mobile terminals to access the services, and wireless carriers do not need to make the infrastructure investments to offer LBS.

Another network-centric method applied in LBS is the software-only approach, where no additional hardware on the cell towers (base stations) or the mobile phones (terminals) is required. The method is based on the signal strength model observed for the area. By merging the information about the actual *received signal strength* (RSS) with the existing (mapped) signal data, the system can predict the user's location. An important feature of this fingerprinting technique is that it can determine the location of any digital cell phone or wireless device without any modification or add-ons or enhancements to the wireless carrier's existing network. By simply analyzing the existing RX (received) signal level (dBm) from multiple base stations to a standard wireless phone or device, the actual location of the phone can be calculated in seconds. The basic mathematical model is the relationship between the signal strength and the distance between the mobile station and the base. As in other ranging techniques, the user is located on a circle around the base with a radius equal to the distance measured using the signal strength. The prototype system based on RSS called GeoMode™, developed by Digital Earth Systems, has been tested in several metropolitan regions where it demonstrated location accuracy of 20 to 50 m for 92% of the time (McGeough, 2002). In more challenging environments such as Manhattan, accuracy better than 100 m for 80% of all stationary tests was obtained.

Several positioning methods presented in this section are based on the time measurement, such as the time of the user's signal arrival recorded at the base stations (TOA), or the time differences between the signal arrival from multiple base stations recorded by the mobile set (E-OTD, CGI-TA), or the signal travel time between GPS satellites and the mobile user. These time measurements can be converted to a range measurement (or the range difference), enabling the user's position determination by a common method of triangulation. Clearly, TOA or CGI-TA have an advantage over TDOA by working with the existing Global System for Mobile communications (GSM) mobiles (discussed in the following section) but may require significant investments in the supporting infrastructure (this is especially true with TOA). CGI-TA is rather inexpensive, as the cell information is already built into the networks. The E-OTD and TDOA methods require an extensive infrastructure support; moreover, E-OTD needs also customized handsets at the users' end (SnapTrack, 2002).

In general, one may argue that the user is less in control when the determination of his or her location is placed entirely within the network, as opposed to the mobile device. Perhaps the most autonomous is the GPS method and A-GPS, which rely on mobile devices that have an integrated GPS receiver. Clearly, GPS is the most accurate method of locating the mobile user; however, its accuracy may be limited by interference; jamming; strong multipath; and losses of signal lock under foliage, overpasses, or in urban canyons as well as other factors. However, with the upcoming GPS modernization bringing the new, stronger civilian signal and providing additional redundancy provisions, it is expected that GPS use in LBS will only increase.

It should be mentioned that the existence of a variety of positioning technologies, without a standardized method, may pose a problem to both the users and the providers, as the user is only covered in the area serviced by his or her provider and may not be covered elsewhere if another provider uses different location-identification method. It is rather difficult to define a single best technology, as each has its own advantages and disadvantages. Perhaps hybrid solutions offer the best choice, as they normally combine highly accurate with highly robust methods, resulting in multiple inputs improving both the robustness and the coverage. One example of a hybrid solution, listed in Table 1.6 and Table 1.7, based on a combination of a handset-based GPS method with a network-based CGI-TA to cover GPS losses of lock, should offer a solution more reliable to the one offered by each technique alone. Other hybrid solutions are, for example, AOA plus TDOA, called enhanced forward link triangulation (E-FLT), also called enhanced forward link time difference; E-OTD plus A-GPS; and AOA plus RSS. Clearly, a selection of the positioning

TABLE 1.6
Cellular Network-Based Techniques Supporting LBS

LBS Technique	Primary Observable	Upgrade of the User Terminal or Network	Location Calculation and Control
GPS, A-GPS	• Time (range) to multiple satellites • 3D location • Minimum of 3 ranges required for 2D positioning	• User terminal (GPS receiver, memory, software) • Nonsynchronized networks may require an enhancement	Mobile terminal
E-OTD	• Signal travel time difference between the user and the base stations • 2D location	• User terminal (memory, software) • Base station time synchronization	Mobile terminal
CGI-TA	• Cell ID • The accuracy does not meet the E-911[a] requirements • 2D location	• None	Network
TOA	• Signal travel time between the user and the base stations • 2D location	• Supports legacy terminals • Monitoring equipment at every base station	Network
TDOA	• Signal travel time difference between the user and the base stations • 2D location	• Network interconnection	Network
AOA	• Time (range) to multiple cell towers (minimum of three measurements is required) • 2D location	• Network interconnection • Antenna arrays to measure angles	Network
RSS	• Received signal strength • 2D location	• None	Network
Location/multipath pattern matching	• Multipath signature at the users location • 2D location	• None	Network
Hybrid system such as A-GPS + CGI	• GPS range • Cell ID • 3D or 2D	• Same as for GPS method	Mobile terminal plus Network

[a] E-911 (Enhanced 911) services. As of 2000, according to a Federal Communication Commission (FCC) mandate, wireless carriers are to provide the location of all emergency calls.

TABLE 1.7

Cellular Network-Based Techniques Supporting LBS: Cost, Latency, and Accuracy

LBS Technique	Total Cost[a]	Latency (TTFF)	Accuracy[b]
A-GPS; GPS	Moderate	<10 s; up to 60 s (cold start)	High (5–10 m)
E-OTD	High	<10 s	Moderate to high
CGI-TA	Low	<10 s	Low; depends on cell size
TOA	High	<10 s	Moderate to high
TDOA	High	<10 s	Moderate to high
AOA	Low	<10 s	Low to moderate
RSS	Low	<10 s	Moderate
Location/multipath pattern matching	Moderate	<10 s	Moderate
Hybrid system (A-GPS + CGI)	Moderate	<10 s	High

[a] Total cost includes handset, infrastructure, and maintenance; for details, see SnapTrack, 2002, Location Technologies for GSM, GPRS, and WCDMA Networks, Snap Track Whitepaper.

[b] The levels of accuracy are defined using the 95% CEP (circular error probable) as follows: high level equals to 95% CEP within 50 m; moderate level equals to 95% CEP within 300 m; low level equals to 95% CEP greater than 300 m (Source: Pietila, S., and Williams, M., 2002, Mobile Location Applications and Enabling Technologies, Proceedings of ION GPS [CD ROM], September 24–27.)

technique supporting LBS should be guided by the following issues documented by Mautz (2012):

- Accuracy/measurement uncertainty (millimeter, centimeter, decimeter, meter, decameter level)
- Coverage area/limitations to certain environments (single room, building, city, global)
- Cost (unique system setup costs, per user device costs, per room costs, maintenance costs)
- Required infrastructure (none, markers, passive tags, active beacons, pre-existing or dedicated, local or global)
- Market maturity (concept, development, product)
- Output data (2D, 3D coordinates, relative, absolute or symbolic position, dynamic)
- Parameters (such as speed, heading, uncertainty, variances)
- Privacy (active or passive devices, mobile or server-based computation)
- Update rate (on-event, on request, or periodically, e.g., 100 Hz or once a week)
- Interface (man–machine interfaces such as text based, graphical display, audio voice, and electrical interfaces such as RS-232, USB, fiber channels, or wireless communications)
- System integrity (operability according to technical specification, alarm in case of malfunction)
- Robustness (physical damage, theft, jamming, unauthorized access)

- Availability (likelihood and maximum duration of outages)
- Scalability (not scalable, scalable with area-proportional node deployment, scalable with accuracy loss)
- Number of users (single user, e.g., total station; unlimited users, e.g., passive mobile sensors)
- Intrusiveness/user acceptance (disturbing, imperceptible)
- Approval (legal system operation, certification of authorities)

In summary, any wireless method of position determination is subject to errors. The major error sources include multipath propagation, non-line of sight (NLOS), and multiple access interference. Multipath affects primarily AOA and RSS but can also affect the time-based methods. Under NLOS the arriving signal is reflected or diffracted, and thus takes a longer path as compared to the direct LOS signal, affecting primarily the time measurement. Co-channel interference is common to all cellular systems, where users share the same frequency band. The multiple access interference can significantly affect the time measurement. In order to obtain satisfactory positioning performance, steps must be taken to mitigate the effects of the error sources (Caffery et al., 1998).

1.3.2.3   GSM and UMTS Ranging Accuracy

GSM and the Universal Mobile Telecommunication System (UMTS) are the current standard supporting mobile communications. GSM, ranging from ~450 MHz to ~2000 MHz, uses time- and frequency-division multiple access (TDMA/FDMA), while UMTS is a code division multiple access (CDMA) system, operating on a carrier frequency of about 2 GHz. TDMA/FDMA means that the signal bandwidth is divided into frequency slots, each one further subdivided into time slots. The FDMA part involves the division by frequency of the (maximum) 25 MHz bandwidth into 124 carrier frequencies spaced 200 KHz apart. One or more carrier frequencies of the radio frequency part of the spectrum designated for mobile communication are assigned to each base station. Each of these carrier frequencies is then divided in time, using a TDMA scheme (Wireless KnowHow, 2002).

CDMA supports synchronized networking, while GSM networks are generally unsynchronized (Pietila and Williams, 2002). Also, CDMA is able to support more calls in the same spectrum, as it dynamically allocates the bandwidth. Wideband CDMA (W-CDMA) has been selected for the third generation (3G) of mobile telephone systems in Europe, Japan, and the United States. CDMA's signal structure is similar to the one used in satellite navigation techniques, that is, channels are allocated on the same frequency and separated with codes. This approach may suffer from near-far effects, where a transmitter close to the receiver will effectively jam the signal from the distant transmitters. A solution to overcome this problem is to implement an idle period down link (IPDL) of the base stations, meaning that a base station will have brief periods of no transmission, allowing all receivers in the vicinity to receive the signal. It should be mentioned that in the W-CDMA standardization, the positioning methods based on time difference are idle period down link–observed time difference of arrival (IPDL-OTDOA) and advanced forward link triangulation (A-FLT, also called advanced forward link time difference).

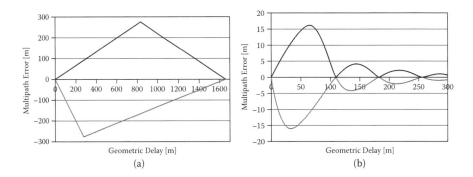

FIGURE 1.14 (See color insert.) Multipath error envelopes for (a) GSM and (b) UMTS. (From Hein, G., 2001, On the Integration of Satellite Navigation and UMTS, CASAN-1 International Congress, Munich, Germany.)

A-FLT and E-FLT mentioned earlier are essentially the same algorithms, according to Djuknic and Richton (2001), except E-FLT covers the legacy handsets.

The signal structure of a mobile communication system is rather complex, and different communication standards, such as GSM or UMTS, will differ with respect to the mode of access, frequency, time slot, power, and so on. Several of these aspects are important also from the positioning standpoint and would affect the final accuracy of the user's location, determined with the ranging signals. For example, GSM and UMTS suffer from different levels of multipath, as shown in Figure 1.14 (Hein, 2001). Multipath and other measurement errors (noise, tropospheric delay, and time synchronization error) accumulate in the user's location accuracy, and their effect is increased by the horizontal dilution of precision (HDOP) factor, which reflects the geometry of the user-base configuration (for more information on DOP factors, see Section 1.2.4.2). Table 1.8 shows examples of the expected positioning accuracy with GSM and UMTS and Table 1.9 lists advantages and disadvantages of stand-alone GSM and UMTS, and their combination with GNSS (Hein et al., 2000b; Hein, 2001).

TABLE 1.8
GSM and UMTS Positioning Error Budget

Error Source	GSM (m)	UMTS (m)
Measurement noise	270	18
Multipath	0–250	0–17
Troposphere	0.3–3	0.3–3
Network/handset synchronization	3–6	3–6
Oscillator error	7.5	7.5
Total error (1 sigma)	270–380	19–26

Source: Hein, G., 2001, On the Integration of Satellite Navigation and UMTS, CASAN-1 International Congress, Munich, Germany.

TABLE 1.7

Advantages and Disadvantages of Stand-Alone GSM and UMTS, and Their Combination with GNSS (Hein et al., 2001).

	GSM	UMTS/CDMA	GSM/UMTS/CDMA and GNSS
Pros	• No upgrade to the existing infrastructure if CGI or RSS is used • Limited increase of accuracy if ranging methods are used	• No additional hardware needed while system design is optimized • Increased accuracy when using ranging methods, as compared to CGI or RSS	• Advanced technology, low integration effort to integrate GNSS into GSM phone • Reduction in TTFF, GNSS receiver uptime reduced (wireless assisted GNSS) • Further reduction in TTFF and signal tracking enabled under bad conditions when common oscillators and microprocessors are used (tight integration)
Cons	• Accuracy might be poor for CGI and RSS • Accuracy strongly depends on location if ranging technique is used • Position availability is always dependent on the access to the network	• Accuracy strongly depends on location • Position availability is always dependent on the access to the network	• Integrated GNSS and GSM units may be relatively large • Network service needed for wireless assisted GNSS • Much development needed to reach the tight integration level

Source: Hein, G., Eissfeller, B., Öhler, V., and Winkel, J. O., 2001, Determining Location Using Wireless Networks, *GPS World,* vol. 12, no. 3, pp. 26–37.

1.3.3 WiFi Positioning Systems

The signals transmitted by access points in 2.4 GHz wireless local area networks (WLANs), or more commonly termed WiFi networks, are an extremely attractive option for positioning, as it is based on the availability of hardware that is already embedded within buildings and mobile devices, which therefore implies that there are no additional cost implications for positioning. The two methods used in WiFi positioning make use of the RSS information from the WiFi access points. Trilateration uses ranges derived from the RSS, and fingerprinting is similar to that used in cellular networks as described in Section 1.3.2.2.

Trilateration works on similar principles to GPS where the access points at a minimum of three known locations are used in combination with ranges derived from the RSS values. Herein lies the biggest difficulty for WiFi trilateration. The WiFi signals themselves are highly sensitive to interference and multipath caused by the operating environment (e.g., walls, people, equipment) and can result in incorrect positioning of the user. The method of fingerprinting is proving to be more successful but requires an additional workload to create the database containing the recorded signal strength data from various access points (APs) at known points spread across the environment in which positioning information is required. Operationally, it is assumed that the interference and multipath signals are accounted for in this process and that the signal's obtained by a mobile user can be matched to the information contained in the database. The location corresponding to the best match using a location estimation algorithm is the estimated position of the user.

Commercial WiFi positioning solutions (e.g., the Real Time Location Service from Ekahau) are claiming 1 to 3 m of indoor positioning accuracy using an enterprise grade WiFi network (Ekahau, 2012). This solution requires a highly detailed survey of WiFi signals in the operating environment as well as other map-based information (e.g., corridors and walkways) in order to assure the accuracy of the fingerprinting technique. Ekahau's current solution is therefore localized and customized to a specific building, organization, or task. The WiFi fingerprinting system offered by Skyhook is mostly global and can be used to determine a user's position in environments where GPS has been recognized to be inadequate. Skyhook maintains a global database of RSS data collected from WiFi access points and cell towers. Vehicles navigating the streets of major cities collect the information that allows them to determine the location of individual access points and cell towers as well as the RSS values at the vehicle's known position. Any WiFi-enabled device using the Skyhook client can access the server and determine a position to an accuracy of 10 to 20 m (Skyhook, 2012).

1.3.4   ULTRA-WIDEBAND POSITIONING

The high bandwidth and accurate pulse timing offered by ultra-wideband (UWB) signals make them extremely attractive for positioning due to its high multipath resistance and penetration and accurate ranging capabilities. Commercial UWB positioning systems such as that offered by Ubisense (Ubisense, 2012) operate within the restricted frequency bands (the Federal Communications Commission [FCC] has restricted the frequency band for unlicensed UWB to 3.1–10.6 GHz with the European Communications Commission restricting it to 6.0–8.5 GHz). UWB systems operate over short distances (typically <100m) and use TDoA, ToA, and signal travel time as measurement techniques to determine ranges between an UWB transmitter and receiver radio. A typical configuration for UWB (Figure 1.12a) tracking could include a number of fixed UWB signal transmitters placed at known locations around the operating environment, and a mobile device that determines the range from three or more transmitters and uses a triangulation process to fix its position. Using UWB techniques submeter or even centimeter level positioning is possible.

1.3.5   RFID

Similar to UWB, radio-frequency identification (RFID) positioning systems require the deployment of RFID scanners across the operational environment. These scanners are then able to interrogate either active or passive tags attached to the object to be tracked. The range between the scanner and the tag is the most important

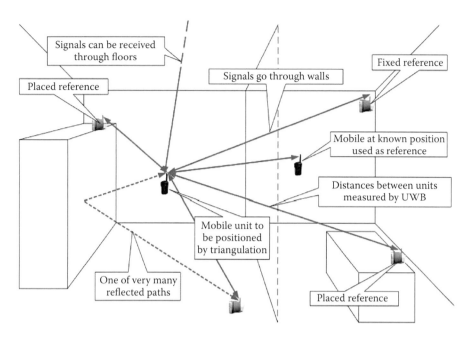

FIGURE 1.15 **(See color insert.)** Ultra-Wideband positioning System (Thales, 2004).

relationship defining the positioning technique used (active tags enable a greater range than passive tags). Cell of origin (CoO) and RSS ranging are the two most popular techniques used in RFID positioning. For CoO the location of the reader is described by a cell identified by the maximum read range to a tag. This technique offers relatively low accuracy and depends on the size of the distinguishable cells (10–20 m). A dense configuration of scanners across an area would improve the granularity of positioning but would incur a significantly higher cost. As a result, it is not practical to use RFID for real-time tracking applications over large areas. To improve the achievable positioning the deduction of ranges to the RFID tags from received signal power levels is used (it can be converted to a distance). Calibration for the signal strength to range conversion is required. The position fix can be obtained using trilateration if range measurements to several tags are performed.

To create a more general localization method than using trilateration, the RFID location fingerprinting is used. The principle of operation of RFID fingerprinting is similar to that used in WiFi and cellular positioning systems.

1.4 OTHER POSITIONING AND TRACKING TECHNIQUES: AN OVERVIEW

1.4.1 INERTIAL AND DEAD RECKONING SYSTEMS

The name "dead reckoning" is an abbreviation of "deduced reckoning," which means that the present location is deduced or extrapolated from a known prior position, modified by the known (observed) direction of motion and the velocity. Thus, in order to find the position coordinates using a dead reckoning system (DRS), the starting location and orientation must be known, since the system is sensitive only to a change in the direction and the apparent distance traveled derived from the continuously measured speed. An example of DRS is an *inertial navigation system* (INS). INS consists of multiple *inertial measurement units* (IMUs), such as *accelerometers* and *gyroscopes*, each one being a separate DRS.

Inertial navigation utilizes the inertial properties of sensors mounted on a moving platform and provides self-contained determination of instantaneous position and other parameters of motion by measuring a specific force, angular velocity, and time. Two primary types of IMUs are accelerometers, which sense linear acceleration in the inertial frame (i.e., fixed nonrotating frame), and gyroscopes, which sense the inertial rotational motion (angular rates, angular increments, or total angular displacements from an initial known orientation). In principle, INS requires no external information except for initial calibration (initialization and alignment), including externally provided 3D position, velocity, and attitude. A stationary self-alignment is routinely performed if no external velocity or attitude data are available. The IMU errors, however, grow with time. Therefore, the INS should be recalibrated periodically (by performing a zero velocity update [ZUPT] or by using an external aid such as GPS) to maintain reliable navigation quality. In the stand-alone mode, INS results are primarily affected by the initial sensor misorientation, accelerometer biases, and gyroscope drifts, causing a time-dependent positioning error.

Accelerometers use a known mass (proof-mass) attached to one end of a damped spring, which is attached to the accelerometer housing. Under no external acceleration condition, the spring is at rest and exhibits zero displacement. An external force applied to the housing will cause its acceleration; however, due to inertia, the proof-mass will lag behind, resulting in a displacement. The displacement of the mass and extension/compression of the spring is proportional to the acceleration of the housing (Allen et al., 2001; Jekeli, 2001). Since, according to Einstein's principle of equivalence, accelerometers do not sense the presence of gravitational field (but can sense the reaction due to the gravitational forces), external gravity information must be provided to obtain navigation information. In inertial navigation, the velocity and position are obtained through real-time integration of the governing differential equations (equations of motion), with accelerometer-measured specific force as an input. More details on inertial navigation can be found in Allen et al. (2001), Grejner-Brzezinska (1999), Grejner-Brzezinska (2001a, 2001b), and Jekeli (2001).

Any spinning object tends to keep its axis pointed in the same direction (so-called gyroscopic inertia, or rigidity in space), and if a force is applied to deflect its orientation, it responds by moving at right angles to the applied force. Gyroscopes (or gyros) are mechanisms displaying strong angular momentum characteristics, capable of maintaining a known spatial direction through an appropriate torque control, since inertially referred rate of angular momentum is proportional to the applied torque. Three mutually orthogonal gyroscopes can facilitate a three-dimensional inertially nonrotating Cartesian frame if they are mounted on a gimbaled platform maintaining the gyros orientation in space (space-stable system). Consequently, the gyroscopes maintain the knowledge of the orientation of the inertial platform, upon which orthogonal triad of accelerometers is mounted. Alternatively, gyros can be mounted directly on the vehicle (strapped down to the vehicle). In this case, since there is no gimbal platform performing the torque to maintain the gyroscopes' orientation, the torque is applied mathematically directly to the gyros. Since the physically or mathematically applied torque is proportional to the gyroscope's inertially referenced angular motion, it can be used to calculate the relative angular orientation between the gyro's initial and present spatial direction. The systems with no moving parts (no gimbaled platform), where the instrumentation of a reference frame is not facilitated physically but rather analytically, are referred to as strapdown INS. Since the lack of gimbaled structure allows for reduction in weight, size, power consumption, and ultimately cost, they are the primary modes of inertial navigation in a number of land and airborne applications.

A gyroscope-based DRS is called a *gyrocompass*. Thus, its directive action is based on the mechanical laws governing the dynamics of rotating bodies. The essential part of a gyrocompass consists of a spinning wheel (gyroscope) mounted in a way allowing freedom of movement about three mutually perpendicular axes. Essentially, a gyroscope becomes a gyrocompass if it can be controlled so that its axis of spin is aligned parallel with the true meridian (north-seeking gyroscope) under the influences of the earth's rotation and gravity. As the earth rotates, gravitational force attempts to change the gyroscope's axis of rotation. The resulting motion of the axis of the gyroscope at right angles to the applied force causes it to move to keep the alignment with the earth's axis of rotation (Navis, 2002).

Another DRS, which can measure the distance traveled, is an *odometer* (wheel counter), based on the concept of counting wheel turns, starting at a known location. The measuring accuracy of odometers may vary, and typically ranges from 0.1 to 0.01%. The main factor that degrades the accuracy of the wheel counter measurement is the error in the scaling factor, which is affected by several conditions, one of them being unevenness of the surface on which the distance is measured (Da and Dedes, 1995). Modern odometers normally have digital displays (control panel), memory storage, built-in calculators for automatic measurements of areas and volumes, instant unit conversion, and can automatically include wheel radius for accurate wall-to-wall measurements. If in addition to the distance traveled the application requires heading and attitude data, an odometer can be coupled with vertical and directional gyroscopes in a self-contained gyroscope package. This kind of DRS is capable of providing full orientation and positioning information.

1.4.1.1   What Are the Errors in Inertial Navigation?

Errors in inertial navigation are functions of the following factors: (1) initial condition errors including the alignment process, (2) errors in gravitational attraction compensation, (3) errors in coordinate transformation, (4) time-dependent accelerometer and gyroscope errors and possibly errors from external navigation aids used, and, finally, (5) errors excited by the dynamics of the vehicle. As a consequence, the INS-determined vehicle trajectory will diverge from the actual path, depending primarily on the quality of the IMU sensors and the mission duration. For example, a high-reliability and medium accuracy strapdown INS, such as Northrop Grumman LN100 (based on Zero-lock™ Laser Gyro [ZLG™] and A-4 accelerometer triad,

TABLE 1.10
INS Performance Error Characteristics
Assuming 4 to 8 min Alignment

Position	1.48 km/h (CEP)
Velocity	0.76 m/s (RMS)
Heading	0.1 deg (RMS)
Pitch and roll	0.05 deg (RMS)
Angular rate	0.04 deg/s (RMS)

Source: May, M. B., 1993, Inertial Navigation and GPS, *GPS World*, September, pp. 56–66

gyro bias of 0.003 deg/h, accelerometer bias of 25 µg), demonstrates the positioning quality of 1.48 km/h CEP (Circular Error Probable rate, at 50% probability level) in the stand-alone navigation mode (Litton Systems, Inc., 1994). Table 1.10 shows representative error characteristics of a medium quality unaided inertial navigator.

1.4.1.2　Microelectromechanical Systems (MEMS) Inertial Sensors

Microelectromechanical systems (MEMS) inertial sensors are a key enabling technology for LBS. These low cost, low profile motion sensors are currently embedded as standard in most modern mobile devices and offer sufficient performance for use in augmenting GPS in difficult environments. MEMS sensors have already demonstrated significantly improved performance from their first-generation configurations with MEMS accelerometers approaching performances close to those of tactical grade IMUs. Table 1.11 shows the performance differences between different grades of IMUs.

1.4.2　Digital Compass

A *digital compass* is another device used for orientation tracking in navigation, guidance, and vehicle compassing. It is a solid-state device capable of detecting the earth's weak magnetic field, whose circuit board includes the basic magnetic sensors and electronics to provide a digital indication of heading. The achievable accuracy in heading measurement is typically at the level of 1 to 5 degrees. The device is sensitive to tilt, and any tilt greater than 10 to 15 degrees will create directional errors. Most advanced systems include compensation for hard iron distortions, ferrous objects, and stray fields. Other applications besides navigation and compassing cover attitude reference, satellite antenna positioning, platform leveling, and integration with other devices, such as GPS or laser range finder.

1.4.3　Additional Location Tracking Systems

The location techniques presented in this section represent other positioning sensors based on various physical media employed, commonly used in location tracking.

TABLE 1.11
INS Performance Characteristics

Sensor Grade	Sensor Name	Type	Characteristics
Navigation	LN100 Gyroscope	Nondithered 18 cm Zero Lock Laser Gyro	bias = 0.003°/h; rw = 0.001°/h$^{1/2}$; sf < 1 ppm
	LN100 Accelerometer	Miniature Accelerometer A4	bias = 25 µg; sf = 40 ppm; ma = 2 arcsec; wh = 5 µg/Hz$^{1/2}$
Navigation	H764G Gyroscope	Dithered GG1320AN RLG	bias = 0.0035°/h; rw = 0.0035°h$^{1/2}$; sf = 5 ppm
	H764G Accelerometer	QA2000	bias = 25 µg; wh = 8.3 µg (100 HZ bw); sf = 100 ppm
Tactical	HG1700 Gyroscope	Dithered GG1308 RLG	bias = 2.0°/h; rw = 0.125~0.3°/h$^{1/2}$; sf = 150 ppm
	HG1700 Accelerometer	RBA500	bias = 1.0 mg; wh = 0.2 mg (100 HZ bw); sf = 300 ppm; ma = 12 m rad
Consumer	IMU400CC Gyroscope	Nondithered Silicon MEMS Gyro	bias = 1°/sec; rw = 2.25°/h$^{1/2}$; sf = 1%
	IMU400CC Accelerometer	Silicon MEMS Accelerometer	bias = 8.5 mg; rw = 0.1 m/s/h$^{1/2}$; sf = 1%

Notes: Manufacturer's specifications for the initial sensors: sf, scale factor; ma, misalignment; wh = white noise; rw = random walk; bw = bandwidth.

These systems are most commonly used for indoor positioning and tracking, but some of them can be also used outdoors.

1.4.3.1 Acoustic (Ultrasonic) Tracking

An *ultrasonic tracker* utilizes high-frequency sound waves (approximately 20,000 Hz) to locate objects either by triangulation of several transmitters, time-of-flight (TOF) method, or by measuring the signal's phase difference between the transmitter and the receiver (phase-coherence method). The TOF method measures the time of travel between a transmitter and the receiver, which multiplied by a speed of sound provides an absolute distance measurement. The phase coherence method provides the phase difference between the sound wave at the receiver and the transmitter, which can be converted to a change in distance if the signal's wavelength (frequency) is known. Since this method is sensitive to a change in distance only, the initial distance to the target must be known. More details on the ultrasonic trackers can be found in Allen et al. (2001). A single transmitter–receiver pair provides a range measurement between the target and the fixed point. Three distance measurements provide two solutions, one of which is normally discarded as impossible. Thus, to estimate the 3D position coordinates, minimum of three range observations between the known locations of the transmitters and the target are needed. This concept is similar to the GPS-based triangulation. An inherent problem of an ultrasonic tracker is a signal travel delay due to slow speed of sound (331 m/s at 0°C, and varies with temperature and pressure).

1.4.3.2 Magnetic Tracking

Magnetic trackers use magnetic fields, such as low frequency alternating current (AC) fields or pulsed direct current (DC) fields, to determine 3D location coordinates, attitude, and heading relative to the transmitter. Typically, three orthogonal triaxial coils generating the source magnetic fields are used at the transmitter and the receiver. A magnetic field is generated when current is applied to the transmitter coil. At the receiving end, a time varying magnetic field induces a voltage with a magnitude proportional to the area bounded by the coil and the rate of change of the field. This voltage varies with a cosine of the angle between the direction of the field lines and the axis of the coil. From the induced voltage level, the information about the distance from the transmitter to the receiver and the axis alignment between them can be extracted. The distance estimation is based on the fact that the magnetic field strength decreases with a third power of distance and with the cosine of the angle between the axis of the receiving coil and the direction of the magnetic field. Subsequently, to find the distance between the receiver and the transmitter, the voltage induced at the receiver is compared to the known voltage of the transmitted signals. The orientation can be found through the comparison of the strength (voltage) of the induced signals (Allen et al., 2001).

Magnetic tracking systems are subject to error, primarily due to distortions of their magnetic fields by conducting objects or due to other electromagnetic fields in the environment, and generally, the error increases with the transmitter–receiver distance. If there is a metal object in the vicinity of the magnetic tracker's transmitter or receiver, the transmitter signals are distorted and the resulting position/orientation measurements will contain errors. One possible solution is the system calibration based on a map of distortions (calibration table) from which a correction term can be derived. More details on magnetic tracking can be found in Allen et al. (2001), Raab et al. (1979), and Livingston (2002).

1.4.3.3 Optical Tracking

Optical tracking systems, also referred to as image-based systems, make use of light to measure angles (ray direction) that are used to find the position location. The essential parts of an optical system are the target and the detector (sensor). These systems rely on a clear LOS between the detector and the target. Detectors can be in the form of charged coupled device (CCD)-based cameras, video cameras, infrared cameras, or lateral-effect photodiodes. Targets can be active, such as a light-emitting diode or infrared-emitting diode; or passive, such as mirrors or other reflective materials; or simply natural objects (Allen et al., 2001). Detectors are used to observe targets and to derive position and orientation of a target from multiple angular observations (multiple detectors). For example, a single point on a 2D detector imaging plane provides a single ray defined by that point and the center of projection; two 2D points allow 3D target positioning and additional points are required for orientation. According to Allen et al. (2001) "in order to determine orientation, multiple targets must be arranged in a rigid configuration. Then the relative positions of the targets can be used to derive orientation." One possible configuration of an optical system is a set of three 1D sensors, each one narrowing the location of a target to a plane. The intersection of three planes provides coordinates of a target in a system-defined reference frame. Alternatively, two 2D sensors, each one determining a line, allow for location of a target in 3D.

In general, the image-based tracking systems provide high positioning accuracy and resolution, but these are a function of the type of sensors used (primarily its angular resolution), distance between the target and the sensor, specific application, and the environment (outdoor versus indoor). Also, the system geometry has an impact on its position location accuracy. For example, fixed sensors can observe moving targets, and moving sensors can observe fixed targets. An important aspect is also the selection of the reference frame, in which the optical measurements are made (Allen et al., 2001). More details on optical target location can be found in Allen et al. (2001), Blais et al. (1996), and Beraldin et al. (2000).

Another type of optical tracking systems is based on laser ranging, which provides range measurements to active or passive targets. To measure a distance, the TOF of a laser beam from a transmitter to the target and back is measured (see, for example, Blais et al., 2000. This method is well suited for measuring distances from several meters to a few hundreds of meters and even considerably longer distances, and thus, it is suitable for both outdoor and indoor applications. The accuracy of the distance measured ranges from micrometers for short-range devices to a decimeter-level for very long-range systems.

In the long-range laser scanning systems placed on moving platforms (airborne or spaceborne), the laser beam is oriented and its projection center located using integrated GPS/INS system (see Section 1.5). This allows for direct estimation of 3D coordinates of a target (Baltsavias, 1999; Grejner-Brzezinska, 2001a, 2001b). Similarly, in the image-based tracking methods, if the imaging sensor can be directly georeferenced by GPS/INS (or ground control points can be used), the absolute location coordinates of the points in the imagery in the selected global or local reference frame can be determined by photogrammetric methods. This approach is normally used in land-based or airborne systems suitable for mapping and GIS data acquisition. Examples of image- or terrain-based navigation techniques can be found in Toth et al. (2009) and Markiel et al. (2009).

1.5 HYBRID SYSTEMS

Virtually all position location techniques presented here display some inherent limitations related either to the associated physical phenomenon, system design specifications, or application environmental constraints. Consequently, no single technique or sensor can provide complete tracking information with continuously high performance and reliability. However, the sensors or techniques can be integrated with each other to provide redundancy, complementarity, or fault-resistance, rendering more robust positioning and tracking systems. One example of a commonly used hybrid or integrated system is an inertial–optical hybrid system, in which the optical system supports calibration of inertial errors during the slow motion of the system, where the optical sensor's performance is the best. During the rapid motion, the inertial system performs better than the optical; thus complementary behavior of both systems renders more reliable and accurate performance.

Another example is a GPS/INS system, which is often combined with additional imaging sensors, such as frame CCD or video. Integrated GPS/INS systems are commonly used in positioning and navigation as well as in direct georeferencing of imaging sensors in mobile mapping and remote sensing. GPS contributes its high accuracy and long-term stability (under no losses of lock), providing means of error estimation of the inertial sensors. GPS-calibrated INS provides reliable bridging during GPS outages and supports the ambiguity resolution after the GPS lock is reestablished. The effective positioning error level depends on systematic and random GPS errors as amplified by satellite geometry. Well-calibrated, GPS-supported INS provides precise position and attitude information between the GPS updates and during GPS losses of lock, facilitating immunity to GPS outages, and continuous attitude solution. In general, using a GPS-calibrated, high to medium accuracy inertial system, attitude accuracy in the range of 10 to 30 arcsec can be achieved (Schwarz and Wei, 1994; Abdullah, 1997; Grejner-Brzezinska, 1998). GPS/INS works well if GPS gaps are not too frequent and not excessively long. However, in the case of urban canyon or indoor navigation, there is usually very limited or no GPS signal. Consequently, a PL array may be used to supplement the satellite signal (Wang et al., 2001; Grejner-Brzezinska, Yi, et al., 2002).

In summary, any combination of GPS and INS functionality into a single integrated navigation system represents a fusion of dissimilar, complementary data, and should be able to provide a superior performance as opposed to either sensor in a stand-alone mode. In fact, integration of these two systems is often the only way to achieve the following goals (Greenspan, 1996):

- Maintaining a specified level of navigation during GPS outages
- Providing a complete set of six navigational parameters (three positional and three attitude components) and high rate (higher than available from conventional GPS, i.e., >20 Hz)
- Reducing random errors in the GPS solution
- Maintaining a GPS solution under high vehicle dynamics and interference

A combination of positioning/orientation and imaging sensors renders a multisensor tracking or imaging system that can be designed for use either in outdoor or indoor environments (Bossler et al., 1991; El-Hakim et al., 1997; Behringer, 1999; El-Sheimy and Schwarz, 1999; You et al., 1999a, 1999b; Grejner-Brzezinska, 2001a, 2001b; Grejner-Brzezinska et al., 2010a; Rizos et al., 2010; Toth and Grejner-Brzezinska, 2010). Table 1.12 lists the example functionality of a multisensor system designed for mobile mapping and LBS. However, the sensor functionality can be generalized for other applications.

The increasing range of positioning technologies that are now available to support LBS or more sophisticated mobile mapping and tracking systems has established a fundamental requirement for a ubiquitous positioning capability. The definition of ubiquitous positioning is based on achieving seamless GPS-like performance in all environments. In achieving a ubiquitous positioning capability there are many challenges that need to be addressed including developing truly interoperable or

TABLE 1.12

Primary Sensors and Information Typically Used in Multisensor Hybrid Tracking/Imaging System and Their Functionality

Primary Sensor	Sensor Functionality
GPS	• Image geopositioning in 3D
	• Time synchronization between GPS and other measurement sensors, e.g., INS
	• Image time-tagging
	• INS error control
	• Furnishes access to the 3D mapping frame through WGS84
INS	• Image orientation in 3D
	• Supports image georeferencing
	• Provides bridging of GPS gaps
	• Provides continuous, up to 400 Hz, trajectory between the GPS measurement epochs
	• Supports ambiguity resolution after losses of lock, and cycle slip detection and fixing
UWB/WiFi/RFID	• Supports positioning in indoor environment
Pseudolite;	• Primary positioning functions identical to GPS
Transmitter/transceiver	• Supports GPS constellation during weak geometry (urban canyons)
CCD camera	• Collects imagery used to derive object position
	• Two (or more) cameras provide 3D coordinates in space
Laser range finder	• Supports feature extraction from the imagery by providing precise distance (typical measuring accuracy is about 2–5 mm)
LIDAR[a] (airborne systems)	• Source of DSM/DTM,[b] also material signature for classification purposes
	• Supports feature extraction from the imagery
Multi-/hyperspectral sensors (airborne systems)	• Spectral responses of the surface materials at each pixel location
	• Wealth of information for classification and image interpretation
Voice recording, touch-screen, barometers, gravity gauges	• Attribute collecting sensors (land-based systems)
Mapping and spatial database information	• Supports map matching and can be used to constrain sensor errors
Platform dynamics (e.g., human locomotion models)	• Supports the development of knowledge based algorithms based on a priori models that describe nontraditional platform motions such as the human body

Source: Grejner-Brzezinska, D., 2001, Mobile Mapping Technology: Ten Years Later, Part I, Surveying and Land Information Systems, vol. 61, no. 2, pp. 79–94.

[a] LIDAR, light detection and ranging.

[b] DSM, digital surface model; DTM, digital terrain model.

compatible geopositioning devices that take into account all signals of opportunity as well as developing computationally efficient measurement fusion algorithms that can undertake real-time signal processing, interference detection, and measurement fusion computations. What is interesting and significant in these developments is the trade-offs between complexity and accuracy and the overall cost of the system (Retscher and Kealy, 2006; Kealy and Scott-Young, 2006).

Typical fusion algorithms for positioning have traditionally revolved around the Kalman filter where both the measurement errors and platform dynamics can be easily combined in a robust estimation process. The emergence of measurements from nontraditional sensors like WiFi or RFID challenge these algorithms due to non-Gaussian error distributions or the requirement for smarter integration techniques to manage the more complex and disparate error sources inherent in these new signals. Techniques that use qualitative information (Hope and Kealy, 2008; Winter and Kealy, 2011), map matching (Kealy and Scott-Young, 2006), artificial intelligence (Grejner-Brzezinska et al., 2008; Moafipoor et al., 2008; Reiterer et al., 2011) or more recently cooperative positioning techniques (Grejner-Brzezinska et al., 2008; Grejner-Brzezinska et al., 2010b; Efatmanshenek et al., 2011, 2012) are creating new fusion architectures that are delivering robust positioning results. It is expected that over the next decade these techniques will be embedded in many LBS algorithms.

1.6   SUMMARY

Location and tracking techniques presented in this chapter play an important role in geoinformatics, which is the foundation component of telegeoinformatics. These techniques are a key component of real-time indoor and outdoor tracking for LBS, such as emergency response, mapping, robot, fleet and personnel tracking, agriculture and environmental protection, or traveler information services. The main emphasis was put on GNSS-based location technology; however, other major techniques used in LBS and in indoor and outdoor tracking were also presented. Moreover, the primary concepts of position determination using different media and observation principles were discussed.

Positions along with the information contained in spatial databases, combined with wireless communication, are crucial in telegeoinformatics where the user's location must be known in real time, enabling a number of applications mentioned earlier. These applications are expected to expand dramatically in the next few years and will require an updated infrastructure, efficient tracking, and communication techniques as well as advanced optimization algorithms. These algorithms should be able to fuse measurements from diverse positioning technologies, integrate the positioning solution with centralized and decentralized spatial information databases, and generate robust navigation and other application outputs. Current generation GNSS seems by far the most accurate and the fastest expanding positioning technology supporting active tracking for outdoor applications, which can be augmented for better performance and reliability by other techniques, such as INS. Future satellite constellation expansions, increasingly improved MEMS inertial sensor performance, and the increasing availability of signals of opportunity that facilitate indoor positioning are creating new paradigms for the development of ubiquitous positioning systems.

REFERENCES

Abdullah, Q., 1997, Evaluation of GPS-Inertial Navigation System for Airborne Photogrammetry, presented at ACSM/ASPRS Annual Convention and Exposition, April 7–10, Seattle, Washington.

Abnizova, I., Cullen, P., and Taherian, S., 2002, Mobile Terminal Location in Indoor Cellular Multi-Path Environment, http://www.wlan01.wpi.edu/proceedings/wlan69d.pdf (accessed July 20, 2012).

Allen, B. D., Bishop, G., and Welch, G., 2001, Tracking: Beyond 15 Minutes of Thought, http://www.cs.unc.edu/~tracker/media/pdf/SIGGRAPH2001_CoursePack_11.pdf (accessed July 20, 2012).

Al-Shaery, A., Lim, S., and Rizos, C., 2011, Investigation of Different Interpolation Models Used in Network-RTK for the Virtual Reference Station Technique, *Journal of Global Positioning Systems*, vol. 10, no. 2, pp. 136–148.

Andersson, C., 2001, Wireless Developer Network web page, http://www.wirelessdevnet.com/channels/lbs/features/mobilepositioning.html (accessed July 20, 2012).

Andersson, C., 2002, Mobile Positioning—Where You Want to Be! http://www.wireless-devnet.com/channels/lbs/features/mobilepositioning.html (accessed July 20, 2012).

Balbach, O., 2000, UMTS-Competing Navigation System and Supplemental Communication System to GNSS, *Proceedings of ION GPS*, Salt Lake City, pp. 519–527.

Baltsavias, E.P., 1999, Airborne Laser Scanning: Basic Relations and Formulas, *ISPRS Journal of Photogrammetry and Remote Sensing*, vol. 54, pp. 199–214.

Behringer, R., 1999, Registration for Outdoor Augmented Reality Applications Using Computer Vision Techniques and Hybrid Sensors, IEEE Virtual Reality, pp. 244–251, Houston, Texas.

Beraldin, J.-A., Blais, F., Cournoyer, L., Godin, G., and Rioux, M., 2000, Active 3D Sensing, http://foto.hut.fi/opetus/295/pg_course2008/beraldin/beraldin.pdf (accessed July 20, 2012).

Blais, F., Beraldin, J.-A., and El-Hakim, S., 2000, Range Error Analysis of an Integrated Time-of-Flight, Triangulation, and Photogrammetric 3D Laser Scanning System, *SPIE Proceedings*, *AeroSense*, vol. 4035, pp. 236–247, April 24–28, Orlando, Florida.

Blais, F., Lecavalier, M., and Bisson, J., 1996, Real-Time Processing and Validation of Optical Ranging in a Cluttered Environment, ICSPAT, pp. 1066–1070, October 7–10, 1996, Boston, Massachusetts.

Bossler, J. D., Goad, C., Johnson, P., and Novak, K., 1991, GPS and GIS Map the Nation's Highway, *GeoInfo System Magazine*, March, pp. 26–37.

Caffery, J. J., and Stuber, G. L., 1998, Overview of Radiolocation in CDMA Cellular Systems, *IEEE Communications Magazine*, vol. 36, no. 4, pp. 38–45.

Da, R., and Dedes, G., 1995, Nonlinear Smoothing of Dead Reckoning Data with GPS Measurements, Proceedings of Mobile Mapping Symposium, The Ohio State University, May 24–26.

Dai, L., Han, S., Wang, J., and Rizos, C., 2003, Comparison of Interpolation Algorithms in Network-Based GPS Techniques, *Journal of Navigation*, vol. 50, no. 4, 277–293.

Deitel, H. M., Deitel, P. J., Nieto, T. R., and Steinbuhler, K., 2002, *Wireless Internet and Mobile Business How to Program*, Prentice Hall, Upper Saddle River, NJ.

Divis, 2012. First GPS III Launch Delayed by Up to a Year, OCX by Two Years, http://www.insidegnss.com/node/3054 (accessed July 20, 2012).

Djuknic, G. M., and Richton, R. E., 2001, Geolocation and Assisted GPS. Computer, vol. 34, no. 2, 123–125.

Ekahau, 2012, Ekahau Real Time Location System (RTLS), http://www.ekahau.com/products/real-time-location-system/overview.html (accessed July 23, 2012).

Efatmanesnek, M., Alam, N., Kealy, A., and Dempster, A. G., 2012, A Fast Multidimensional Scaling Based Filter for Vehicular Cooperative Positioning, *Journal of Navigation*, vol. 65, no. 2, 223–243.

Efatmanesnek, M., Kealy, A., and Dempster, A., 2011, Information Fusion for Localization within Vehicular Networks, *Journal of Navigation* 64(3): 401–416.

El-Hakim, S. F., Boulanger, P., Blais, F., Beraldin, J.-A., and Roth, G., 1997, A Mobile System for Indoors 3-D Mapping and Positioning, *Proceedings of the Optical 3-D Measurement Techniques IV*, pp. 275–282, Zurich, September 29–October 2.

Elrod, B. D., and Van Dierendonck, A. J., 1996, Pseudolites, in: *Global Positioning System: Theory and Applications*, Vol. II, B. W. Parkinson and J. J. Spilker (eds.), pp. 51–79, American Institute of Astronautics and Astronautics, Washington, DC.

El-Sheimy, N., and Schwarz, K. P., 1999, Navigating Urban Areas by VISAT—A Mobile Mapping System Integrating GPS/INS/Digital Cameras for GIS Application, *Navigation*, vol. 45, no. 4, pp. 275–286.

European Space Agency (ESA), 2012, Galileo, http://www.esa.int/esaNA/galileo.html (accessed July 23, 2012).

Federal Aviation Administration, Global Navigation Satellite Systems, 2012, http://www.faa.gov/about/office_org/headquarters_offices/ato/service_units/techops/navservices/gnss/ (accessed July 20, 2012).

Francica, J., 2002, Location-Based Services: Where Wireless Meets GIS, *GEOWorld*, December.

Greenspan, R. L., 1996, GPS and Inertial Integration, in: *Global Positioning System: Theory and Applications*, Vol. II, B. W. Parkinson and J. J. Spilker (eds.), pp. 187–220, American Institute of Aeronautics and Astronautics, Washington, DC.

Grejner-Brzezinska, D. A., 1998, High Accuracy Airborne Integrated Mapping System, in: *Advances in Positioning and Reference Frames: IAG Scientific Assembly, Rio De Janeiro, Brazil, September 3–9, 1997*, F. K. Brunner (ed.), pp. 337–342, Springer, Berlin.

Grejner-Brzezinska, D. A., 1999, Direct Exterior Orientation of Airborne Imagery with GPS/INS System: Performance Analysis, Navigation, vol. 46, no. 4, pp. 261–270.

Grejner-Brzezinska, D., 2001a, Mobile Mapping Technology: Ten Years Later, Part I, Surveying and Land Information Systems, vol. 61, no. 2, pp. 79–94.

Grejner-Brzezinska D., 2001b, Mobile Mapping Technology: Ten Years Later, Part II, Surveying and Land Information Systems, vol. 61, no. 3, pp. 83–100.

Grejner-Brzezinska, D. A., 2002, GPS Instrumentation Issues, in: *Manual of Geospatial Science and Technology*, J. Bossler, J. Jensen, R. McMaster, and C. Rizos (eds.), Chapter 10, Taylor & Francis, New York.

Grejner-Brzezinska, D., Toth, C., Gupta, J., Lei, L., and Wang, X., 2010a, Challenged Positions: Dynamic Sensor Network, Distributed GPS Aperture, and Inter-Nodal Ranging Signals, *GPS World*, September, pp. 35–42.

Grejner-Brzezinska, D. A., Toth, C. K., Li, L., Park, J., Wang, X., Sun, H., Gupta, I. J., Huggins, K., and Zheng, Y. F., 2009, Positioning in GPS-challenged Environments: Dynamic Sensor Network with Distributed GPS Aperture and Inter-nodal Ranging Signals, *Proceedings of the 22nd International Technical Meeting of The Satellite Division of the Institute of Navigation* (ION GNSS 2009), Savannah, Georgia, September 2009, pp. 111–123.

Grejner-Brzezinska, D., Toth, C. K., and Moafipoor, S., 2008, Performance Assessment of a Multi-Sensor Personal Navigator Supported by an Adaptive Knowledge Based System, *International Archives of Photogrammetry and Remote Sensing*, vol. XXXVII, Part B5, pp. 857–867.

Grejner-Brzezinska, D., Toth, C., Sun, H., Wang, X., and Rizos, C., 2010b, A Robust Solution to High-Accuracy Geolocation: Quadruple Integration of GPS, IMU, Pseudolite and Terrestrial Laser Scanning, *IEEE Transactions on Instrumentation and Measurement*, DOI 10.1109/TIM.2010.2050981.

Grejner-Brzezinska, D. A., Yi, Y., and Wang J., 2002, Design and Navigation Performance Analysis of an Experimental GPS/INS/PL System, Proceedings of 2nd Symposium on Geodesy for Geotechnical and Structural Engineering, Berlin, Germany, May 21–24, pp. 452–461.

Heggarty C., 2008, Evolution of the Global Navigation Satellite System (GNSS), http://ieeexplore.ieee.org/ieee_pilot/articles/96jproc12/jproc-CHegarty-2006090/article.html (accessed July 20, 2012).

Hein, G., 2001, On the Integration of Satellite Navigation and UMTS, CASAN-1 International Congress, Munich, Germany.

Hein, G., Eissfeller, B., Öhler, V., and Winkel, J. O., 2000, Synergies Between Satellite Navigation and Location Services of Terrestrial Mobile Communication, *Proceedings of ION GPS*, Salt Lake City, Utah, pp. 535–544.

Hein, G., Eissfeller, B., Öhler, V., and Winkel, J. O., 2001, Determining Location Using Wireless Networks, *GPS World,* vol. 12, no. 3, pp. 26–37.

Hellebrandt, M., and Mathar, R., 1999, Location Tracking of Mobiles in Cellular Radio Networks, *IEEE Transactions on Vehicular Technology*, vol. 48, no. 5, pp. 1558 –1562.

Hofman-Wellenhof, B., Lichtenegger, H., and Collins, J., 2001, *Global Positioning System: Theory and Practice*, 5th ed., Springer-Verlag, Wien.

Hope, S., and Kealy, A., 2008, Using topological relationships to inform a data integration process, *Transactions in GIS,* vol. 12, no. 2, pp. 267–283.

Inside GNSS, 2012, Japan Aims at 4-Satellite QZSS by Decade's End, http://www.insidegnss.com/node/3014 (accessed July 20, 2012).

Jekeli, C., 2001, *Inertial Navigation Systems with Geodetic Applications*, Walter de Gruyter, Berlin.

Kealy, A., and Scott-Young, S., 2006, Augmented Reality: Technology Fusion for an Innovative Application, *Transactions in GIS*, vol. 10, no. 2, pp. 279–300.

Kleusnerg, A., 2009, GPS and GLONASS, http://gauss.gge.unb.ca/gpsworld/EarlyInnovation-Columns/Innov.1990.11-12.pdf (accessed July 20, 2012).

Lachapelle, G., 1990, GPS Observables and Error Sources for Kinematic Positioning, in: *Kinematic Systems in Geodesy: Surveying, and Remote Sensing*, K. P. Schwarz and G. Lachapelle (eds.), pp. 17–26, Springer-Verlag, New York.

Lachapelle, G., Ryan, S., and Rizos, C., 2002, Servicing the GPS User, in: *Manual of Geospatial Science and Technology*, J. Bossler, J. Jensen, R. McMaster, and C. Rizos (eds.), pp. 201–215, Taylor & Francis, London.

Langley, R. B., 1991, The GPS Receiver: An Introduction, *GPS World*, January, pp. 50–53.

Langley, R. B., 2008, A New, Timely Service from Natural Resources Canada, http://www.gpsworld.com/gnss-system/innovation-online-precise-point-positioning-4252 (accessed July 20, 2012).

Litton Systems, Inc., 1994, LN-100G EGI Description, Litton Systems, Inc., September.

Livingston, M., 2002, UNC Magnetic Tracker Calibration Research, http://www.cs.unc.edu/~us/magtrack.html (accessed July 20, 2012).

Locata Corp., 2012, LocataTech Explained, http://www.locatacorp.com/technology/locata-tech-explained/ (accessed 23 July 23, 2012).

Markiel, J. N., Grejner-Brzezinska, D., and Toth, C., 2009, Flash LADAR Navigation: Locating Features with Moving Acquisition, Joint Navigation Conference 2009 (CD ROM), June 1–4, Orlando, Florida,.

Mautz, R., 2012, Indoor Positioning Technologies, Habilitation thesis, ETH Zurich.

May, M. B., 1993, Inertial Navigation and GPS, *GPS World*, September, pp. 56–66.

McGeough, J., 2002, Wireless Location Positioning Based on Signal Propagation Data, http://www.wirelessdevnet.com/library/geomode1.pdf (accessed July 20, 2012).

Moafipoor, S., Grejner-Brzezinska, D. A., and Toth, C. K., 2008, A Fuzzy Dead Reckoning Algorithm for a Personal Navigator, *Navigation*, vol. 55, no. 4, pp. 241–254.

National Coordination Office for Space-Based Positioning, Navigation, and Timing, 2012, GPS Control Segment, http://www.gps.gov/systems/gps/control/ (accessed July 20, 2012).

Navis, 2002, Aids to Navigation, Gyrocompass, http://www.navis.gr/navaids/gyro.htm (accessed July 20, 2012).

OmniSTAR, 2012, How it Works, http://www.omnistar.com/ (accessed July 20, 2012).

Parkinson, B. W., and Spilker J. J. (ed.), 1996, *Global Positioning System: Theory and Applications*, American Institute of Aeronautics and Astronautics, Washington, DC.

Pietila, S., and Williams, M., 2002, Mobile Location Applications and Enabling Technologies, Proceedings of ION GPS (CD ROM), September 24–27.

Progri, I. F., and Michalson, W. R., 2001, An Alternative Approach to Multipath and Near-Far Problem for Indoor Geolocation Systems, Proceedings of ION GPS (CD ROM), September 11–14.

Raab, F., Blood, E., Steioner, T., and Jones, H., 1979, Magnetic Position and Orientation Tracking System, *IEEE Transactions on Aerospace and Electronic Systems*, vol. 15, no. 5, pp. 709–718.

Raquet, J., and Lachapelle, G., 2001, RTK Positioning with Multiple Reference Stations, *GPS World*, vol. 12, no. 4, pp. 48–53.

Reiterer, A., Egly, U., Vicovac, T., Mai, E., Moafipoor, S., Grejner-Brzezinska, D., and Toth, C., 2011, Application of Artificial Intelligence in Geodesy: A Review of Theoretical Foundations and Practical Examples, *Journal of Applied Geodesy*, vol. 4, no. 4, pp. 201–217.

Retscher, G., and Kealy, A., 2006, Ubiquitous Positioning Technologies for Modern Intelligent Navigation Systems, *Journal of Navigation*, vol. 59, no. 1, pp. 91–103.

Rizos, C., 2002a, Making Sense of the GPS Technique, in: *Manual of Geospatial Science and Technology*, J. Bossler, J. Jensen, R. McMaster, and C. Rizos (eds.), Taylor & Francis, London.

Rizos, C., 2002b, Introducing the Global Positioning System, in: *Manual of Geospatial Science and Technology*, J. Bossler, J. Jensen, R. McMaster, and C. Rizos (eds.), Taylor & Francis, London.

Rizos, C., 2002c, Where Do We Go from Here? in: *Manual of Geospatial Science and Technology*, J. Bossler, J. Jensen, R. McMaster, and C. Rizos (eds.), Taylor & Francis, London.

Rizos, C., Grejner-Brzezinska, D., Toth, C., Dempster, A., Li, Y., Politi, N., Barnes, J., Sun, J., and Li, L., 2010, Hybrid Positioning: A Prototype System for Navigation in GPS-Challenged Environments, *GPS World*, March, pp. 42–47.

Schwarz, K. P., and Wei, M., 1994, Aided Versus Embedded A Comparison of Two Approaches to GPS/INS Integration, *Proceedings of IEEE Position Location and Navigation Symposium*, Las Vegas, Nevada, pp. 314–321.

SnapTrack, 2002, Location Technologies for GSM, GPRS and WCDMA Networks. Snap Track Whitepaper.

Thales, 2004, Ultra Wide Band Indoor Positioning, http://esamultimedia.esa.int/docs/ NavigationProjects/UWB_tech_Proj_Overview.pdf (accessed July 20, 2012).

Toth, C., and Grejner-Brzezinska, D., 2010, Error Analysis of Airborne Multisensory Systems, ISPRS Technical Commission IV & AutoCarto/ASPRS/CaGIS Specialty Conference (CD ROM), Orlando, Florida, November 15–19.

Toth, C., Grejner-Brzezinska, D., Oh, J., and Markiel, J. N., 2009, Terrain-Based Navigation: A Tool to Improve Navigation and Feature Extraction Performance of Mobile Mapping Systems, *Boletim de Ciências Geodésicas*, vol. 15, no. 5, pp. 807–823.

Trimble Corporation, 2012, Trimble R7: Trimble R-Track Technology for GPS Modernization, http://www.trimble.com/trimbler7.shtml (accessed July 23, 2012).

Turner, D., 2010, Update on the US GNSS International Cooperation Activities, 50th Meeting of the Civil GPS Service Interface Committee, Institute of Navigation GNSS 2010 Conference, Portland, Oregon, September 20–21.

Ubisense, 2012, Real Time Location Solutions, http://www.ubisense.net/en/rtls-solutions/ (accessed July 20, 2012).

U.S. Coast Guard (USCG) Navigation Center, 2012, GPS General Information, http://www.navcen.uscg.gov/?Do=constellationStatus (accessed July 23, 2012).

U.S. Wireless Corp., 2002, Location Pattern Matching and The RadioCamera™ Network, http://www.uswcorp.com/USWCMainPages/our.htm (accessed July 22, 2012).

Wang J., Tsujii, T., Rizos, C., Dai, L., and Moore, M., 2001, GPS and Pseudo-Satellites Integration for Precise Positioning, *Geomatics Research Australasia*, vol. 74, pp. 103–117.

Winter, S., and Kealy, A., 2011, An Alternative View of Positioning Observations from Low Cost Sensors, *Computers, Environment, and Urban Systems*, vol. 36, no. 2, pp. 109–117.

Wireless KnowHow, 2002, GSM, http://www.m-indya.com/mwap/gsm/whatis.htm (accessed July 20, 2012).

You, S., Neumann, U., and Azuma, R. T., 1999a, Hybrid Inertial and Vision Tracking for Augmented Reality Registration, *Proceedings of IEEE Virtual Reality*, pp. 260–267.

You, S., Neumann, U., and Azuma, R. T., 1999b, Orientation Tracking for Outdoor Augmented Reality Registration, *IEEE Computer Graphics and Applications*, vol. 19, pp. 36–42.

2 WiFi Location Fingerprinting

Prashant Krishnamurthy

CONTENTS

ABSTRACT

Positioning in indoor areas poses several challenges that cannot be easily overcome using GPS. Although the near-ubiquitous availability of WiFi in most indoor and campus areas of interest makes it an inexpensive choice (in terms of hardware) for positioning, WiFi was not designed with positioning as the application. Hence, traditional time or direction-of-arrival techniques are not suitable for positioning with WiFi. Alternatively, over the last decade, location fingerprinting has been widely suggested and investigated for positioning with WiFi indoors. This chapter provides an overview of WiFi location fingerprinting, the challenges therein, and emerging research and practical directions in the area.

2.1   INTRODUCTION

Positioning in indoor areas has always been problematic since commonly used positioning infrastructures for outdoors, such as the Global Positioning System (GPS), do not work very well in buildings. The reasons why GPS does not work very well in indoor areas include significant attenuation of the signal as it propagates into buildings and multipath effects. Accuracy, which is further impacted by the signal propagation effects, is an additional problem. Although GPS may indicate that a mobile device is *at* a building, the limited accuracy and available map information cannot pinpoint the exact location of the device; whether it is in the building or outside the building, and if inside the building, where exactly it is located.

It may be possible to build an entirely new infrastructure for positioning in indoors. This will however incur substantial cost in a variety of ways. The cost would include spectrum for positioning, embedding hardware capabilities in mobile devices to sense signals for positioning, and installing anchor devices similar to base stations at known locations to transmit signals that can be used for positioning. This substantial cost of new infrastructure has led to the investigation of commonly employed wireless technologies for positioning purposes in indoors.

Radio frequency-based approaches for positioning in indoors have primarily considered the use of wireless local area networks (WLANs) based on the IEEE 802.11 standard (also called WiFi) as a suitable infrastructure for positioning. Most mobile communication devices of today are WiFi enabled including iPods and tablets. In fact, WiFi is widely used for outdoor positioning and navigation with smartphones through databases that are maintained by companies such as Google, Apple, and Microsoft, or positioning service providers like Skyhook. Other technologies such as Bluetooth, radio-frequency identifiers (RFIDs), and cellular telephone signals have been considered as possibilities for positioning in indoors but are not as popular since they are not as widely available in as many devices as WiFi. Also, cell phone signals do not propagate well in all indoor areas. Further, some of these approaches, such as RFIDs, incur additional expense for hardware and for installation. Techniques based on ultrasound for positioning have also been considered in experimental work, but again have not been popular due to the lack of commercial devices that can exploit ultrasound.

The widespread use of WiFi in homes, hotels, coffee shops, airports, malls, and other large and small buildings makes WiFi an attractive technology for positioning.

Typically, a WiFi system consists of fixed access points (APs) that are installed in convenient locations in indoors and campus. The locations of APs are usually known to the system or network administrator. Mobile devices, such as laptops, cell phones, and cameras that are WiFi enabled, communicate with other devices or the Internet through these APs. Thus, this makes WiFi suitable for positioning in addition to communication for mobile devices. However, WiFi signals are not designed for positioning as the main application. Further, the harsh radio propagation environment in indoors makes it difficult and challenging to employ well-known algorithms based on time or time-difference of arrival (TOA/TDOA) to determine distances to known APs. Using direction of signals for positioning is also difficult for the same reasons in indoors. Moreover, installing directional antennas in every WiFi network is also costly. Consequently, the approach that has been considered in recent years and has been investigated at length is *location fingerprinting*.

This chapter provides a tutorial on indoor and campus positioning using location fingerprinting with WiFi access points and mobile devices. The chapter includes the basic idea of location fingerprinting, WiFi and its use for positioning, location fingerprinting algorithms that have been proposed for use with WiFi, challenges and performance issues, and miscellaneous issues in WiFi location fingerprinting.

2.2 BASICS OF LOCATION FINGERPRINTING

Location fingerprinting refers to the idea of associating locations in an environment with some sort of *fingerprint* that is unique to that location. The fingerprint could be single or multidimensional, and it is typically a *feature* of a signal, or information that is transmitted by a mobile device at the location or received by a mobile device at a location. In the former case, fixed receivers that are part of the positioning system sense the transmitted signal or information and estimate the mobile's position based on the detected features. This is often called *remote positioning* or *network* positioning. In the latter case, the mobile device senses signals that are transmitted by fixed transmitters that are part of the positioning system and estimates its own position based on the detected features. This is often called *self positioning*. The mobile device may also communicate the detected features to entities such as servers in the network, which would then be responsible for estimating the location of the mobile device. This is often called *hybrid positioning*. In all cases, it is necessary for some entity to *match* the sensed features of the signal with features stored in a database. In essence, the process can be viewed as a pattern recognition problem.

2.2.1 WHAT COMPRISES A LOCATION FINGERPRINT?

Fingerprinting can be of several types. Any unique characteristic that differentiates locations from one another can be used as a location fingerprint. The multipath structure of the transmitted signal at a location, the visible access points or base stations at a location, the received signal strength (RSS) from the visible access points or base stations at a location, and the round-trip time or delays of the signal could all form part of the fingerprint or the entire fingerprint itself. Here we describe two signal features that have been suggested (Bahl and Padmanabhan, 2000; Pahlavan and Krishnamurthy 2002) or employed for location fingerprinting: the multipath structure and the RSS.

2.2.1.1 Multipath Structure

Propagation of radio signals that are at carrier frequencies larger than 500 MHz or so can be approximated using rays (as in optics) (Pahlavan and Krishnamurthy 2002). When such a radio signal is transmitted, the "rays" can be reflected off fairly smooth surfaces (smooth compared to the wavelength) such as walls of buildings or the ground, transmitted through objects, diffracted by sharp edges or roofs of buildings, and scattered by smaller objects such as foliage on trees. As a result, there are several rays that arrive at the receiver through multiple paths. Each ray that arrives has a different delay and power. The delay depends on the distance traveled by the ray and the power depends on the distance and the propagation phenomenon (e.g., the ray may have been reflected off multiple objects or diffracted off edges). Each ray that arrives at the receiver is called a multipath component and the multipath structure of the channel corresponds to the set of power and delays in the rays that are received. This multipath structure is also called the *power-delay profile*. An example of the power delay profile is shown in Figure 2.1. Here, six significant multipath components have been received with powers and delays corresponding to the sets $\{\beta_1, \beta_2, \beta_3, \beta_4, \beta_5, \beta_6\}$ and $\{\tau_1, \tau_2, \tau_3, \tau_4, \tau_5, \tau_6\}$, respectively.

Assuming that a sufficiently wideband signal has been transmitted (e.g., using direct sequence spread spectrum or ultra-wideband), it is possible to *resolve* the multipath components at the receiver. Depending on the environment, the resulting multipath structure may be unique at various locations in the area and may be used as the fingerprint associated with specific locations. This has been proposed, for example, by Ahonen and Eskelinen (2003) for positioning cell phones in 3G UMTS networks. This work reports that the accuracy of the fingerprinting scheme that employs the multipath structure is within 25 m in 67% of the estimates and within 188 m in 95%

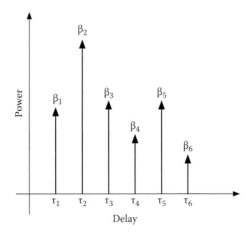

FIGURE 2.1   Power delay profile.

of the estimates. Such a performance is suitable for the E-911 requirements that the Federal Communications Commission has imposed for cell phones.

2.2.1.2   Received Signal Strength

The RSS or received power in a signal also depends on the location. It is easier to compute the received signal strength, as it is a requirement for most wireless communications for operational purposes. Most wireless systems need the RSS information to assess the quality of the link, to make handoffs, to adjust the transmission rates, and other operational reasons. The RSS is not influenced by the bandwidth of the signal. It is not necessary to have a wide bandwidth as in the case of the multipath structure to determine the RSS. Thus, the RSS is an attractive signal feature and has been widely considered for positioning purposes.

If there is exactly one transmitter, the average RSS from that particular transmitter in decibels falls linearly with the log of the distance d from the transmitter. That is, in the simplest case, it is possible to express the RSS as follows:

$$RSS = P_t - K - 10\alpha \log_{10} d \qquad (2.1)$$

The factor affecting the slope, α, of this linear drop is called the path-loss exponent. The transmit power is P_t and K is a constant that depends on the frequency and environment. The RSS can be used to extract the distance of the mobile device from the access point or base station in the network. This begs the question as to whether the extracted distance can be used for trilateration or triangulation of the mobile device. Unfortunately, there are significant variations around this average due to environmental effects (this is called shadow fading). Consequently, the errors in positioning using the extracted distance are likely to be very high, precluding this as a good solution for indoor positioning.

If, however, the mobile device can sense signals from multiple transmitters or multiple fixed receivers can sense the signal from a mobile device, it may be possible to employ the *vector* of RSS values from the multiple transmitters or at multiple receivers as the fingerprint associated with a location. This is the typical situation with WiFi location fingerprinting as described next. Most WiFi network interface cards (NICs) are able to measure the RSS from multiple APs, albeit one at a time. Today, in most indoor, urban, and campus areas (e.g., universities and large organizations), it is possible for a WiFi-enabled mobile device to detect multiple APs at most locations. Thus, employing the RSS from these APs as the unique fingerprint of a location makes sense and this will be the focus of the rest of this chapter.

We note here that the RSS is itself an average value computed over a period of time, so a single sample is not unreasonable. In WiFi networks, APs are usually configured to transmit a beacon packet, which contains information about the network, the service set ID (which is a name for the wireless network), transmission rates supported, and other types of system information. The beacon packet is one of many management packets used in WiFi. The RSS is typically measured using the beacon packet, which is transmitted by APs every 100 ms or so. The beacon is not encrypted, so that even closed networks can be often used for positioning. The transmission of the beacon packet is not completely periodic, although it is close. The reason why it is not periodic is that transmissions in WiFi have to be deferred if the medium of transmission is sensed to be busy. If the medium is busy, the beacon will be transmitted as soon as it becomes free and repeated at the next time at which it has to be transmitted, without waiting for the full 100 ms period. Further, if the APs are transmitting over different channels, a scenario that is likely, to avoid interference the mobile device has to spend some time scanning for these channels before measuring the RSS from the APs. The WiFi standard (also called IEEE 802.11) specifies 11 channels in the 2.4 GHz bands and more in the 5 GHz bands. Usually, only 3 non-overlapping channels are used in the 2.4 GHz bands, although it is not uncommon to find others being used in a geographical area. The reader is referred to (Perahia and Stacey 2008) for details about WiFi and the IEEE 802.11 standards.

2.2.2 ESTIMATING THE POSITION FROM MEASUREMENTS AND THE FINGERPRINT DATABASE

The determination of the position of the mobile device using location fingerprinting typically has two phases. In the *offline phase*, a laborious survey of the area to be covered by the positioning system is performed to collect the fingerprints at various positions to populate a database. Sometimes this is also referred to as the training set. In the *online phase*, the unknown position of a mobile device is estimated. We describe the two phases in more detail next. We also note here that the coordinates of the position of a mobile device are usually a local coordinate associated with the indoor rather than a geographic coordinate such as a latitude or longitude.

2.2.2.1 Offline Phase

The association of a fingerprint with the position is typically performed offline. In the most common scenario, the geographical area that needs to be covered is filled with a rectangular grid of points as shown in Figure 2.2. In this figure, an indoor office area has a grid of four rows and eight columns for a total of 32 *grid points*. There are two APs that cover the entire office area. These APs were probably deployed for communication purposes but are also being exploited for positioning. At each of the grid points, the average RSS from each AP is determined by sampling the RSS for a period of time (ranging from 5 to 15 minutes, with samples taken every second or so). The RSS may be collected with different orientations or directions of the mobile device. The fingerprint, in this example, at a given grid point is a two-dimensional (2D) vector $\rho = [\rho_1 \ \rho_2]$, where ρ_i is the average RSS from AP i. As we will see later, it is also possible to record the *distribution* of the RSS samples as the fingerprint or record other statistics of the RSS samples, such as the sample standard deviation. For simplicity, we will assume, unless otherwise mentioned, that the fingerprint is simply the average of the RSS samples.

The 2D fingerprint is collected at each of the grid points in the area shown in Figure 2.2 and a database of grid point coordinates and the corresponding fingerprint is created. This process is sometimes called the *calibration phase*. The fingerprint database is sometimes called the *radiomap*. In this simple example, a partial

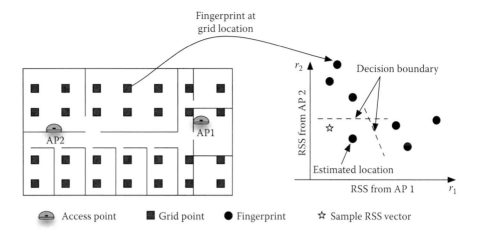

FIGURE 2.2 **(See color insert.)** WiFi RSS-based location fingerprinting and Euclidean distance for estimating location.

TABLE 2.1
Example Fingerprint Database

Coordinates of Grid Point	ρ_1 in dBm	ρ_2 in dBm
(0,0)	−65	−50
(0,1)	−64	−45
(0,2)	−60	−51

database may look like what is shown in Table 2.1. The right side of Figure 2.2 shows the fingerprints in a 2D vector space (called signal space hereafter). In the general case, N APs are visible at each grid point and the fingerprint ρ is an N-dimensional vector, which is harder to pictorially represent.

We also note that while the RSS measurements were taken on a *physical rectangular grid* of points, in the signal space, the location fingerprints are no longer arranged in any regular manner. As we will see later due to the vagaries of radio propagation, the rectangular grid of locations gets converted into an irregular pattern of signal vectors in signal space. Some of the signal vectors, while far apart in physical space, end up being closer in signal space increasing the chance for errors. Consequently, a reasonable part of the labor for fingerprint collection turns out to be not very useful and in some cases harmful for positioning.

2.2.2.2 Online Phase

In the online phase, a mobile device is in the geographical area under consideration but at an unknown location. It is even unlikely that the mobile device is on a grid point. Assuming hybrid positioning, the mobile device measures the RSS from the visible APs (in the example in Figure 2.2, only two APs are visible in the office area). One or more RSS samples may be collected but typically far fewer than the number collected to create fingerprints. In here we assume that a single sample is collected. When the RSS from all of the visible APs is obtained, the *vector* of RSS measurements is transmitted to the network. Let the vector of RSS measurements in the example related to Figure 2.2 be $\mathbf{r} = [r_1 \ r_2]$. The problem of determining the position of the mobile device now becomes one of finding the best match for $\mathbf{r}$ from the entries for ρ in the database of fingerprints. Once the best match is found, the mobile device position is estimated at the corresponding coordinates. For example, if $\mathbf{r} = [-65 \ -49]$, the best match is the first entry in Table 2.1 and the mobile device is located at (0,1). In the general case, the vector $\mathbf{r}$ is also N-dimensional.

   This discussion makes many simplifying assumptions about the coordinates, the entries for the fingerprints, the measurement of the RSS values, and matching the vector $\mathbf{r}$ to the vector ρ. In the following sections, we consider some issues that arise in more detail. We first start with the algorithms for matching $\mathbf{r}$ to ρ.

2.3 POSITIONING ALGORITHMS USING LOCATION FINGERPRINTING

There are two types of algorithms typically employed for determining the location of a mobile device using the fingerprint database as described in Section 2.2. They are *deterministic algorithms* that compare measured signal features (e.g., the vector $\mathbf{r}$) with precomputed statistics that are stored in the fingerprint database (e.g., the vector ρ) and *probabilistic algorithms* that employ a likelihood of a signal feature belonging to a fingerprint distribution (stored in the fingerprint database). We consider examples of these algorithms next. The objective here is not to provide a comprehensive survey but to sample some of the available algorithms.

2.3.1 DETERMINISTIC POSITIONING ALGORITHMS

The earliest work on using location fingerprints with WiFi for positioning and tracking in indoor areas was done by Microsoft (Bahl and Padmanabhan 2000). The authors of this work employed the Euclidean distance between the measured RSS vector $\mathbf{r}$ and the fingerprint vector ρ to determine the location of the mobile device. Let us suppose that the location fingerprint is N-dimensional. That is, there are N APs visible at each grid point in the area under consideration. Let us also suppose that there are M grid points in the given area and that the corresponding M fingerprints populate the database. The Euclidean distance between the measured RSS vector $\mathbf{r}$ and a given fingerprint vector ρ is defined as

$$D = \sqrt{\sum_{i=1}^{N} |r_i - \rho_i|^2} \qquad (2.2)$$

   With this, the simplest algorithm for estimating the position of the mobile device can be stated as "of the M location fingerprints in the database, find the location fingerprint that has the *smallest* Euclidean distance in signal space to the measured RSS vector, and associate its coordinates with the position of the mobile device." This approach that makes use of the Euclidean distance is also called finding the nearest neighbor in signal space, since the objective here is to determine the closest fingerprint in signal space to the measured RSS vector. On the right side of Figure 2.2, the general idea is illustrated, where the star represents the measured RSS vector and the brown circles represent the location fingerprints in signal space. Decision boundaries can be drawn separating the location fingerprints into Voronoi regions that contain areas or volumes in the signal space that are closest to a location fingerprint. The coordinates of the location fingerprint associated with the Voronoi region where the measured RSS vector lies become the estimated position.

As we will see in Section 2.4, not all fingerprints are reliable and a more complex database (compared to that shown in Table 2.1) may contain entries for each location fingerprint corresponding to the standard deviations of the RSS elements in ρ or some weight factor for each element in ρ. In such cases, it is possible to have a *weighted* Euclidean distance as the metric for determining the position of the mobile device. There may be a single weight for each fingerprint or separate weights for each element in the fingerprint vector. Other distance metrics such as the Manhattan distance and the Mahalanobis distance have also been employed in the literature for position estimation.

2.3.2 Probabilistic Positioning Algorithms

One of the earliest probabilistic positioning algorithms with WiFi fingerprinting was proposed by Youssef et al. (2003). The idea here is that using simply a statistic of the RSS samples (such as the sample mean) may cause errors because of the actual distribution of the RSS values (see more on this in Section 2.4). Hence, a *joint distribution* of the RSS samples from multiple access points collected at each grid point is used as the fingerprint. Clearly, this is not a trivial problem since the correlations between the RSS values from multiple access points are not obvious. The authors assume independence (which is reasonable) and simply use the product of the marginal distributions of the RSS at given grid points as an approximation of the joint distribution of the RSS vector. Let us suppose that the measured RSS vector is $\mathbf{r} = [r_1\, r_2\, r_3 \ldots r_N]$. Then the position is estimated as the grid point, which has the maximum probability of resulting in the observed RSS vector $\mathbf{r}$. Bayes rule is used to estimate the position of a mobile device given $\mathbf{r}$. That is, first the probability that a certain grid point would have produced $\mathbf{r}$ is calculated as follows:

$$P\{\text{Grid point}|\mathbf{r}\} = \frac{P\{\mathbf{r}|\text{Grid Point}\} \times P\{\text{Grid Point}\}}{P\{\mathbf{r}\}} \qquad (2.3)$$

Then the grid point that has the *maximum* such probability is used as the estimate of the position of the mobile device.

2.3.3   CLUSTERING OF FINGERPRINTS

One of the problems not considered so far is that not all grid points may see the same set of APs all the time. The work by Youssef et al. (2003) described earlier and the work by Swangmuang and Krishnamurthy (2008b) consider *clustering* in different ways. The work by Youssef et al. (2003) groups grid points based on the commonality of APs. The grid points that share a common set of k APs are referred to as a cluster. The cluster is determined based on the probability of seeing the set of k APs at the given grid point. Hence this approach has been called the *joint clustering* or JC technique by the authors. Swangmuang and Krishnamurthy cluster fingerprints based on how *close* they are in signal space to reduce the complexity of searching through the fingerprints. Here, it is assumed that all of the locations see the same set of APs.

2.3.4   OTHERS

Other pattern-matching algorithms have been suggested for use with WiFi location fingerprinting. These include Bayesian inference, statistical learning theory, support vector machines, and neural networks (see, for example, Battiti et al. 2002).

2.4   PERFORMANCE OF LOCATION FINGERPRINTING

In this section, we consider the positioning performance (accuracy and precision) with location fingerprinting and analyze the labor associated with location finger-printing and how it may perhaps be reduced. We first look at the reasons why there are errors in the estimation of the position using location fingerprinting. We then describe the results of the error analyses and performance of location fingerprinting reported in the research literature.

2.4.1   CAUSES FOR ERRORS

In the ideal case, the measured RSS vector **r** should very closely match the finger-print in the database that is associated with the closest *physical position* of the mobile device. Unfortunately, this is not the case and there are several reasons for this that may cause significant errors in positioning with WiFi fingerprinting.

2.4.1.1   Radio Propagation Vagaries

Radio propagation is quite harsh, and even more so in indoor areas and urban canyons. Figure 2.3 shows a time series of RSS samples measured from an AP when a user is sitting and working with the mobile device (in this case a laptop) over a period of several minutes. Clearly, the RSS samples vary with time and there are occasions where there are significant changes in the RSS samples. In fact, over a period of 5 minutes, it is not impossible for the RSS values to change by over 20 dB. The *orientation* of the human being in relation to the mobile device has a significant effect on what the device measures. Network interface cards that are made by dif-ferent vendors compute the RSS in different ways leading to inconsistencies. It is

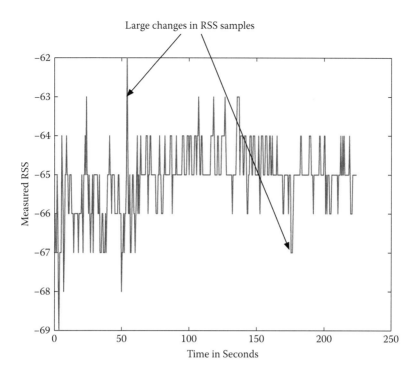

FIGURE 2.3 **(See color insert.)** Time series of RSS samples (in dBm) measured on a laptop.

possible that the RSS distribution is not even stationary in time. Consequently, it is possible that when the vector **r** is measured, it may actually match a fingerprint corresponding to a position that is very different from the actual position. It is beyond the scope of this chapter to describe all of the characteristics of the RSS samples and how they may impact positioning. The reader is referred to Kaemarungsi and Krishnamurthy (2004b, 2011) for an exhaustive analysis of the WiFi RSS characteristics for indoor positioning.

2.4.1.2 Censored Data

As mentioned in Section 2.3.3, not all APs are seen at all positions all of the time. For example, when the fingerprint is collected at a grid point, it may be the case that only three APs are visible. So the fingerprint corresponds to those three APs. However, in the online phase, when a user is trying to determine her position, the mobile device may actually see four or even five APs. Increasing the fingerprint dimensionality is actually beneficial in the sense that it makes the grid points separated in signal space from one another. However, the unreliable nature of some signals makes it challenging to implement position determination algorithms correctly. The fingerprint database can of course include information about the APs (such as their 48 bit MAC addresses), but handling of censored data is not easy. The work in Youssef et al. (2003) tries to always use a set of k APs and ignores any others that are seen. The k APs that are used are selected based on how reliable the signal is, from each of the APs. If the wrong set of k APs is used, the position that is estimated may be very far away from the actual position.

2.4.2 ERROR ANALYSES

First we summarize a few of the error performance results reported in the literature. Most of these results have been determined through simulations or experiments. In the seminal work by Bahl and Padmanabhan (2000), the authors determined the median error to vary between 3 m and 6 m depending on the number of grid points employed with three APs visible (i.e., the dimension of ρ is 3) and using the deterministic nearest-neighbor scheme. With a similar deterministic algorithm, the cumulative distribution function of errors in Swangmuang and Krishnamurthy (2008a) indicated that the probability that the positioning error is smaller than 4 m is 0.9 in an office area with 25 grid points and three visible APs. In Youssef et al. (2003), the authors employ the joint clustering probabilistic position determination algorithm described in Section 3.2. The experiments were conducted in an officelike building that measured roughly 225 feet by 85 feet. A total of 110 grid points were used to create the fingerprint database. Most of the fingerprints were collected along long corridors. The cumulative distribution function of positioning errors indicates that the errors are smaller than 7 feet with a probability of 0.9.

There are few analytical considerations of the positioning error with location fingerprinting for problems discussed next. The earliest work on analyzing the error performance of location fingerprinting was that in Kaemarungsi and Krishnamurthy (2004a). Here, the authors assumed that the RSS samples had a Gaussian or normal distribution, the sample mean was the actual mean, and that the variance of the normal distribution of the RSS from all of the APs was identical. These assumptions make the problem of evaluating the error performance somewhat tractable. In reality, the distribution of the RSS is not Gaussian (Kaemarungsi and Krishnamurthy 2004b, 2011). Figure 2.4 shows two histograms of RSS, one close to an AP and the other far away from an AP. The first histogram shows a *left skew*, while the second is more symmetric and could perhaps be modeled as a normal distribution. The variances of the RSS samples are not constant across APs. The work in (Kaemarungsi and Krishnamurthy 2011) shows that the variance is smaller at larger physical distances from an AP. Thus, the simplifying assumptions are not necessarily valid in reality. The analyses in (Kaemarungsi and Krishnamurthy 2004a), extended further in (Swangmuang and Krishnamurthy 2008a), match simulations/experiments in terms of the cumulative distribution of errors, especially for positioning errors larger than a few meters. Thus this analysis is potentially useful even under the simplifying assumptions. The analysis was used to evaluate the marginal benefits of increasing the dimensionality of the location fingerprint and the impact of the path-loss exponent in simple scenarios in (Kaemarungsi and Krishnamurthy 2004a).

Assuming that the RSS has a normal distribution, the error analysis becomes similar to the error analysis that has been traditionally performed for computing bit error rates in digital communications. An increasing variance of the RSS increases the error probabilities. However, there is a twist here. The RSS fingerprints in signal

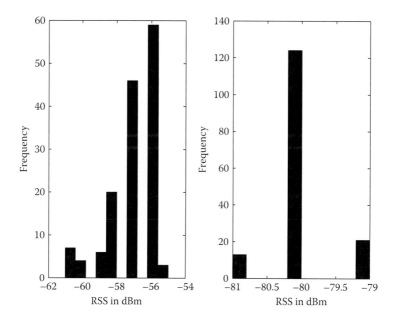

FIGURE 2.4 **(See color insert.)** Histograms of RSS samples.

space are not arranged in any regular fashion and the variance of the RSS distributions can be quite different along different dimensions making it extremely difficult to derive analytically closed form solutions for the positioning errors. If only two fingerprints are considered (as shown in Figure 2.5), each with only two dimensions, it is possible to derive expressions for the probability of mistaking one fingerprint as the closest to a measured RSS vector **r** compared to the correct fingerprint using the decision line shown in Figure 2.5. As more fingerprints are added and they create highly irregular Voronoi regions, the problem becomes intractable without approximations. In Swangmuang and Krishnamurthy (2008a), proximity graphs are employed to approximate the analysis with reasonable results.

2.4.3 LABOR ASSOCIATED WITH LOCATION FINGERPRINTING

Although it is certainly not as expensive compared to an entirely new positioning system, WiFi positioning using location fingerprints has its own costs as well. Most of the research literature assumes that location fingerprints are collected offline on a virtual grid of locations in the desired area as described in Section 4.2. For instance, in a 100 m × 100 m area, location fingerprints could be collected on a grid with a spacing of 2 m for a total of approximately 50 × 50 = 2500 fingerprints. Each fingerprint consists of RSS measurements from the multiple access points visible at a grid point, over a certain time frame ranging from 5 minutes to 15 minutes, and sometimes with several different orientations and several measurement devices (laptops or phones). This process is extremely labor intensive and may have to be repeated periodically to account for changes in the environment. In the aforementioned example, with a

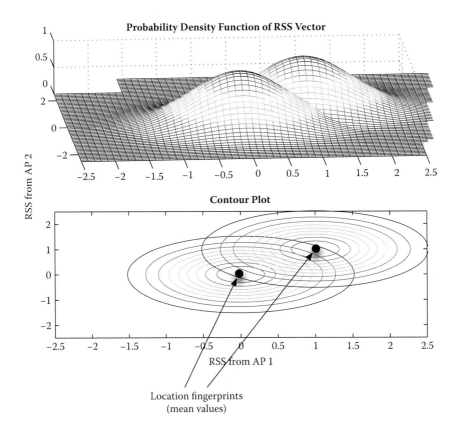

FIGURE 2.5 **(See color insert.)** Gaussian RSS distributions in 2D with two location fingerprints.

single orientation and device, collecting the location fingerprints takes $2500 \times 5 = 12500$ minutes or almost 9 days of RSS measurements. Measurements can be done in parallel speeding up the process, but this is still labor intensive. In what follows, we examine a few attempts and approaches for reducing the labor associated with the offline phase.

2.4.3.1 Reducing the Number of Fingerprints Collected

It would be a good investment of labor and time if all of the RSS location fingerprints were useful and improved the system performance. That is often not the case. As demonstrated by the work of Youssef et al. (2003) and Swangmuang and Krishnamurthy (2008b), this is not necessarily the case. A "good" fingerprint should not cause errors in positioning, at least more often than not. From an RSS standpoint, the variance of the RSS should be small and there should be no other location fingerprint that is "close" enough in signal space. However, the Euclidean distance between some pairs or sets of location fingerprints can be very small compared to the variation of the RSSs at the corresponding locations. The labor invested in collecting such fingerprints with small "signal distance" may not improve performance

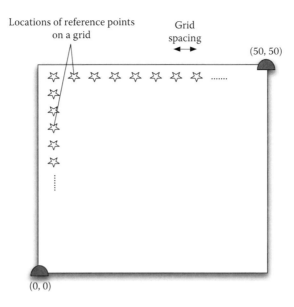

FIGURE 2.6 Simple consideration of labor versus useful location fingerprints.

and also cause extra computational effort while estimating the location of a MS. Including such fingerprints in the radio map may even reduce location precision.

As a simple example, consider the following scenario of a square 50 m × 50 m area where location fingerprints are collected at reference points as shown in Figure 2.6. We use the path-loss model "D" for IEEE 802.11n with a transmit power of 10 dBm with two APs located at (0, 0, 10) and (50,50,10). The path-loss model is given by (Perahia and Stacey 2008)

$$L_p = 20\log_{10}(f) - 127.5 + 35\log_{10}\left(\frac{d}{d_{bk}}\right), d > d_{bk} \tag{2.4}$$

Here, f is the frequency and d_{bk}, the breakpoint distance, is 10 m. The grid points are placed regularly in the square area as shown in Figure 2.6. The location fingerprints that result in this scenario for various grid spacing values are shown in Figure 2.7. Clearly, collecting fingerprints every meter (or even 2 m) may not make much sense since the fingerprints are really close in such cases. When a mobile device reports the measured RSS vector **r**, the system may make errors in associating it with the closest fingerprint, especially if the variance of the RSS is high. Since this is the case, it may be reasonable to simply collect fingerprints on a grid with a spacing of 4 m or even 8 m since the accuracy obtained will be unlikely to change.

In Swangmuang and Krishnamurthy (2008a), the authors used the probability (assuming that the RSS had close to a normal distribution) that a location fingerprint is likely to be identified as closest to a measured RSS vector to eliminate "bad" fingerprints in the radio map. The work showed that up to 20% of the location fingerprints can be eliminated from the radio map without any significant impact on the

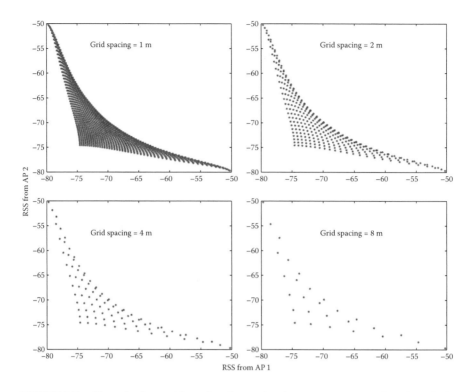

FIGURE 2.7   **(See color insert.)** Location fingerprints in a square area as a function of grid spacing.

accuracy or precision. In this work, the "bad fingerprints" were detected after their laborious collection. However, the authors also presented qualitative guidelines for reducing the labor of deployment.

2.4.3.2   Subarea Localization

Other approaches to reducing the offline calibration efforts have been reported that include online measurements between access points rather than laborious finger-printing (e.g., Gwon and Jain 2004). Work done by Aksu and Krishnamurthy (2010) considers using only information about *which* APs can "hear" a mobile device rather than using fine-grained location fingerprinting. That is, instead of recording the average RSS values or the distribution of the RSS samples in the location finger-print ρ, the positioning system only records a 1 or 0 indicating whether an AP can be "heard" for various APs in the area. For example, let us suppose that there are four APs located at the corners of a square geographical area. If the coverage of these access points can be tuned, it is possible to create 13 unique subareas that are covered by *unique* subsets of these APs. Based on this information, the mobile device can be located in a subarea that is covered by exactly a given set of APs. This subarea can be small or large depending on the placement and coverage of APs impacting the accuracy and precision of positioning. Note that there are 16 unique fingerprints that are possible in this case, namely, [0 0 0 0] through [1 1 1 1], but it is not possible for certain fingerprints to exist, making the maximum number of unique fingerprints as 13. This number could be smaller; for instance, if all four APs cover the square area, the only fingerprint is [1 1 1 1], which indicates that the mobile device is *somewhere* in the square area.

2.4.3.3   Organic Creation of Fingerprint Database

Recently, the use of user-based RSS collection has been suggested to reduce the labor of collecting location fingerprints (e.g., Park et al. 2010). In this work, users are prompted to supply the fingerprints, but the challenges are in the issues that arise for incrementally building the radio map. Park et al. (2010) used clustering and Voronoi regions to organically develop the radio map. Park et al. focused on spatially separating indoor areas (e.g., rooms) into Voronoi regions to facilitate the organic creation of the fingerprint database.

2.5   MISCELLANEOUS ISSUES

In this section we briefly consider some miscellaneous issues related to WiFi fingerprinting-based positioning. In particular, we look at reducing energy consumption, throughput problems, latency, and security.

2.5.1   USE OF MULTIPLE TECHNOLOGIES

With many mobile devices, especially smartphones, being equipped with more than one wireless technology (e.g., some of these devices have Bluetooth, RFIDs, and near-field communication capability), the use of multiple technologies for location fingerprinting becomes viable and possible. There are papers that have considered the use of GSM and WiFi in indoors, but no comprehensive work exists on fusing location fingerprinting from various technologies together.

2.5.2   REDUCING ENERGY CONSUMPTION

Reducing energy consumption in mobile devices through efficient communications protocols has been an area of research for several years now (Pahlavan and Krishnamurthy 2002). Only recently, there have been efforts to make positioning functions energy efficient as well, primarily with GPS that can consume a significant amount of power in smartphones. Paek et al. (2010) present rate adaptive positioning with GPS for smartphones. The idea here is that by reducing the duty cycle of GPS, it is possible to save energy at the cost of reducing accuracy. However, GPS itself does not provide uniform accuracy everywhere (e.g., reduced accuracy in urban areas) and may be unavailable in indoors. By exploiting this knowledge, GPS is switched off or operated at a low duty cycle to increase the battery life in a phone by a factor of 3.8. The use of lower power positioning modes (e.g., using WiFi instead of GPS) is suggested by Lin et al. (2010) to reduce the energy consumption, again perhaps at the cost of reduced accuracy.

2.5.3 THROUGHPUT

When WiFi is used for positioning, a mobile device can spend a lot of time scanning for WiFi signals, and this interrupts data flow and thus the throughput. By reducing the scanning intervals when a user is not moving, King and Kjaergaard (2008) show that throughput can increase by up to 122% and the percentage of dropped packets by 73%. The reduction of scanning in King and Kjaergaard's study is related to user movement. When motion is detected, the scanning rate is increased. Otherwise, the device does not scan, as the assumption is that the user is not really changing his position.

2.5.4 LATENCY

To the best of our knowledge, no work exists on relating the capacity or latency of a WiFi positioning system to provide positioning information if there are large numbers of requests. Metrics such as time-to-first-fix do not exist for WiFi fingerprinting. However, when the number of requests is small, coarse-grained position estimates are much faster to obtain than with GPS that typically requires several seconds and sometimes minutes to obtain a fix.

2.5.5 SECURITY

With a large number of devices now being WiFi enabled, monitoring of WiFi signals is very easy. If a network of such devices was created for malicious purposes, it could be possible to track a large number of devices based on their MAC addresses and RSS values. This would result in a serious attack on user privacy. The work by Husted and Myers (2010) considers the creation of such *malnets* for tracking mobile devices.

2.6 SUMMARY

This chapter provides a tutorial overview of WiFi fingerprinting and describes the work in some of the important research papers in this area. The basics of fingerprinting for position determination, algorithms for positioning using fingerprinting, positioning error performance, and miscellaneous issues were considered. With the increasing emphasis on indoor navigation and other indoor activities, positioning in indoors can only be expected to grow, and WiFi location fingerprinting will likely be a cornerstone for these efforts.

WiFi is nearly ubiquitous in availability in most major indoor and campus areas making it an attractive option for positioning with no additional hardware costs. However, time and direction-of-arrival approaches do not work very well with WiFi signals, making location fingerprinting the primary choice for positioning. However, location fingerprinting is labor intensive and may need constant tuning with changes in the environment. Further, collection of location fingerprints is not an intuitively easy problem due to the vagaries of radio propagation. Measurements, analyses, and simulations have shown that some rules-of-thumb approaches can be taken to reduce the labor-intensive efforts for collecting location fingerprints. When accuracy can be somewhat sacrificed, other approaches such as subarea localization or organic collection of location fingerprints have been suggested to reduce the labor of location fingerprinting.

Although location fingerprinting with only WiFi has been demonstrated to be a viable solution for accurate positioning in indoor areas, the emergence of new technologies and the spread of smartphones with additional sensing capabilities are likely to result in positioning solutions that integrate and fuse sensing and positioning using multiple approaches and technologies. It may be possible in the next few years to use near-field communications capabilities, the accelerometer in smartphones, and perhaps signals from Bluetooth as additional parameters to improve accuracy, while keeping the cost reasonable.

REFERENCES

Ahonen, S., and P. Eskelinen. 2003. Mobile terminal location for UMTS. *IEEE Aerospace and Electronic Systems Magazine*, vol.18, no. 2, pp. 23–27.

Aksu, A., and P. Krishnamurthy. 2010. Sub-area localization: A simple calibration free approach. ACM MSWiM'10.

Bahl, P., and V. N. Padmanabhan. 2000. Radar: An in-building RF based user location and tracking system. *IEEE INFOCOM 2000,* pp. 775–784.

Battiti, R., M. Brunato, and A. Villani. 2002. Statistical learning theory for location fingerprinting in wireless LANs. Technical Report (October 2002). http://rtm.science.unitn.it/~battiti/archive/86.pdf.

Gwon, Y., and R. Jain. 2004. Error characteristics and calibration-free techniques for wireless LAN-based location estimation. Proceedings of the Second International Workshop on Mobility Management & Wireless Access Protocols, October 1, Philadelphia, Pennsylvania.

Husted, N., and S. Myers. 2010. Mobile location tracking in metro areas: Malnets and others. ACM CCS.

Kaemarungsi, K. and P. Krishnamurthy. 2004a. Modeling of indoor positioning systems based on location fingerprinting. IEEE INFOCOM, Hong Kong, China.

Kaemarungsi, K., and P. Krishnamurthy. 2004b. Properties of indoor received signal strength for WLAN location fingerprinting. IEEE/ACM Mobiquitous.

Kaemarungsi, K., and P. Krishnamurthy. 2011. Analysis of WLAN's received signal strength indication for indoor location fingerprinting. *Pervasive and Mobile Computing*, DOI: 10.10.16/j.pmcj.2011.09.003.

King, T., and M. B. Kjaergaard. 2008. Composcan: Adaptive scanning for efficient concurrent communications and positioning with 802.11. ACM Mobisys.

Lin, K., A. Kansal, D. Lymberopolous, and F. Zhao. 2010. Energy-accuracy aware localization for mobile devices. ACM Mobisys.

Paek, J., J. Kim, and R. Govindan. 2010. Energy-efficient rate-adaptive GPS-based positioning for smartphones. ACM Mobisys.

Pahlavan, K., and P. Krishnamurthy. 2002. *Principles of Wireless Networks: A Unified Approach*. Prentice Hall PTR.

Park, J.-G., B. Charrow, D. Curtis, et al. 2010. Growing an organic indoor location system. ACM Mobisys.

Perahia, E., and R. Stacey. 2008. *Next Generation Wireless LANs*. Cambridge University Press.

Swangmuang, N., and P. Krishnamurthy. 2008a. Location fingerprint analyses toward efficient indoor positioning. PerCom'08, IEEE, pp. 100–109.

Swangmuang, N., and P. Krishnamurthy. 2008b. On clustering RSS fingerprints for improving scalability of performance prediction of indoor positioning systems. ACM MELT.

Youssef, M. A., A. Agrawala, and A. U. Shankar. 2003. WLAN location determination via clustering and probability distributions. *Proceedings of IEEE International Conference on Pervasive Computing and Communications (PerCom '03)*, Dallas-Fort Worth, Texas, pp. 23–26.

3 Geocoding Techniques and Technologies for Location-Based Services

Daniel W. Goldberg

CONTENTS

ABSTRACT

Geocoding, the process of converting textual information into a geographic representation, is used to convert postal mailing addresses into geographic coordinates for a number of purposes. This technology is critical to location-based services (LBS) because it is used to (a) generate the geographic layers that LBS search to answer queries; and (b) associate a geographical context with an input query to an LBS. As such, geocoding presents specific challenges and opportunities for LBS research and practice. This chapter explores these issues through an examination of the geocoding process and its relationship to LBS. Opportunities and exemplar cases of the use of geocoding in LBS are presented as guidance for what is currently possible. Similarly, challenges and hurdles in merging these two interrelated technologies are described to warn against common mistakes and shortcomings in both. The chapter concludes with a series of research and implementation tasks that would serve to bring the combination of geocoding and LBS to even higher levels in order to plant the seeds for future work in both domains.

3.1   INTRODUCTION

Geocoding is the process of converting textual locational information into one or more geographic representations (Boscoe 2008). Most commonly this process is used to convert postal mailing addresses into geographic coordinates for mapping, visualization, or analysis purposes. Geocoding plays several critical roles within the context of location-based services (LBS). Among many others, these include (a) the translation of an input query location into a geographic context that can be used as input to an LBS query such as "find golf stores nearby 123 Main Street"; and (b) the generation of the location databases that serve as the foundational layers upon which an LBS can search for answers to queries such as "find all grocery stores nearby my current location."

Because geocoding is critical to both the generation of the input query as well as the reference data layers that are used to compute an answer, this topic presents specific challenges and opportunities for LBS research and practice. This chapter explores these issues through an examination of the geocoding process and its relationship to LBS. Opportunities and exemplar cases of the use of geocoding in LBS are presented as guidance for what is currently possible. Similarly, challenges and hurdles in merging these two interrelated technologies are described to warn against common mistakes and shortcomings in both. The chapter concludes with a series of research and implementation tasks that would serve to bring the combination of geocoding and LBS to even higher levels in order to plant the seeds for future work in both domains.

3.2   THE RELATIONS BETWEEN GEOCODING AND LOCATION-BASED SERVICES (LBS)

3.2.1   CONTEXTUALIZING USER QUERIES

At their core, LBS rely on two primary input data sources to provide services or information that are geographically and contextually relevant to an end user (Dao,

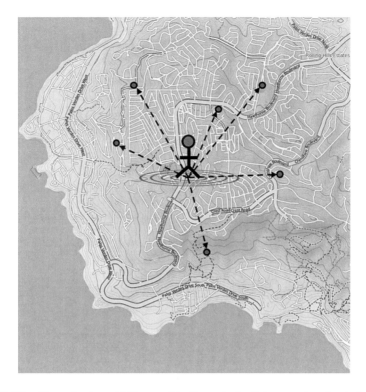

FIGURE 3.1 **(See color insert.)** Geographically contextualizing user location information in an LBS.

Rizos, and Wang 2002). The first is a geographic position of interest that serves as the context within which LBS are to be provided. For an LBS to be useful to an end user, the location of the LBS end user must be spatially referenced and known to the LBS. This ensures that the LBS can deliver information or services that are relevant to the current geographic context of the end user (Jiang and Yao 2006) (Figure 3.1).

This information could be the current geographic location of an individual, group, or other geographic object in the form of a latitude–longitude pair; for example, "In-car navigation, tell me how far the next all night diner is from where we are." As discussed elsewhere in this book, the recent and accelerating proliferation of GPS-enabled devices is helping to provide much of the data necessary to place the end user of an LBS within a geographic context. Devices such as smart phones, iPads and other tablets, and in-car navigation systems now provide LBS applications and developers with the ability to determine a reasonably accurate location for an LBS end user. With such detailed information regarding an end user's location in hand, it becomes possible to expand the utilization of LBS technology, enhance the ability of service providers to deliver contextually relevant services, and drive the development of ever more complex LBS and applications (Rao and Minakakis 2003).

GPS-enabled devices are not the only means by which an end user can express his or her geographic context to a LBS, however. Similar context-defining information could be in many diverse forms and formats beyond latitude and longitude determined through the use of GPS-enabled technologies. For example, an end user may submit a textual or spoken query that describes a location to an LBS to request information or services relevant to that location (Adams, Ashwell, and Baxter 2003). Such queries could include data in the form of full postal addresses: "I'm interested in moving to Texas A&M University. Show me crime statistics around 123 Main Street, College Station, Texas." Alternatively, these locational descriptions may refer to named places and relative locations: "What is the parking availability in Lot 5 across from Kyle Field?" In each of these cases, an LBS must be able to translate these textual forms of user input into one or more digital geographic representations for geographically contextualizing the scope of the services to deliver to the end user of the LBS (Amitay et al. 2004). The process of geocoding performs these tasks.

3.2.2 CONTEXTUALIZING REFERENCE DATA SOURCES

The other side of the coin necessary for an LBS to provide contextually relevant data to an end user is the geographic data layer or layers that contain information about the items of interest to the end user (Rao and Minakakis 2003; Steiniger, Neun, and Edwardes 2006). For example, an LBS that recommends coffee shops based primarily on proximity to a user's current location, as in answering the query "I need caffeine right now, real bad; give me directions to the closest place I can get an espresso," relies on the fact that somewhere there exists a database of coffee shops, gas stations, convenience stores, and other such caffeine purveyors. This database should contain a number of attributes about the location that could help the end user filter possible caffeine refueling options by preference. These might include the type of establishment (i.e., national coffee house chain, mom-and-pop cafe, gas station rest stop), the phone number, the types of coffee that are sold, and the hours of operation.

However, in addition to these nice-to-have attributes that help an end user make an informed decision based on criteria he or she finds important, the database queried by an LBS absolutely must contain one or more spatial footprints in the form of geographic points, lines, or polygons (Steiniger, Neun, and Edwardes 2006). These spatial footprints provide the geographic context that an LBS can exploit to identify relevant results to return to an end user. When these databases of reference data, that is, the information a user of an LBS is interested in, are spatially referenced, it becomes a relatively simple and straightforward task to select the subset that could or should be of interest to the user based on his or her location. Further, the geographic context offered by spatially enabling reference data layers provides the ability to rank potential results based on geographic criteria, such as proximity to current user location as in the caffeine example discussed earlier (Zhou et al. 2005; Mountain and Macfarlane 2007). Without such a spatially enabled database of geographic objects, it would be nearly impossible for an LBS to provide contextually relevant information to an end user because the domain of potential solutions could not be filtered using a geographically relevant context.

Thus, the generation and availability of these reference data layers is a major issue that affects the quality of services delivered by an LBS (D'Roza and Bilchev 2003). For example, if a reference data layer containing the information that a user is interested in is not available for a particular region relevant to the current location of the user, the service is effectively rendered useless from the perspective of that end user for that particular region. Similarly, if the information contained in the reference data sets used by an LBS is incomplete, inaccurate, or of otherwise questionable quality (such as in the case of reference data with little to no metadata or provenance information), the quality of the results provided to the end user may be suspect at best and completely inaccurate at worst. Two of the major factors that influence the quality of the reference data layers utilized by LBS are (a) the data sources used to collect, coordinate, and integrate the underlying reference data sources; and (b) the processing techniques used to massage these data into a form and format useful for an LBS.

With regard to the first, the data in LBS reference layers may come from many different types of sources including administrative lists such as parcel information drawn from local tax assessor databases, business listings such as yellow page directories of businesses, and, increasingly, information gathered from the public at large, a phenomenon recently termed volunteered geographic information (VGI) (Amitay et al. 2004; Mountain and Macfarlane 2007). Each of these sources of data present opportunities and challenges for LBS systems, as described later in this chapter. Issues related to the second stem directly from the processes used to transform the locational attributes associated with the reference data layers into a digital geographic representation. These locational attributes could include a postal address (123 Main Street), a name (Vons Grocery Store), or a relative description of the location associated with the business (northeast corner of Vermont Ave and Exposition Blvd). Again, geocoding performs this task.

3.3 GEOCODING SYSTEMS

Geocoding is the computational process responsible for translating the description of a location into one or more digital geographic representations (Boscoe 2008). This process is used to support the input queries provided to an LBS as well as to generate the reference data layers used within an LBS. Although relatively straightforward in terms of input (locational descriptions) and output (geographic representations), the internal workings of geocoding systems can differ dramatically (Goldberg, Wilson, and Knoblock 2007). This situation may result in highly variable results across geocoding systems. In turn, these quality-related artifacts resulting from the geocoding process can directly impact the quality of the services or information provided by an LBS. The following subsections detail the internal components of modern geocoding systems with a particular focus on how the choices made in a geocoding system may impact the use of geocoded data in LBS. The primary components of a geocoding system are (a) input data; (b) address standardization and normalization algorithms; (c) feature matching algorithms; (d) reference data files; (e) feature interpolation algorithms; and (f) output data (Goldberg 2008) (Figure 3.2).

3.3.1 INPUT DATA

The input data provided to a geocoding system can be in several different formats. These include but are not limited to postal addresses, named places, and relative locational descriptions.

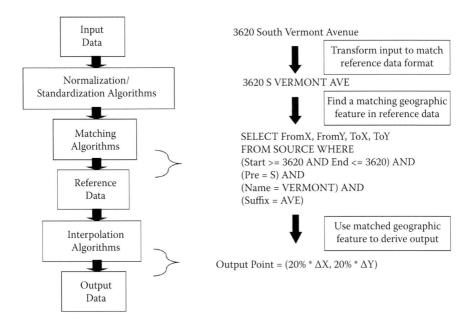

FIGURE 3.2   Components of the geocoding process.

3.3.1.1   Postal Address Data

Postal address data are the most typical form of information processed by a geocoding system. However, postal address data themselves can take one of many forms. For example, the most commonly encountered form of postal address data in the United States is a city-style address. This format of address includes the street address, city, state, and U.S. Postal Service (USPS) ZIP code of a location. One example of a well-defined postal address is "3616 Trousdale Parkway, Room B57G, Los Angeles, California 90089-0374." This example contains the street address that is composed of the street number (3616), the street name (Trousdale), the street suffix (Parkway), the unit type (Room), and the unit number (B57G), along with the city (Los Angeles), the state (California), the USPS ZIP code (90089), and the four digits that extend the ZIP code (0374). This particular example address format follows the USPS Publication 28 Address Standard (U.S. Postal Service 2012), but other address formats are possible and often encountered. The major difference between city-style postal address formats is usually the granularity of the attributes maintained, for example, separating "Old Main Street" into the street pretype (Old), the street name (Main), and the street suffix (Street), rather than consolidating "Old Main" as the single name attribute. More details on the challenges related to handling different address formats are covered in several places in this section.

Alternatives to city-style postal addresses that are seen in rural areas of the United States are formats that describe mail delivery routes instead of the locations of premises along a street. In these cases, the address describes the location of a mail delivery point, which is a shared box along a particular street. A postal worker or delivery service would place the mail for multiple people in the same location.

Each person or household has a particular box number assigned within the drop-off location. The recipient of mail items will travel to these shared locations to retrieve his or her mail. This is in contrast to city-style addresses where mail is delivered to a particular address (Beyer, Schultz, and Rushton 2008).

Addressing systems in rural areas that fall into this category are rural routes, star routes, and highway contract routes among other possibilities. In addition to these is the Post Office box (PO box), a rented mailbox within a USPS facility or other mail delivery location. These can be found in rural and urban areas, and are the most common form of nonpremise location information encountered by geocoding systems. The PO box address describes the location of the mail facility rather than the physical location of the object of interest. As such, the amount of useful information that a PO box address can provide to an LBS is limited at best if the intent is to be able to use the location of the addressee in a geographically relevant context (Oliver et al. 2005).

3.3.1.2   Named Places

Named places are a second common form of input data seen by LBS. Examples include searches for the names of built-environment features like businesses (Apple headquarters), hospitals (Los Angeles County USC Medical Center), and universities (Texas A&M University). Other named places include administratively or colloquially defined locations or areas such as the Greater Los Angeles Metropolitan Area and Pioneer Square. Similarly, natural features such as the Gulf of Mexico, the Amazon Rain Forest, and the Grand Canyon often comprise many of the queries seen by LBS. Each of these types of named places could be used as both the query to an LBS ("Give me metro directions from my current location to Pioneer Square") as well as within the underlying reference data that provide information and services to LBS users (e.g., the case of a database of hospital locations used in an emergency response application).

When input data to a geocoding system are in the form of a named place rather than a postal address, the process of converting them to spatial locations is sometimes known as georeferencing (Hill 2006). The distinction between the use of the terms "geocoding" (typically referring to locating postal address data) and "georeferencing" (typically referring to locating named places) is mostly academic rather than practical. Many authors have defined the process of geocoding as a superset of tasks that include georeferencing (Goldberg, Wilson, and Knoblock 2007). Others have reversed the order with georeferencing being the superset and geocoding being the subset (Hill 2006). In practice, however, these two terms are often used synonymously. The main distinction between the two is that the georeferencing community has developed a specialized reference data format, termed a gazetteer, which is composed of reference features with data along three axes: (a) the toponym, which is the geographic name for the reference data feature; (b) the type of the geographic reference feature, which is drawn from large ontologies of natural and built environment features; and (c) the geographic footprint of the reference feature (Hill 2000).

The reference data sources used by geocoding systems contain each of these types of data about the reference features they contain; however, it is often the case that only a single feature type is contained in the reference data layers instead of multiple geographic data types. An example of this would be an address point data-base, which contains just a single geographic feature type—points representing the location of structures. A more subtle distinction between the two is the primary uses of the data stored within geocoding and georeferencing systems. The primary purpose of geocoding systems is to translate from a name in the form of a postal address or named place to a geographic representation (Goldberg, Wilson, and Knoblock 2007). The community of researchers and practitioners who perform georeferencing and maintain gazetteer databases would argue that the other two axes are just as important. These include (a) the ability to find all geographic objects of a particular type, that is, searching based on the type axis; and (b) the ability to identify all geographic objects based on location, that is, searching based on location (Hill 2000).

Regardless of which term (geocoding or georeferencing) is used to describe the process of associating a named place with a geographic location, named places present unique challenges for these systems not present in postal address data. The primary challenge is that in many instances, the named place for a location may have indeterminate boundaries as well as no official postal address. For example, a built environment feature such as a university may have specific buildings on its campus that are addressable using a postal addressing system. One example is the address "University of Southern California, Allan Hancock Foundation Building," which would equate to the postal address "3616 Trousdale Parkway, Los Angeles, CA 90089." However, the more general query "Give me directions to USC" would not be capable of making use of a postal address because the University of Southern California (USC) is a large campus that covers nearly a square mile. Although an LBS could assume any particular address within the USC campus and use it to compute directions, it could be the case that the user simply wants to find his or her way relatively close to the campus. After arriving in the vicinity, he or she may then issue a more refined query for the particular building or consult other sources to reach the ultimate destination.

These issues are even more pronounced in the case of built environment features and colloquially defined terms (Wieczorek, Guo, and Hijmans 2004). For example, the Grand Canyon is a destination that many people may seek to visit. An LBS may be responsible for providing directions to or from, information about, or services within this location. The challenge is that the spatial footprint of this geographic feature can vary from over 275 linear miles if measured in terms of river miles (from the Lees Ferry to Grand Wash Cliffs) to over 1,900 square miles if measured in terms of the area of the national park within which the actual canyon is situated. Although there are certainly postal addresses within Grand Canyon National Park, these may not be representative of how the user of an LBS would issue a query to visit the Grand Canyon unless he or she had a specific museum, ranger post, or information booth in mind. Similarly a user wishing to find apartment rentals near Koreatown, Los Angeles, would be equally challenged to provide a postal address for the colloquially defined region of Los Angeles.

3.3.1.3 Relative Directions

The final type of input data that the user of an LBS may wish to issue as a query or may be associated with reference data used by an LBS are locations defined in terms of relative geographies. An example of this type of data would be "the crime occurred 2 miles northwest of the park entrance." Although less common than named places and postal addresses, these types of locational descriptions are encountered in many application domains. These could include field notes indicating the locations of settlements, the discovery of artifacts, or the reports of crimes that occur in large areas such as parks and campuses (Wieczorek, Guo, and Hijmans 2004). These types of input data are more commonly found in the reference data layers used by LBS rather than in queries issued by a user, but use-cases for both could be constructed. In either, a geocoding system that seeks to associate a spatial footprint with the location description must be capable of parsing and geographically referencing the origin of the location description ("the park entrance") as well as the relative or absolute distance ("2 miles" or "halfway between") and the relative or absolute direction ("northwest" or "toward Mojave") in order to compute a final output geographic location.

3.3.2 Reference Data

The reference data used by a geocoding system consist of databases of records that have one or more textual locational descriptions associated with them in addition to one or more digital geographic footprints. The location descriptions can be any of the types just discussed—postal addresses, named places, or relative locations—among any other identifiable terms such as a geographically defined coordinate system. The geographic attributes associated with these objects could be any valid geographic data type including points, lines, and polygons, which represent the spatial footprint of the reference feature (Figure 3.3).

The information contained within the reference data sets available to a geocoding system represents all of the knowledge a geocoding system can use to return a geographic location as output to a user in response to a textual location description query (Zandbergen 2008). Although inference systems do exist to produce new knowledge for answering queries about data that are not contained in a reference

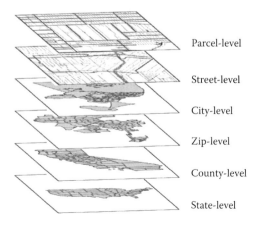

Parcel-level

Street-level

City-level

Zip-level

County-level

State-level

FIGURE 3.3   (See color insert.) Geocoding reference data layers.

data set, typically, if an addressable geographic object is not contained within a geocoder's reference data sets, the geocoding system will fail to return an accurate output to the user.

Each of the aforementioned input data types can be supported by different types of reference data layers. For example, city-style postal address data ("123 Main Street") are often supported by point, line, and areal-unit reference data sources. Point data sets might include GPS locations for postal addresses that have been collected by a local government to support emergency response applications. One example is the E-911 databases that are increasingly becoming available as local government agencies transition rural-route-type addressing systems to city-style addresses for the purposes of routing emergency vehicles in response to 911 calls. Other examples include building centroid databases that have been digitized from aerial imagery.

Linear reference data include databases of street segments that have street address attributes associated with each segment. These attributes include those previously discussed such as the street name, suffix, and pre- and postdirectional as well as the valid address ranges for structures on both sides of the street. The TIGER/Line files provided at no cost by the U.S. Census Bureau are one example of this type of reference data file. Another is the commercial versions provided by for-profit organizations such as TomTom (formerly sold under the brand Tele Atlas) (Tele Atlas Inc. 2012) and Nokia (formerly sold under the brand NAVTEQ) (NAVTEQ 2012). Linear reference data sets are far more common than point-based versions, primarily due to the extreme costs associated with field visits to acquire GPS points or the digitization efforts to derive point locations from imagery. Linear reference data files typically have nationwide coverage, whereas point-based files can be found intermittently across large geographic regions.

Areal-unit reference data files contain geographic objects made up of polygon data. These can range from parcel databases created and maintained by local tax assessor offices to city boundaries provided by government organizations such as the U.S. Census Bureau. Again commercial data providers such as TomTom, Nokia, and Esri also provide enhanced versions of these data layers at an increased cost. The level of geography represented by areal-unit reference data spans the full spectrum of locational specificity from the polygon footprints of individual buildings to the boundaries of entire countries. As with GPS data, polygon boundaries representing highly precise geographic regions such as individual buildings are often difficult to obtain, again due to the cost of their production. However, automated techniques that derive building footprints from aerial imagery or from lidar clouds are becoming more readily available as consortiums of cities, regions, and counties pool resources to pay the cost of obtaining these data. One such example is the Los Angeles Region Imagery Acquisition Consortium (LAR-IAC), which pools resources from various Los Angeles County agencies, local city governments and agencies within the county, and educational institutions to fund the acquisition of these data by private firms (Los Angeles County Department of Regional Planning 2012). These data are used to derive detailed building footprint layers that are then provided to consortium participants.

In all cases, the reference data sets used to support queries for city-style postal address data must store the address data associated with points, lines, or polygons in some address format. The address format used will often vary by the source of data, with each company or government agency using a data format that fits the operational needs of the organization. For example, the parcel data sets that can be obtained from the Los Angeles County Assessor's Office separate the prearticle of a street name from the name of the street name as in "La" being the prearticle and "Brea" being the name for the street known as "La Brea" in Los Angeles (Los Angeles County Assessor's Office 2012). In comparison, the TIGER/Line files provided by the U.S. Census Bureau do not maintain the prearticle as a separate field and thus the street name field contains "La Brea" in this data set (U.S. Census Bureau 2012). These types of issues have a great impact on the quality of the results provided by geocoding systems if not properly accounted for during address parsing, normalization, and feature matching operations of the geocoding system. More details on these issues are discussed in the following sections.

3.3.3 ADDRESS PARSING AND NORMALIZATION ALGORITHMS

As noted, city-style postal address data contain many different address attributes within a single address. These include the address number and the street name, suffix, predirectional, postdirectional along with the suite type, suite number, city, state, and USPS ZIP code. The role of the address parsing component of a geocoding system is to identify which pieces of the input address text are which address attributes. The role of the address normalization component is to transform the values of the address attributes identified by the address parser into standardized values and structures consistent with those maintained by any particular reference data set (Goldberg, Wilson, and Knoblock 2007).

Address parsing systems break an input address into its constituent pieces based on a desired address standard. In many cases, this activity is accomplished by processing an input address as a string of tokens using tables of synonyms to identify a likely type for each set of letters that makes up a single token. Here, "type" is used to refer to the domain of attributes that an address standard supports—street predirectional, suffixes, suite types, and so on. This process usually begins by using whitespace between consecutive letters to separate the original input string into a series of tokens. The values of these tokens are then queried against known synonyms to associate potential types. One token could be assigned multiple potential types. For example the "Trl" component of the address "123 Skyline Drive Trl 456" could represent both the suffix "Trail" if the name is "Skyline Drive" or the suite type "Trailer" if the name is "Skyline" and the suffix is "Drive." Once each token is assigned one or more types, the most likely combination of tokens and types is used to associate a final type to each. This could be based on heuristics, such as "Trl" most likely being "Trailer" because a series of numbers follows it indicating a suite number; determined empirically from available reference data (does the combination exist in the reference data?); or determined probabilistically from the frequency of occurrence and co-occurrence of terms in available reference data (does something that looks like the combination exist in the reference data?).

One of the challenges in this process is that the address schemas used to store the postal address attributes of reference features can vary by reference data source. In practice, this means that the granularity of address components supported by

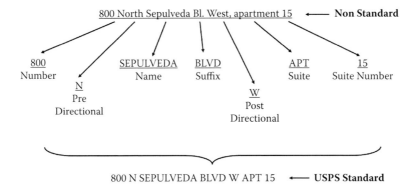

FIGURE 3.4   Address normalization.

one reference data file may not be the same as that of another. The example of the street "La Brea" discussed in the previous section is illustrative of this scenario. In the Los Angeles County parcel files, the street "La Brea" is broken into "La" as a prearticle and "Brea" as the name, whereas in the Census TIGER/Line files the name is stored as "La Brea" with the type values collapsed into one. The first role of the normalization component of the geocoding system is to tackle this problem by mapping the address attributes of the input address data item onto the schema of the reference data source. This capability would separate "La" and "Brea" to match the expected attributes for the LA County Assessor files, and would combine them into "La Brea" to match the Census Bureau files.

The second role of the normalization component is to transform the values of the input postal address attributes into the standardized forms of the address system used by the reference data set (Figure 3.4). Here, nonstandard street suffix values such as "Street," "Str," or "St." would be rewritten as the standardized form "ST" or an alternative official version used by the address standard of a reference data layer. This approach is applied to each component of the input address attribute values including the pre- and postdirectional, that is, changing "North" to "N," "Sth" to "S," and so on; the suffix as in the "Street" to "ST" example, and the suite type "Trailer" to "TRL," "ROOM" to "RM," and so on. However, these processes are not usually applied to the street number or name field as these do not usually have standardized values. The exception here is for street names that are numbers like "1st Street." These are often normalized to either all numeric as in "1st" or all textual as in "First."

3.3.4   Feature Matching Algorithms

The feature matching algorithms used by a geocoding system try to identify one or more candidate reference features (individual geographic objects) within the reference data layers available to the geocoding system that match the attributes associated with an input address (Boscoe 2008). These processes can either be deterministic or probabilistic. The former refers to systems that search for matches and either find one or do not; the latter refers to systems that return a set of matches each with a

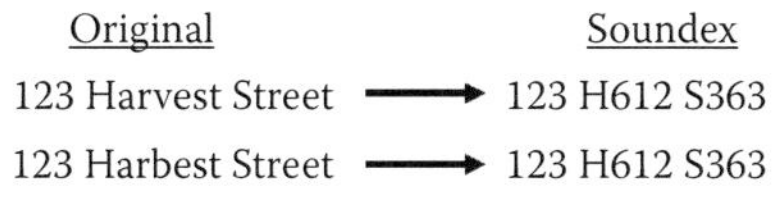

Original		Soundex
123 Harvest Street	⟶	123 H612 S363
123 Harbest Street	⟶	123 H612 S363

FIGURE 3.5 Ambiguous Soundex queries.

probability of being correct. Both systems have their benefits: deterministic methods always return the same results given the same input data; probabilistic methods may return different results based on the frequency of occurrence and co-occurrence of terms in a specific reference data set.

In either case, the matching algorithm will return zero, one, or more than one reference feature that meet the selection criteria as being a match candidate. These criteria could be an exact match where all of the attributes of the input address are present on the reference feature, a relaxed match where some of the attributes are present, or another heuristic match where some quality of the attributes match. An example of a heuristic matching approach is a phonetic match where the SOUNDEX value of the input and reference attributes are compared rather than the full text of the value (string equivalence) (Figure 3.5).

Each match candidate selected by the matching algorithm is assigned a score that indicates the quality of the match between the input address attributes requested by the user and the attributes associated with the match candidate (Boscoe 2008). These scores are determined by adding or subtracting a value from the total score based on the existence of an attribute in both the input and the reference feature as well as the similarity between the values of the attributes in each. To compute an overall match score for a particular match candidate, the geocoding system uses a weighting scheme that assigns a weighted value (proportion of the total score) to each available attribute. If the attribute values in both the candidate and reference feature are equivalent, the weight for that attribute is added to the overall score. If they differ, the weighted score is subtracted by a penalty indicating the amount by which they differ. Exactly how this penalty is calculated may vary between geocoding systems. Edit distance, or the number of operations it would take to add, remove, or change letters to make the two attribute values equivalent, is one common approach used to quantify the difference between input and candidate values.

Most geocoding systems provide the user with the ability to set a minimum candidate matching score below which candidates are excluded as non-matches. This minimum match score filters the results of the matching algorithm by excluding non-exact matches that differ by a user-definable sensitivity metric. Setting the match score high (99% match) will reduce the amount of false positive matches (matches that were returned but should not have been) but potentially increase the number of false negatives (matches that were not returned but should have been). Setting the score too low (75%) will achieve the opposite result, increasing the false positive rate and decreasing the false negative rate (Figure 3.6).

Which is the better option depends on the application scenario that is using the geocoded results and if any additional filtering is performed following the matching algorithm execution. If additional processing takes place at the application level (following the matching algorithms), it may be beneficial to lower the match

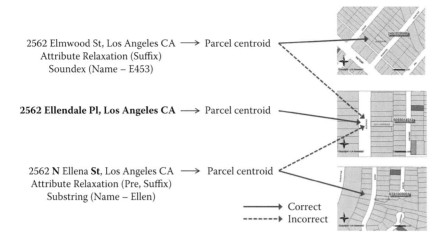

2562 Elmwood St, Los Angeles CA ⟶ Parcel centroid
        Attribute Relaxation (Suffix)
        Soundex (Name – E453)

2562 Ellendale Pl, Los Angeles CA ⟶ Parcel centroid

2562 **N** Ellena **St**, Los Angeles CA ⟶ Parcel centroid
        Attribute Relaxation (Pre, Suffix)
        Substring (Name – Ellen)

⟶ Correct
------▶ Incorrect

FIGURE 3.6   (See color insert.) Example geocoder mismatches.

rate to provide the application with the largest set of candidate features possible. However, this increases execution time as these candidates must be postprocessed in the application layer, which may be an issue for scenarios requiring real-time or near-real-time responses.

3.3.5   INTERPOLATION ALGORITHMS

If the feature matching algorithms used by a geocoding system are successful in identifying one or more candidate reference features in the reference data layers of sufficient match quality, the next step of the geocoding process is to determine where within or along the geography associated with the reference feature the ultimate output should be placed. The interpolation algorithms used by a geocoding system perform this task. These functions use the attributes of the input data, the attributes and geography of the reference feature, and any ancillary information that may provide additional knowledge that could improve the position of the output result to compute the ultimate output location for a user query.

The type of interpolation algorithm used by a geocoding system depends primarily on the type of geography maintained within its reference data layers. For example, reference data layers that contain geographic points— building centroids, GPS locations, and so forth—do not typically require any form of interpolation because the point associated with the reference data object can be directly returned. Linear and area-unit reference data layers require some form of interpolation, however, because (typically) one single point must be chosen to be returned which could be anywhere along a linear geographic object or within a polygon object (Bakshi, Knoblock, and Thakkar 2004). It should be noted here that an LBS may require or request the full geography of the reference feature instead of an interpolated output point. This is currently an infrequent request to geocoding systems, but LBS applications may increase the use of the full geography as LBS-provided information and services begin to be consumed in novel and innovative ways.

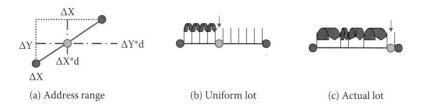

(a) Address range   (b) Uniform lot   (c) Actual lot

FIGURE 3.7   (See color insert.) Linear interpolation methods.

Linear reference data layers (i.e., street segment databases) use linear interpolation techniques to return a single output point derived from the geography of the reference data feature. Here, the address number portion of the input postal address is used to compute a location at a proportional distance from the starting node of the street segment to the ending node of the street segment. For example, if the input address is "123 Main Street" and the candidate reference feature is the "100–200 Block of Main Street" the output point would be computed to as the position that is 23/50 of the way down the street segment, starting from the from-node heading toward the to-node. Once the position along the street is computed, an offset is applied orthogonal to the direction of the street segment to move the output away from the centerline of the street toward the side of the street that corresponds to the parity of the input address (odd or even) (Figure 3.7).

This process, known as address-range interpolation, assumes in this example case that there are 50 potential addresses on each side of the street (one side even and one side odd) and uses a simple proportion to compute where along the linear geography of the reference feature the output should be placed (Bakshi, Knoblock, and Thakkar 2004). The assumption that all addresses associated with the full address range of the street segment exist is usually an overestimation of the true number of addresses present on a street segment that results in the output data points bunching up at the low end of the street segment. Approaches such as the uniform-lot method that account for the actual number of houses on a street can improve the quality of the interpolation results because a more realistic denominator is used in the proportion equation. Similarly, approaches that include the size and orientation of parcels on the street can improve this calculation further. However, if detailed information about parcel data is known, one may achieve better output results by simply using a parcel data layer instead of a linear interpolation approach.

Areal-unit reference data layers have a similarly diverse set of methods that can be used to compute an output point from the geography associated with candidate reference feature matches (Figure 3.8). The simplest and quickest is the bounding

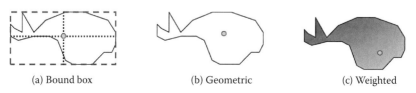

(a) Bound box   (b) Geometric   (c) Weighted

FIGURE 3.8   Areal-unit interpolation methods.

box centroid method. Here, a bounding box that encompasses the full geography of the object is first computed. Next, the minimum and maximum extents of the box are determined, for example, the top left and bottom right corners. Computing the output X and Y locations then becomes the simple task of taking the average of the height and width of the box: $outputX = (minX + maxX)/2$, $outputY = (minY + MaxY)/2$. This approach has the benefit that it is extremely quick to compute but has the drawback that the output point may fall outside of the actual boundaries of the reference feature match if the geography is an odd shape.

A center of mass (centroid) equation can be used to avoid the scenario where the output location falls outside of the geography of the reference feature. This approach ensures that the output is contained within the reference feature but has the drawback that it is much more computationally intensive than a simple division equation as in the bounding box approach (Thomas et al. 2006). Nonetheless, the centroid method is typically used by most geocoding systems because the Open Geospatial Consortium STCentroid() method is now supported by most modern database systems as well as natively in programming languages that support spatial data types.

The centroid approach is useful and relatively accurate when applied to high-specificity geographic reference data layers like parcels in an urban area. However the centroid approach becomes much less useful when applied to large geographic regions like city or USPS ZIP code boundaries. In these cases, it often makes more sense to apply a weighted centroid algorithm that biases the location of the output based on the density of some other attribute of interest. This approach is commonly employed when input address data match to the ZIP code level. Here, additional geographic data layers describing the distribution of population within the ZIP code can be incorporated into the centroid computation method to bias the location of the output toward areas with higher population density. Because the center of mass approach is a best guess approximation anyway, biasing the result using additional data layers has the effect of producing a higher quality geocode because it moves the output location towards a location that has a higher probability of being correct based on the distribution of a variable of interest (Krieger, Waterman, et al. 2002).

3.4 OUTPUT DATA

As noted earlier, the output of most geocoding systems in use today is a single geographic point (latitude and longitude) computed from the geography of a reference data feature that is selected as a candidate match. In addition to this spatial representation, there are often a series of nonspatial metadata attributes associated with a particular geocode output as well as with a geocoding system as a whole.

At the per-record level, these attributes include the level of geography that the output geocode is matched to. Effectively, this metadata item simply lists the type of geographic objects included in the reference data layer from which the matching candidate was selected: GPS point, building centroid, parcel, street segment, street intersection, USPS ZIP code, city, county, and so on. One commonly used hierarchy is the North American Association of Central Cancer Registries' GIS Coordinate Quality Code metadata scheme for describing geocode quality (Table 3.1).

TABLE 3.1
NAACCR GIS Coordinate Quality Codes

Code	Description
1	GPS
2	Parcel centroid
3	Complete street address
4	Street intersection
5	Mid-point on street segment
6	USPS ZIP5+4 centroid
7	USPS ZIP5+2 centroid
8	Assigned manually
9	USPS ZIP5 centroid
10	USPS ZIP5 centroid of PO Box or RR
11	City centroid
12	County centroid

Although informative, these qualitative descriptions of "geocode accuracy" do not define a quantitative metric that can be used to determine a spatial confidence interval for a particular geocode (Goldberg and Cockburn 2010). What users and developers of LBS are more likely to be interested in is the spatial accuracy of the output point, as in "the output geographic point is ±20 m from the true location." Users need to know this information to judge how far the geocoded point provided to them by a geocoding system is from the true location they are seeking. Similarly, LBS providers need to know the spatial accuracy of a query location that is translated into a geographic position by geocoding. Knowledge of the spatial accuracy of the items contained in the underlying reference data layers used by the LBS is equally important. New approaches to representing spatial uncertainty in geocoded data are making strides toward true quantitative error metrics, but geocoding systems that support such error reporting metrics are currently limited in number and often seen only in research scenarios (Goldberg and Cockburn 2010).

The other record-level metadata item typically returned for an output geocode is the match score. As detailed earlier, this metric captures the similarity between the attributes of the reference feature selected as a candidate match to those of the input address requested of the system. Although a nonspatial method of describing the accuracy of a geocode, this metric does provide some guidance on how likely it is that the output geocode returned is in fact the one requested by the query (Boscoe 2008). The challenge here is that multiple matches may be possible during the geocoding process, and typically only one is presented back to the user as the final output. These situations often occur when the input data are incomplete or inaccurate. For example, if a user were to search for the address "123 Main Street, Los Angeles, CA" they would be presented with the situation where two candidate reference features are equally likely because the input query was incompletely specified: "123 N Main Street" and "123 S Main Street" (Figure 3.9).

Which of these two should be returned is certainly context dependent on what the user intended by the search. But without any further information both are equally likely

FIGURE 3.9   Ambiguous feature matching result.

and thus equally valid to return as a candidate matching feature, use for interpolation, and provide as a geocoded output. Different geocoding systems handle this ambiguity scenario in different ways—some flip a coin and return one or the other, some pick one or the other based on frequency of searches, and some report a failure and do not return either (Goldberg 2011). Similar cases occur where the match scores of multiple candidates are very close, for example, two candidates with scores of 95.6% and 95.5%. Without any knowledge of what the input query was or what the reference features are, it is clear that the two candidates are quite similar and perhaps both equally valid.

To accommodate these scenarios, geocoding systems are beginning to return all available matches (and resulting interpolated outputs) and allow the application to determine what the most appropriate result should be. This shifts the burden of determining the applicable context to the LBS developers but perhaps this is appropriate given that a LBS may have access to other contextual clues about intent of query that a geocoding system may not.

In comparison to the per-record metadata elements just described, geocoding systems are also characterized in terms of match rates (Zhan et al. 2006). The match rate identifies the proportion of total input data items a geocoding system was capable of computing a geocode for. For example, if a user submitted 100 records to be geocoded and the system was able to compute geocodes for 88 of them, the match rate for this geocoding attempt would be 88%. This quality metric does little to provide accuracy measures that would be useful to an LBS in terms of spatial distance from the real location, but does provide an indication of the coverage of a geocoding system. This may be important if multiple geocoding systems are available that can be used opportunistically as a user transfers from one location to another.

3.5   OPPORTUNITIES AND CHALLENGES FOR GEOCODING IN LBS

The previous two sections attempted to provide information about the role of geocoding systems in an LBS as well as detail the inner workings of the multiple components of the geocoding process. Together, these two sections set the context

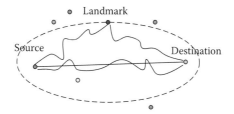

FIGURE 3.10   Geocoding for travel routing in an LBS.

for this current section which is tailored toward describing many of the opportunities and challenges inherent to employing geocoding technology in LBS. Following from the two sources of input data used by a LBS, this section walks through a series of examples that illustrate how LBS and geocoding systems interact to provide users of LBS with contextually relevant information and services. This discussion will be cast from the two complementary views of how a geocoding system facilitates: (a) processing an input query for services; and (b) generating the reference data layers an LBS uses to answer a user query. In each case, the opportunities for additional capabilities that geocoding services provide to LBS are discussed, as are the challenges and limitations that LBS users and developers should be aware of and take into consideration as LBS are utilized and developed in the future.

3.5.1   Travel Routing

One of the most common forms of LBS in use today are travel routing applications provided by online Web services like Google Maps,[*] Yahoo Maps,[†] Bing Map,[‡] and MapQuest.[§] This same service is also now routinely provided by personal navigation devices like in-car navigation systems and GPS-enabled smartphones. These services capture the location of a user-defined set of origin and destination points then compute a travel route that will take the user from an origin location to the single or multiple set of destination points of interest (D'Roza and Bilchev 2003) (Figure 3.10).

The origin of a user's location is increasingly becoming his or her current location as determined by the GPS readings of the in-car navigation system or mobile device in his or her pocket. This is especially true when the user is already in one's vehicle or when already in transit via walking, biking, or riding public transportation. However, many queries of this type still begin in the predeparture stage so that the user may generate a map of the route, investigate alternative routes, or identify points of interest (POIs) along the route at which to stop and enjoy dinner, the natural scenery, or other interesting aspects of the journey. In these instances the LBS that provides the service of computing a route require a starting location defined in terms of a postal address, named place, or some other form of locational description.

[*] http://maps.google.com/
[†] http://maps.yahoo.com/
[‡] http://www.bing.com/maps/
[§] http://www.mapquest.com/

In contrast to the origin location, which may be determined by GPS, the destination locations of routes are nearly always defined in terms of a postal address, named place, or other locational description.

Regardless of the mode of transportation route computed—driving directions along streets and highways; walking directions along sidewalks; biking directions along bike paths, streets, and trails; transit networks that include busses, trains, and ferries; or any combination of transportation modes as in multimodal transit routing—the very first fundamental step of an LBS must be to geographically situate the start, end, and intermediate points of the route (Raubal, Miller, and Bridwell 2004). When the origin or destination of the route is specified by the user in one of the many forms of locational descriptions, an LBS must thus employ a geocoding system to determine one or more geographic representations of the origin, destination, and intermediate points of interest in order to compute a route. These geographic locations can then intersect with the travel networks of choice to enable a graph-searching algorithm for routing the user from the origin, to the set of points of interest along the route, and ultimately to the destination.

The challenges for geocoding systems in these cases, and the LBS which rely on them, are twofold. First, the locations must be expressed in such a way that the geocoding system will be capable of parsing and processing the input data into an output geocoded location for each destination identified. When the input locations are defined in terms of postal addresses, this is commonly a simple task as described in the previous sections detailing the inner workings of geocoding systems. However, when the origin, destination, or intermediate points are described as named places this situation becomes more difficult. In order for a named place to be located using a geocoding system, it is typically the case that the specific name must be present in the reference data files used by the geocoding (or georeferencing) system.

Being primarily deterministic systems that query data sets of known geographic objects, geocoding systems have a difficult time processing data that are not contained in the knowledge base available to the system. The result of this scenario is that if a named location is not contained in any reference data sets, the geocoding system may completely fail to locate the named place (if false negatives are preferred) or provide an inaccurate guess (if false positives are preferred) (Brimicombe and Li 2006). This challenge still occurs when the geographic object of interest is contained in the reference layers, but the colloquial, nonofficial, or abbreviated name for the object specified by the user is not. One example would be issuing a query "give me directions to TAMU," when in reality the user wanted to find "give me directions to Texas A&M University."

3.5.2 K-NEAREST POINTS OF INTEREST

A second extremely common form of an LBS in use today is services that allow a user to search for POIs in relation to his or her current location or some other user-specific location. The prevalence of these services is due partly to the increasingly ubiquitous nature of Web-mapping software such as Google Maps that provide relatively straightforward and intuitive mapping interfaces for browsing and searching in a geographic space. As the population at large has become ever more facile with using these technologies, it is a fair statement to say that nearly every person with an Internet-capable computer or device has at one time or another searched for locations of interest using an online mapping interface. Each time that this process is performed, the LBS underneath the online mapping interface queries a number of geographically referenced POI data sets to determine what results would be "of interest" to the user (Zheng et al. 2006). Here "of interest" means contextually relevant, and the context that determines relevancy in a mapping and LBS context is typically geographic proximity. This notion of relevance obviously assumes that the LBS is capable of filtering the domain of all possible POI to just the subset that match the search intent of the user, that is, not returning eating establishments when the user requested the locations of nearby car repair shops.

This type of LBS is an example of a classic k-nearest neighbor search where the intent of the search is the find the closest k objects to an input geographic location from a spatial database (Zheng et al. 2006). As discussed earlier in this chapter, this type of search is becoming nearly rudimentary when the underlying reference data are stored in a modern spatial database. In fact, a single query can often accomplish this task as expressed in the following SQL pseudocode example: "SELECT TOP 5 ID, DISTANCE(InputPoint, FeaturePoint) AS D FROM TABLE ORDER BY D ASC". This query computes the distance from an input point to the geographic points associated with each object in a database table and returns the first $k = 5$ results in ascending order of distance between the input point and the feature point. More efficient methods exist to accomplish the same task, particularly those that compute the distance only for relevant records instead of the all records in the entire database table (Lee, Zhu, and Hu 2005). Regardless of the method used to determine the result, the main goal of the approach is the same in all cases—find the k closest objects to an input location (Figure 3.11).

Although Volunteered Geographic Information (VGI) approaches to gathering data are increasing in number, quality, and user participation, it is the case that most reference data sets used in LBS are still derived from lists of locations obtained from some nonvolunteered source. The geographic component of the items in these lists is most commonly in the form of postal address information. Thus, the geocoding process remains one of the primary tools used to create these POI databases upon which LBS rely to return meaningful data to a user. Whether it is a list of hotel locations obtained from a hotel chain, a list of coffee places screen-scraped from a national coffee chain's Web site, a list of houses for rent or sale, or a list of public schools, any administrative or business listing that uses an address as the primary way to denote the site location must be geocoded for it to be useable as a reference data layer within an LBS.

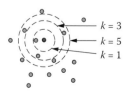

FIGURE 3.11   Example k-nearest computation for $k = 1$, 3 and 5.

Like the travel routing example described earlier, the primary challenge in *k*-nearest POI searching is the availability and quality of reference data layers that can be queried by a geocoding system. Despite the deluge of locationally relevant data with which LBS developers and service providers now find themselves inundated, thanks to GPS-enabled devices, the geographic data sets that describe the location of services or information for which the end user of an LBS may have an interest have, unfortunately, not witnessed such marked increases in availability, completeness, or spatial accuracy. In the majority of cases, the same methods that were used to generate lists of POI in the early days of LBS development are still in use today. These include purchasing business and property listings from commercial vendors or government agencies, geographically enabling proprietary database listings drawn from customer interactions or queries, using official lists of geographic places such as the Geographic Names Information System (GNIS) and GeoNames gazetteer databases provided by U.S. government agencies (U.S. Board on Geographic Names 2012), or scraping data from publicly available internet sources. VGI is beginning to play a larger and more prominent role in how these databases are gathered, but it remains the case that large commercial or government lists form the backbone for the majority of the LBS POI systems.

A second challenge that *k*-nearest POI searching LBS must account for (or at least acknowledge) is the spatial accuracy of the geocode returned to the LBS or end user. Here again, spatial accuracy refers to the geographic distance between the output location generated by a geocoding system and the true location of the object in the real world. It is important to note that geographic accuracy represents error in both two and three dimensions (2D and 3D); a greater discussion of this follows later in this chapter. A higher degree of spatial accuracy would place the output geocode closer to the true location, whereas a lower degree of spatial accuracy would place the geocode farther away from the true location. This issue is relevant to *k*-nearest searching because of the high degree of spatial accuracy required by existing and emerging LBS applications. For example, consider an LBS where a participating user earns points automatically by visiting a business establishment. After accruing a certain number of points, the establishment may provide the user with a coupon for free or reduced-cost services. One method to assign points to the user for entering the establishment would be to actively monitor the user's location through a smartphone application and provide a point when the user entered a fixed boundary around the geographic location of the establishment. In this case, the spatial accuracy of the geographic location of the establishment is of utmost importance because the business would not want to give away points to users who did not enter the store, and the users would want to make sure that they earned points every time they entered the premises. The accuracy of the geocoding process that generated the location for the store plays a key role in how well this system would work.

Because systems that provide subpremise geocoding, that is, producing output data to a greater spatial accuracy than a building centroid, are currently uncommon mainly due to the lack of reference data sources that could provide these details, other strategies are beginning to emerge to increase the initial spatial accuracy of the locations of business establishments. One such method is to make use of the GPS-enabled devices carried by the population of interest to gather more accurate spatial locations than are possible with geocoding systems. One such approach is provided by social media sites and services that allow users to "check in" at locations. When a user clicks the check in command on his or her smartphone, the GPS location derived from the device could be sent back to the server containing the reference data layers to update the location for the geographic reference feature they are purportedly checking in at. As more users check into the same location, they collaboratively increase the degree of spatial accuracy associated with the footprint of the geographic feature (Cho, Myers, and Leskovec 2011). Unfortunately, the GPS locations for geographic reference features captured by these means are not often released by the companies that collect these data because they are in many cases proprietary or subject to data use and licensing agreements. However, using these methods, improved versions of reference data layers are being created and will undoubtedly become more readily available as business cases are made for their release.

3.5.3 Disease Surveillance and Disaster Response

The previous two examples, LBS for travel directions and *k*-nearest searches, are classic examples of LBS and geocoding interaction that have been around since the earliest days of LBS development. Thanks to the proliferation of GPS-enabled smartphones, Web 2.0 technologies, and the explosion of public participation in social media services in recent years, far more advanced combinations of geocoding and LBS are rapidly evolving. Two such related examples include LBS for disease surveillance (Boulos 2003) and LBS for disaster response (George et al. 2010). The nature of these types of LBS relies heavily on data submitted by the public from and about a location. In both cases, messages transmitted in social media platforms can be monitored by researchers, officials, and members of the public to obtain and consolidate information about situations of concern, be they disease outbreaks or the location of downed power lines, flood inundation, gas leaks, or structures on fire. The benefit of these approaches is that information can be gathered from a large number of people in a short amount of time. This type of activity has been termed "citizens as sensors" (Goodchild 2007) and allows for the rapid production of situational awareness for a region that would not be possible using official sources of data or data streams alone. These approaches are known to bring with them data quality issues related to completeness, validity, reliability, and privacy; however they have been successfully used in response to recent natural disasters including the earthquakes in Haiti and Japan (Zook et al. 2010).

The common approach in both disease surveillance and disaster response LBS is to leverage existing social media platforms as a source of intelligence about an area. Systems described in the literature to date have mined Twitter feeds for posts that describe conditions that may be related to a particular disease or to a specific region where a disaster is unfolding (Davis et al. 2011). In the case of disease surveillance, researchers have mined social media streams for message posts that describe disease symptoms such as nausea, vomiting, and diarrhea. Using this information, it is possible to identify hot spots where the number of posts that describe disease symptoms of interest exceeds what would be considered normal background occurrence rates. Areas of elevated symptom reporting might indicate a region where a disease outbreak is unfolding and resources can be dispatched to the region to determine if such is the case and take appropriate action if so.

Disaster response LBS applications operate in a similar manner except for the main distinction that individuals submitting data are most often describing the environment around them in order to alert government officials, agencies, and the general public of areas and instances where immediate help is needed or, alternatively, where affected citizens can go for supplies, housing, or safety (Goodchild and Glennon 2010). Often, these types of posts include digital multimedia such as images, voice recordings, and video that can help provide additional context about the environment in addition to text supplied by the user (Zook et al. 2010).

In both the disease surveillance and emergency response cases, the posts that are used to provide information about the event of interest must include some form of spatial context in order to be of the most relevance to those who would make use of the data. In the case of Twitter messages specifically, posts can be spatially enabled using one of three options: (a) the GPS coordinates associated with the user's phone (if enabled); (b) the location the user has identified him- or herself as being located at (if provided); or (c) from the location listed in the text of the message. Recent studies have found that options (a) and (b) provide the highest level of accuracy and specificity, but are not enabled or provided in the majority of messages that are transmitted via the Twitter system (Davis et al. 2011).

As such, geocoding is often required to infer location context from the text of Twitter messages. However, geocoding in the context of Twitter messages is exceptionally challenging due to the technical nature of the messages and the means by which users of this system communicate location. First, Twitter messages can only be 140 characters long, which necessitate a high degree of abbreviation in terms of location descriptions. It is not uncommon to see messages of the form "fdtrk dwntn @ 1 & Grnd" representing the statement "I am eating at the food truck downtown on the corner of First and Grand." The deterministic address parsing systems described earlier as part of the geocoding process may have extremely difficult times handling locational descriptions of this nature because they do not follow traditional address structuring, contain a high level of abbreviation, and use colloquial terminology.

In these instances, probabilistic natural language processing (NLP) systems may be required to infer both the components of the location description as well as likely values for each of the attributes (Corvey et al. 2010). These NLP systems use large training corpuses and machine learning techniques to develop probabilistic models that can tag tokens based on training data and apply these models to unseen examples. The result provides a means by which the content of an unstructured query can be translated into a structured and annotated format that a geocoding system can then use as the basis for beginning a deterministic search process (Mahmud, Nichols, and Drews 2012). An NLP system would transform the unstructured input text of the food truck example into a query that a geocoding services could use similar to: region = "downtown," intersection = "First and Grand."

3.6 THE ROAD AHEAD

Despite the challenges outlined in the previous section with regard to the three example application areas, geocoding remains closely tied to the operation and development of LBS. It is anticipated that this connection will only become stronger as both geocoding and LBS technology advance, as well as the availability and quality of reference data available to both systems. Several research areas have emerged in recent years that, if progress is made, will greatly enhance the quality of both geocoding and LBS. This section discusses three such research areas.

3.6.1 Volunteered Geographic Information

As discussed several times in this chapter, volunteered geographic information (VGI) has the potential to be a game-changer with respect to geocoding output accuracy and completeness, which will in turn affect the quality of services and information that can be provided by LBS (Goodchild 2007). As two examples, the community driven efforts of Open Street Map* and Wikimapia† are constructing some of the most comprehensive reference data layers available. Although nonofficial in many cases, these databases are often highly vetted by members of the community to ensure the spatial accuracy of the geographic footprints associated with the reference features. Similar quality controls are used to ensure the accuracy of the non-spatial attributes including the names and types associated with the geographic features. These approaches have the advantage that the local community is charged with creating and maintaining the reference data, which results in greater levels of colloquial knowledge being included. This means that multiple names for features that are non-official versions could be available to geocoding systems, improving the quality of geocoded results for non-standardized queries, such as those required to process social media messages like the Twitter examples discussed earlier. Similarly, updates to community-created reference data layers can occur more frequently using a VGI approach, resulting in reference data layers that more closely resemble the present state of a local region (Figure 3.12).

Despite the great promise that VGI brings to the geocoding and LBS communities, several research challenges exist that must be addressed before these data can be exploited to their fullest potential. Primary among these are methods to ensure minimum levels of accuracy and reliability of the data gathered or created by VGI systems (Flanagin and Metzger 2008). Specifically, many LBS require that specific

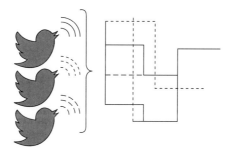

FIGURE 3.12   Crowdsourcing to build geocoding reference data layers.

---

* http://openstreetmap.org/
† http://wikimapia.org/

data quality thresholds be met before data are integrated into an LBS and used to provide services to end users. The rationale for this is simple: if an LBS provides the wrong answer to a user, it is highly likely that the user has just become an ex-user and may go so far as to provide negative reviews of the LBS. The marketplace of LBS is rapidly heating up, with new companies launching new services every day. The social media landscape with which LBS users are familiar means that negative software reviews can spell the end of a fledgling LBS provider.

Similar research challenges exist about user trust: who is authorized to submit: what data and how are these data vetted? It is unfortunately the case that competing LBS may have an incentive to falsify or otherwise corrupt VGI databases if a competitive advantage can be gained. Because the reputation of an LBS provider or developer depends on the quality of the results provided, mechanisms must be in place to ensure that actors cannot collude to invalidate VGI resources for nefarious purposes. User ratings systems have made some progress in these areas (Maue 2007), as have sites like Wikipedia and movements like Open Street Map, which may provide some guidance on how these issues can be appropriately addressed.

3.6.2   3D AND INDOOR GEOCODING

Currently, geocoding system support for producing 3D and indoor spatial locations is limited at best. Although some prototype systems exist for small areas that primarily support indoor routing, to the best of the author's knowledge, no publicly or commercially available geocoding Web service has the ability to provide 3D or indoor geocoding for large-scale regions (Lee 2004; Lee 2009; Worboys 2011; Vanclooster and Maeyer 2012). The main limitations that are hindering the creation and adoption of these tools are two interrelated aspects: (a) a simple lack of data describing indoor environments; and (b) limited integration of geographic information systems (GIS) and the application platforms used by architects and facilities management services, which include computer-aided design (CAD) and building information management (BIM) systems.

With regard to the first, geocoding systems must have access to floor plans or other architectural drawings in order to compute positions in 3D indoor spaces (Vanclooster and Maeyer 2012). If these data were widely available, the computational methods used to produce an output using these data sources would not be dramatically different from those used to currently compute geocoded locations for postal street address data. Examples of successful geocoding applications using these data can be found at large shopping complexes and convention centers that allow a user to search for and obtain directions to particular stores or meeting rooms (Zeimpekis, Giaglis, and Lekakos 2002; di Flora et al. 2005; Worboys 2011). However, for the majority of structures in the world, floorplan data are simply not available at all and even less so in a format that would be useful to a geocoding system.

This may seem curious because in order to obtain construction permits in most areas that follow modern construction regulations, these documents must have been available at one time because they often must be provided to an inspector to approve a permit to begin construction. This means that these data must have existed in some format at some time prior to the construction of the structure. The challenge here is

FIGURE 3.13 **(See color insert.)** 3D indoor geocoding.

that once approved for construction, copies of these documents, while certainly used by the construction firm to erect the structure, may or may not be retained by the regulating public agency. If they are retained, they are typically kept in paper form. Further, access to these documents by members of the public or commercial data providers may be protected for security or confidentiality, and costs associated with obtaining them one by one may be prohibitive.

The second problem relates to the different software systems used by two professional communities. GIS professionals have used mapping software and database management systems on a regular basis for the better part of the last 40 to 50 years. These systems typically see the world as a series of vector and raster data layers, and are extremely efficient at storing, querying, visualizing, and analyzing these types of data. Example systems include the ArcGIS* suite of tools provided by the Environmental Systems Research Institute (Esri) and, more recently, free applications like Google Earth.† These applications are data oriented and enforce strict constraints to maintain the validity and consistency of geographic data. In contrast, the software used by the architectural community is often seen as more design oriented rather than data oriented. The AutoDesk AutoCAD‡ product used by architects worldwide is a primary example—it allows an author to create extremely creative designs for structures but has few constraints that enforce data validity from a geographic or geometric perspective. The result of this situation is that CAD data are

---

* http://esri.com/
† http://earth.google.com/
‡ http:// autodesk.com/autocad/

not easily imported into a GIS system without a great deal of quality assurance and quality control (QA/QC) in many instances. Therefore, even when CAD diagrams of the interiors of buildings are available, they are often not immediately useable within a geocoding system without a considerable amount of up-front effort, time, and expense (Isikdag and Zlatanova 2009).

The emerging field of Building Information Modeling (BIM), sometimes also referred to as building information science (BIS), is making great strides at narrowing the gap between the GIS and CAD communities (Döllner and Hagedorn 2007). Software systems such as Revit*, Rhino†, and 3D Studio Max‡ provide truly sophisticated object-oriented platforms for designing structures like buildings as well as more complex 3D scenes and full-fledged computer graphics suitable for feature films. These platforms enable architects to creatively design structures, while at the same time maintaining the rigid constraints that would be familiar to GIS professionals. The output of these systems can be easily and natively integrated within GIS systems such as the topic of this chapter, geocoding. Although presently on the cutting edge of research and practice in geocoding and LBS, the intersection of GIS, CAD, and BIM is an active area of research that has the potential to form the glue that enables true indoor and 3D geocoding.

3.6.3 SPATIOTEMPORAL GEOCODING

A third emerging area of geocoding and LBS research is the development of spatiotemporal geocoding algorithms. Currently, geocoding systems return an output representing a single snapshot in time (Murray et al. 2011). A user asks a geocoding system for the geographic position of some geographic feature either by postal address or name, and the geocoding system consults its reference databases and returns an output spatial location. What is implicit here is that the temporal extent of the output geocode—the time period which is applicable—is intimately related to the time period covered by the reference data layers used to compute the output (Krieger, Chen, et al. 2002). If the reference data layers were created in the year 2010, the geocode returned using these layers is applicable to the state of the world in 2010 when the reference data layer was generated or last updated.

Short of returning the vintage associated with the reference data layers used to compute the geocode, current geocoding systems do not capture the fact that the world changes over time. The end user of an LBS may be interested in the relationships of where something used to be, currently is, or is planned to be in the future (McElroy et al. 2003). One can imagine LBS applications where a user could enter the name of historic landmarks that have been removed from the landscape in order to create maps or virtual walking tours of locations that no longer exist. Similarly, one could create applications that allow the citizenry to perform virtual routing and searching in areas that are planned for development in the future. Multiple potential scenarios could be instantiated representing different views of how a future structure

---

* http://autodesk.com/revit/
† http://www.rhino3d.com/
‡ http://autodesk.com/3ds-max/

or region should be built to allow citizenry to evaluate which option is the most appealing from a variety of viewpoints—business, social, governance, and so on.

Although the current geocoding technologies of today do not typically differentiate between the time periods requested by a user, these enhancements are on the horizon. For example, it is common for the reference data utilized by a geocoding system to be updated to the latest versions available as data providers release updates on yearly or quarterly bases. These updates usually include enhanced geographic representations for features in the reference data layers that have been gathered opportunistically or systematically as well as updates to the nonspatial attributes as street names change, ZIP codes are added, removed, or merged, and so on. Similarly new real estate developments are added as they are surveyed and integrated into the reference data layer.

One immediate method that can be employed to take the first steps toward temporally enabling geocoding systems is simply to keep historical reference data sets instead of purging them from the system. With these multiple versions of historical reference data layers available, a user could request a geocode relevant to a specific time period, and LBS developers and service providers could begin to make use of spatiotemporal geocoding in novel and creative ways. One drawback to this practice could be that these databases typically take large amounts of disk space for storage. However, storage has become a low-cost commodity in recent years and the potential of storing large amounts of historical data is often no longer a concern for data managers and IT professionals. In addition, techniques can be used to avoid having to store the full data of every version. For example, an organization could store the full version of major releases and maintain only the changes from these base versions for intermediate releases.

3.7 CONCLUSIONS

This chapter has attempted to shed light on the relationship between geocoding systems and LBS in order to provide guidance and warning to LBS end users and developers in terms of the opportunities and challenges that geocoding systems offer to the LBS community. The various roles that geocoding systems play in providing context to LBS queries as well as generating the underlying reference data sets have been examined to emphasize the tight connections between these two interrelated technologies. The inner workings of modern geocoding systems have been discussed in detail to provide LBS developers and end users with an understanding of what happens under the hood as data are translated from text to geographic representations.

Particular attention has been paid to the various options that are present at each level of the geocoding process to highlight the fact that the potential for a great deal of variability exists when comparing two geocoding systems. No two geocoding systems perform all underlying tasks the same, utilize the same algorithms, or even use the same reference data layers for that matter. LBS developers, providers, and even end users must understand the intricacies of the geocoding systems they rely on to convert text to geography if they hope to be able to judge the quality or reliability of the data produced. As previously noted, the reputation of an LBS system rests completely in the quality of the results it provides to its end users. The process of geocoding is responsible for several critical tasks that contribute to overall LBS quality, and care must be paid when employing these technologies in the LBS query–data–service pipeline.

REFERENCES

Adams, P. M., G. W. B. Ashwell, and R. Baxter. 2003. Location-based services—an overview of the standards. *BT Technology Journal* 21 (1):34–43.

Amitay, E., N. Har'El, R. Sivan, and A. Soffer. 2004. Web-a-Where: Geotagging Web Content. In *SIGIR'04*. Sheffield, South Yorkshire, UK.

Bakshi, R., C. A. Knoblock, and S. Thakkar. 2004. Exploiting online sources to accurately geocode addresses. In *Proceedings of the 12th Annual ACM International Workshop on Geographic Information Systems*, 194–203. Washington, DC: ACM Press.

Beyer, K. M. M., A. F. Schultz, and G. Rushton. 2008. Using ZIP codes as geocodes in cancer research. In *Geocoding Health Data: The Use of Geographic Codes in Cancer Prevention and Control, Research, and Practice*, eds. G. Rushton, M. P. Armstrong, J. Gittler, B. R. Greene, C. E. Pavlik, M. M. West, and D. L. Zimmerman, 3768. Boca Raton, FL: CRC Press.

Boscoe, F. P. 2008. The science and art of geocoding. In *Geocoding Health Data: The Use of Geographic Codes in Cancer Prevention and Control, Research, and Practice*, eds. G. Rushton, M. P. Armstrong, J. Gittler, B. R. Greene, C. E. Pavlik, M. M. West, and D. L. Zimmerman, 95–109. Boca Raton, FL: CRC Press.

Boulos, M. N. K. 2003. Location-based health information services: A new paradigm in personalised information delivery. *International Journal of Health Geographics* 2 (2).

Brimicombe, A., and Y. Li. 2006. Mobile space-time envelopes for location-based services. *Transactions in GIS* 10 (1):5–23.

Cho, E., S. A. Myers, and J. Leskovec. 2011. Friendship and mobility: User movement in location-based social networks. In *The 17th ACM SIGKDD International Conference on Knowledge Discovery and Data Mining*, 1082–1090.

Corvey, W. J., S. Vieweg, T. Rood, and M. Palmer. 2010. Twitter in mass emergency: What NLP techniques can contribute. In *The NAACL HLT 2010 Workshop on Computational Linguistics in a World of Social Media*, 23–24.

D'Roza, T., and G. Bilchev. 2003. An overview of location-based services. *BT Technology Journal* 21 (1):20–27.

Dao, D., C. Rizos, and J. Wang. 2002. Location-based services: Technical and business issues. *GPS Solutions* 6 (3):169–178.

Davis Jr, C. A., G. L. Pappa, D. R. R. de Oliveira, and F. de L Arcanjo. 2011. Inferring the location of Twitter messages based on user relationships. *Transactions in GIS* 15 (6):735–751.

di Flora, C., M. Ficco, S. Russo, and V. Vecchio. 2005. Indoor and outdoor location based services for portable wireless devices. In *25th IEEE International Conference on Distributed Computing Systems Workshops*, 244–250.

Döllner, J., and B. Hagedorn. 2007. Integrating urban GIS, CAD, and BIM data by service-based virtual 3D city models. In *Urban and Regional Data Management*, eds. M. Rumor, V. Coors, E. M. Fendel, and S. Zlatanova, 157–170. London: Taylor & Francis.

Flanagin, A. J., and M. J. Metzger. 2008. The credibility of volunteered geographic information. *GeoJournal* 72 (3):137–148.

George, S. M., W. Zhou, H. Chenji, M. G. Won, Y. O. Lee, A. Pazarloglou, R. Stoleru, and P. Barooah. 2010. DistressNet: A wireless ad hoc and sensor network architecture for situation management in disaster response. *IEEE Communications Magazine* 48 (3):128–136.

Goldberg, D. W. 2008. *A Geocoding Best Practices Guide*. Springfield, IL: North American Association of Central Cancer Registries.

Goldberg, D. W. 2011. Improving geocoding match rates with spatially-varying block metrics. *Transactions in GIS* 15 (6):829–850.

Goldberg, D. W., and M. Cockburn. 2010. Toward quantitative geocode accuracy metrics. In *Accuracy 2010*, 329–332. Leicester, UK.

Goldberg, D. W., J. P. Wilson, and C. A. Knoblock. 2007. From text to geographic coordinates: The current state of geocoding. *Urisa Journal* 19 (1):33–47.

Goodchild, M. F. 2007. Citizens as sensors: The world of volunteered geography. *GeoJournal* 69 (4):211–221.

Goodchild, M. F., and J. A. Glennon. 2010. Crowdsourcing geographic information for disaster response: A research frontier. *International Journal of Digital Earth* 3 (3):231–241.

Hill, L. L. 2000. Core elements of digital gazetteers: Placenames, categories, and footprints. In *ECDL '00: Research and Advanced Technology for Digital Libraries. 4th European Conference*, eds. J. L. Borbinha and T. Baker, 280–290. Lisbon, Portugal: Springer.

Hill, L. L. 2006. *Georeferencing: The Geographic Associations of Information*. Cambridge, MA: MIT Press.

Isikdag, U., and S. Zlatanova. 2009. A SWOT analysis on the implementation of building information models within the geospatial environment. In *Urban and Regional Data Management*, eds. A. Krek, M. Rumor, S. Zlatanova, and E. Fendel, 15–30. Netherlands: Taylor & Francis.

Jiang, B., and X. Yao. 2006. Location-based services and GIS in perspective. *Computers, Environment and Urban Systems* 30 (6):712–725.

Krieger, N., J. T. Chen, P. D. Waterman, M. J. Soobader, S. V. Subramanian, and R. Carson. 2002. Geocoding and monitoring of US socioeconomic inequalities in mortality and cancer incidence: Does the choice of area-based measure and geographic level matter? *American Journal of Epidemiology* 156 (5):471–482.

Krieger, N., P. D. Waterman, J. T. Chen, M. J. Soobader, S. V. Subramanian, and R. Carson. 2002. ZIP code caveat: Bias due to spatiotemporal mismatches between ZIP codes and US census-defined areas: The public health disparities geocoding project. *American Journal of Public Health* 92 (7):1100–1102.

Lee, D. L., M. Zhu., and H. Hu. 2005. When location-based services meet databases. *Mobile Information Systems* 1 (2):81–90.

Lee, J. 2004. 3D GIS for geo-coding human activity in micro-scale urban environments. In *Third International Conference on Geographic Information Science, GIScience 2004*, eds. M. J. Egenhofer, C. Freksa and H. J. Miller, 162–178. College Park, MD.

Lee, J. 2009. GIS-based geocoding methods for area-based addresses and 3 D addresses in urban areas. *Environment and Planning B: Planning and Design* 36 (1):86–106.

Los Angeles County Assessor's Office. 2012. *GIS ready map base data*. Los Angeles, CA: Los Angeles County Assessor's Office.

Los Angeles County Department of Regional Planning. 2012. *Los Angeles Region Image Acquisition Consortium*. Los Angeles, CA: Los Angeles County Department of Regional Planning.

Mahmud, J., J. Nichols, and C. Drews. 2012. Where Is This Tweet From? Inferring Home Locations of Twitter Users. In *Sixth International AAAI Conference on Weblogs and Social Media*, 511–514.

Maue, P. 2007. Reputation as tool to ensure validity of VGI. In *Workshop on Volunteered Geographic Information*.

McElroy, J. A., P. L. Remington, A. Trentham-Dietz, S. A. Roberts, and P. A. Newcomber. 2003. Geocoding addresses from a large population based study: Lessons learned. *Epidemiology* 14 (4):399–407.

Mountain, D., and A. Macfarlane. 2007. Geographic information retrieval in a mobile environment: Evaluating the needs of mobile individuals. *Journal of Information Science* 33 (4):515–530.

Murray, A. T., T. H. Grubesic, R. Wei, and E. A. Mack. 2011. A hybrid geocoding methodology for spatio-temporal data. *Transactions in GIS* 15 (6):795–809.

NAVTEQ. *NAVTEQ Maps and Traffic* 2012. Available from http://www.navteq.com.

Oliver, M. N., K. A. Matthews, M. Siadaty, F. R. Hauck, and L. W. Pickle. 2005. Geographic bias related to geocoding in epidemiologic studies. *International Journal of Health Geographics* 4 (29).

Rao, B., and L. Minakakis. 2003. Evolution of mobile location-based services. *Communications of the ACM* 46 (12):61–65.

Raubal, M., H. J. Miller, and S. Bridwell. 2004. User-centred time geography for location-based services. *Geografiska Annaler: Series B, Human Geography* 86 (4):245–265.

Steiniger, S., M. Neun, and A. Edwardes. 2006. Foundations of location based services. *Lecture Notes on LBS* 1.

Tele Atlas Inc. 2010. *Mapping & navigation solutions: TeleAtlas* 2012. Available from http://www.teleatlas.com/index.htm.

Thomas, S., T. Fusco, A. Tokovinin, M. Nicolle, V. Michau, and G. Rousset. 2006. Comparison of centroid computation algorithms in a Shack–Hartmann sensor. *Monthly Notices of the Royal Astronomical Society* 371 (1):323–336.

U.S. Board on Geographic Names. 2012. *Geographic Names Information System* 2012. Available from http://geonames.usgs.gov.

U.S. Census Bureau. 2012. *U.S. Census Bureau TIGER/Line*. U.S. Census Bureau 2012. Available from http://www.census.gov/geo/www/tiger.

U.S. Postal Service. 2012. *Publication 28: Postal Addressing Standards*. Washington, DC: U.S. Postal Service.

Vanclooster, A., and P. Maeyer. 2012. Combining indoor and outdoor navigation: The current approach of route planners. *Advances in Location-Based Services,* 283–303.

Wieczorek, J., Q. Guo, and R. J. Hijmans. 2004. The point-radius method for georeferencing locality descriptions and calculating associated uncertainty. *International Journal of Geographical Information Science* 18 (8):745–767.

Worboys, M. 2011. Modeling indoor space. In *The 3rd ACM SIGSPATIAL International Workshop on Indoor Spatial Awareness*, 1–6.

Zandbergen, P. A. 2008. A comparison of address point, parcel and street geocoding techniques. *Computers, Environment and Urban Systems* 32:214–232.

Zeimpekis, V., G. M. Giaglis, and G. Lekakos. 2002. A taxonomy of indoor and outdoor positioning techniques for mobile location services. *ACM SIGecom Exchanges* 3 (4):19–27.

Zhan, F. B., J. D. Brender, I. De Lima, L. Suarez, and P. H. Langlois. 2006. Match rate and positional accuracy of two geocoding methods for epidemiologic research. *Annals of Epidemiology* 16 (11):842–849.

Zheng, B., J. Xu, W. C. Lee, and L. Lee. 2006. Grid-partition index: A hybrid method for nearest-neighbor queries in wireless location-based services. *The VLDB Journal—The International Journal on Very Large Data Bases* 15 (1):21–39.

Zhou, Y., X. Xie, C. Wang, Y. Gong, and W. Y. Ma. 2005. Hybrid index structures for location-based web search. In *Proceedings of the 14th ACM International Conference on Information and Knowledge Management*, 155–162.

Zook, M., M. Graham, T. Shelton, and S. Gorman. 2010. Volunteered geographic information and crowdsourcing disaster relief: A case study of the Haitian earthquake. *World Medical & Health Policy* 2 (2).

4 Multimodal Route Planning
Its Modeling and Computational Concerns

Lu Liu and Liqiu Meng

CONTENTS

ABSTRACT

This chapter deals with the modeling of the multimodal route planning problem as well as the development of optimal path-finding algorithms. The weighted digraph structure can well represent the fundamental static networks. The diverse and case-specific mode-switching actions in the real world, however, require some extra treatments. The authors put forward an efficient approach to describe such actions as switch points, which are somewhat analog to plugs and sockets between different mode graphs. Based on the well-defined data model, the multimodal routing problem can be formalized as a variant of the classical shortest-path problem. Consequently, the multimodal path-finding algorithms are designed as extensions of the shortest-path algorithms and verified with experiments.

4.1   INTRODUCTION

With the development of infrastructure construction for the purpose of facilitating mobility in a modern society, transportation networks become more and more complex. However, high-complexity, high-density, multilayer and multimodal transportation networks do not automatically bring convenience to users. They may rather confuse users' planning to go from one place to another. Transportation modes are the means by which people and freight achieve mobility (Rodrigue et al. 2009). In practice, the term *mode* is usually defined as a transportation means or a tool, for example, private car, bicycle, and underground train. Besides transportation means, a mode also refers to the type or functional class of a physical transportation network, for example, motorized road, pedestrian way, and bus line. Throughout this chapter, both implications are adopted.

Personal navigation services have become ubiquitous due to the rapid development of supporting such technologies as high-accuracy positioning, high-speed data communication, powerful mobile computation, efficient data processing and analysis, and openness and standardization of navigation data. Although a personal navigation service or application contains multiple navigation-related functions, and has a complex underlying workflow, it basically answers three questions: *where I am, where the destination is,* and *how to get there.* Route planning as an important function of the service is responsible for a part of the third question based on the results of the first two.

People may encounter difficulties when they try to plan routes in a complex transportation network across multiple different transport means with the existing navigation applications. For instance, most car navigation systems do not work in pedestrian-only areas. And a public transit route planner cannot tell a user, who has a bicycle available on departure, how to ride to a feasible station and then take a public transit line. Although some online route planning systems support different transportation modes, for example, Google Maps added "Walking," "By public transit," and "Bicycling" options in addition to "By car" in its "Get Directions" function for some areas as shown in Figure 4.1, each mode is performed totally independent from other modes, that is, one mode at a time.

In fact, more reasonable solutions can usually be yielded if two or more modes are taken into consideration when planning a route. Furthermore, some routing problems that are unsolvable in monomodal situations do have solutions in multimodal situations. For example, if a traveler wants to find a path in Munich from Albrechtstraße 37 to a spot near Scholss Nymphenburg 205, which is in a pedestrian-only area, there is definitely no direct motor way in between (Figure 4.2). Although a pedestrian route is possible between this pair of addresses, due to the length of the route, the travel time will be long. However, a double-modal and faster route is easily found by driving to the parking lot near the east gate of Schloßpark Nymphenburg first, parking the car there, then walking to the destination as demonstrated in Figure 4.3.

There are of course more complicated multimodal route planning cases in our everyday life. Unfortunately, automatic multimodal routing solutions are not available in conventional route-planning systems. With the increasing availability and integration of various types of road networks corresponding to different transportation

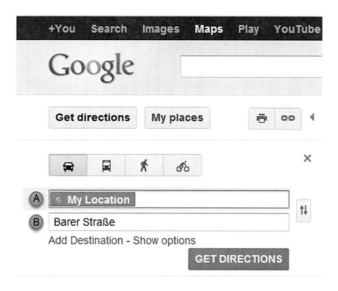

FIGURE 4.1   Four independent routing options for "Get Directions" in Google Maps.

FIGURE 4.2   **(See color insert.)** No direct path provided by Google Maps.

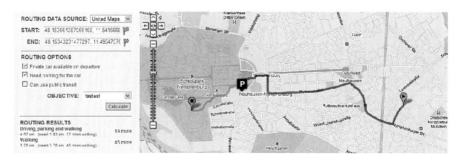

FIGURE 4.3   **(See color insert.)** A feasible double-modal route provided by multimodal route planner. (After Liu, L., 2011, Data model and algorithms for multimodal route planning with transportation networks, PhD thesis, Technische Universitaet Muenchen.)

modes, it can be anticipated that the multimodal route planning will soon become a popular service.

In a nutshell, *multimodal route planning* refers to the problem of route planning involving different transportation modes. The multimodal network modeling and optimal path-finding approaches are addressed in this chapter. They are intended to provide theoretical and technical support to the next generation of personal navigation services.

Behind all routing or route planning problems is the concept of the *shortest path*—the path(s) through the network from a known starting point to an optional ending point that minimizes distance, or some other measure based on distance, such as travel time. This is a classic problem in *graph theory* and has been extensively investigated for the past half a century. Here we provide a brief introduction to graph theory and the shortest path problem because they form the methodological foundation of this chapter. More details about the theory can be found in such textbooks as *Graph Theory* (Diestel 2006) and *Introduction to Algorithms* (Cormen et al. 2009) or in the up-to-date electronic book under GNU Free Documentation License* *Algorithmic Graph Theory* (Joyner et al. 2012).

A graph $G = (V, E)$ is an ordered pair of sets. Elements of V and $E \subseteq V \times V$ are called *vertices* and *edges*, respectively. V refers to the vertex set of G, and E the edge set in G. In the case that any direction of the edges is disregarded, G is referred to as an *undirected graph*.

One can label a graph by attaching labels to its vertices. If $(u, v) \in E$ is an edge of a graph $G = (V, E)$, u and v are said to be *adjacent* vertices. The edge (u, v) is also said to be *incident* with the vertices u and v. A *directed edge* is an edge such that one vertex incident with it is designated as the head vertex and the other as the tail vertex. A directed edge (u, v) is said to be directed from its tail u to its head v. A *directed graph* or *digraph G* is a graph with directed edges. The *indegree/outdegree* of a vertex $v \in V(G)$ counts the number of edges such that v is the head/tail of those edges. Similarly, the *incoming/outgoing* edges of a vertex $v \in V(G)$ are a set of edges such that v is the head/tail of those edges. Furthermore, the set of *in-neighbors/out-neighbors* of $v \in V(G)$ consists of all those vertices that contribute to the indegree/outdegree of v. A *multigraph* is a graph in which there are multiple edges between a pair of vertices. A *multiundirected* graph is an undirected multigraph. Similarly, a *multidigraph* is a directed multigraph.

A graph is said to be *weighted* when a numerical label or weight is assigned to each of its edges. There might be a cost involved in traveling from a vertex to one of its neighbors, in which case the weight assigned to the corresponding edge can represent such a cost. Sometimes, the weight is expressed by a cost function $c: E \rightarrow R$ mapping edges to real-valued weights. The term *cost* instead of weight is adopted in this chapter.

Based on the concept of weighted graphs, a *shortest-path problem* can be defined. Being given a weighted digraph $G = (V, E)$ with the edge cost function $c: E \rightarrow R$, the cost of path p from v to w is the sum of the edge costs for edges in p, denoted by $c(p)$. The minimal value of $c(p)$ for all paths from v to w is the *shortest-path cost* denoted by $\delta(v,w)$. A *shortest path* from vertex v to vertex w is then defined as any path p with cost $c(p) = \delta(v,w)$.

---

* http://www.gnu.org/copyleft/fdl.html

4.2 MULTIMODAL TRANSPORTATION NETWORKS

To represent the multimodal transportation networks, the real-world facilities including road networks and intermodal facilities should be modeled as data structures in computers. The typical modeling approaches can be categorized into two types:

1. *One planar graph with multilabeled edges/vertices*—The multimodal transportation networks are modeled as one graph according to the physical linkages. The multiple mode functions on one edge are distinguished by assigning different labels, for example, w for pedestrian ways, d for motorized roads, and so on (Barrett et al. 2000, 2008; Hoel et al. 2005; Pajor 2009).
2. *Multiple graphs connected with linking edges*—For each transit mode, a directed weighted graph is built to represent the topology. The discrete graphs can be connected via some static or dynamic linking edges. The representation of the spots where people can change from one mode to another, which is called switch point in this chapter, plays a significant role as bridges between different modal networks (Fohl et al. 1996; Goodchild 1998; David and Michael 2008; Ayed et al. 2009; Liu and Meng 2009).

This chapter is focused on the second modeling method. The transportation modes relevant in multimodal route planning within an urban area usually include a motorized road network for private car driving; a pedestrian way network for walking; and public transit networks including underground, suburban, and tram lines (see Table 4.1).

It can be observed from Table 4.1 that there is a one-to-one relationship between the transportation mode and the functional type of transportation network. So it turns out that given a specific mode, a network with identical functional type can be obtained. Such a network is termed as a mode graph according to the following definition.

Definition 4.1: Mode Graph

For a given mode m_i, the corresponding network is modeled as graph G_i. G_i is called the mode graph of m_i. ∎

Specifically, the road segments allowed for cars are modeled as a motorized graph, whereas those allowed for pedestrians are modeled as a pedestrian graph. In the implementation of a routing system, one edge is created for each street line in one direction. Therefore, it is not necessary to support multigraph in the data models and routing algorithms, which makes it easier to apply the classic non-multigraph-oriented path-finding algorithms on the data model.

For a public transit network, the timetable is of particular interest. The timetable is usually taken as the anchor point for public transit network models, and great efforts have been made to explore how the time-dependent schedule information

TABLE 4.1

Transportation Modes and Networks

Mode Type	Mode (abbr.)	Transportation Network	
		Functional Type	**Carrier Type**
Private	Walking (W)	Pedestrian-allowed	Road
	Car driving (D)	Private car-allowed	
Public	By underground train (U)	Underground line	Railway
	By suburban train (S)	Suburban line	
	By tram train (T)	Tram line	

Source: Liu, L., 2011, Data model and algorithms for multimodal route planning with transportation networks, PhD thesis, Technische Universitaet Muenchen.

can be best integrated with the time-independent physical networks. Keeping timetable information separated, the public transit network can be modeled in a similarly straightforward and concise way to the road network. However, instead of modeling each public transit link between two stations as an edge in the corresponding mode graph directly, preprocessing is necessary due to the fact that a typical public transit network dataset is insufficient to construct the corresponding mode graph directly. The networks should be modeled as detailed as the platform level to guide a pedestrian from one platform to another in public transit networks.

On the basis of mode graphs, the mode-switching action can be modeled. In modern urban transportation systems, many intermodal facilities, such as parking places, park-and-ride lots, transit hubs, and trailheads, serve as "bridges" linking different mode graphs and allow the easy transfer between different transportation modes. With the digitalization of intermodal facilities and their incorporation into navigation databases, automatic multimodal route planning is made possible. The intermodal facilities can be modeled as *switch points*.

Liu and Meng (2009) defined a switch point matrix (SPM) where all special conditions to be fulfilled by a switching action between two different modes can be recorded. The *special conditions* under which a mode is allowed to switch to another were illustrated by the example of SPM in transportation (SPM-T). By looking up in the SPM, the answer to the question "What kinds of facilities can I use when changing from mode A to mode B?" can be given. In other words, the item in the matrix at row m and column n indicates the feasible types of switching actions from mode m to mode n. Thus, the special conditions are defined as switch types.

Definition 4.2: Switch Type

A switch type λ is a class corresponding to one type of possible mode-switching actions in a specific application domain. $\Lambda = \left\{\lambda_i | i \in \mathbb{N}\right\}$ consists of all switch types in that domain. ∎

For an arbitrary mode pair, a subset $\Lambda' \subseteq \Lambda$ can be determined based on domain-specific knowledge. The original SPM can be scrutinized to a more general form switch type matrix (STM), see Equation (4.1), where N is the number of modes. The STM-T (originally called SPM-T) is an instance in the domain of transportation.

$$\text{STM} = \left[\Lambda'_{i,j}\right]_{N \times N} \tag{4.1}$$

Note that Λ' can be $\varnothing$, meaning the nil-elements in the SPM. Λ' can also contain more than one element, which means there are different possible switch types between two modes. Taking the Λ' in STM-T between the modes of car driving and walking, for example, there are at least two switch types. If the user is a driver, an available parking lot is necessary for this double-modal route and thus all the vertex pairs meeting this condition are eligible for switching. However, if the user is only a passenger, which means somebody else drives the car (e.g., a taxi driver), any position allowed for temporary parking in the road network can be used for mode switching in this case.

Let $M = \left\{m_i \big| i \in \mathbb{N}\right\}$ be a set of modes, Λ a set of switch types, $G_M = \{G_i = \{V_i, E_i\}\}$ the multimodal graph set (MMGS) consisting of the graphs with each corresponding to a mode in M, $V = \bigcup V_i$, $E = \bigcup E_i$, and C a set of switching-cost functions. The following concepts can be defined.

Definition 4.3: Switch Relation

A switch relation Γ is a 6-ary relation on $M \times M \times \Lambda \times V \times V \times C$ defined as follows:

$\left(m_f, m_t, \lambda, v_f, v_t, c\right)$ a switching action from vertex v_f in the graph of mode m_f to vertex v_t in the graph of mode m_t with the type of λ and cost c ∎

Definition 4.4: Switch Point

Given a switch relation Γ, a switch point γ is identified by a 6-tuple in Γ.   ∎

Definition 4.5: Switch Condition

Given a switch relation Γ, a switch condition φ is a set of criteria expressed by a propositional formula applying on Γ.   ∎

From the point of view of a relational database system, Γ can be implemented as a table consisting of a set of 6-tuples expressed by γ, and φ can correspond to the predicates in a WHERE clause of SQL selection operation.

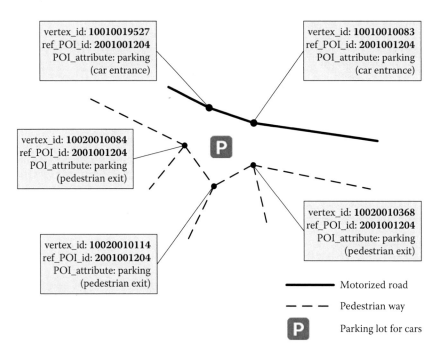

FIGURE 4.4   Attributes of a parking lot for cars. (After Liu, L., 2011, Data model and algorithms for multimodal route planning with transportation networks, PhD thesis, Technische Universitaet Muenchen.)

TABLE 4.2
Eligible Switch Points for the Attributes of the Parking Lot in Figure 4.4

from_mode (m_f)	to_mode (m_t)	type (λ)	from_vertex (v_f)	to_vertex (v_t)	cost* (c)
car_driving	walking	car_parking	10010019527	10020010084	2.5
car_driving	walking	car_parking	10010019527	10020010114	4
car_driving	walking	car_parking	10010019527	10020010368	3.5
car_driving	walking	car_parking	10010010083	10020010084	3.5
car_driving	walking	car_parking	10010010083	10020010114	4.5
car_driving	walking	car_parking	10010010083	10020010368	2

Source: Liu, L., 2011, Data model and algorithms for multimodal route planning with transportation networks, PhD thesis, Technische Universitaet Muenchen.
* The metric of the cost in this table is an estimated time (in minutes) of the corresponding switching.

If we take the parking lot for cars as an example, an attributed expression can be given as shown in Figure 4.4, and based on the given definitions, six eligible switch-point-tuples in this case are listed in Table 4.2 expressing a subset of Γ.

Furthermore, a switch point can be time dependent and assigned with some case-specific constraints. Such characteristics make it easy to express the dynamics in multimodal route planning, for example, the timetable of public transit systems.

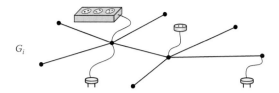

FIGURE 4.5   A mode graph with plugs and sockets. (From Liu, L., 2011, Data model algorithms for multimodal route planning with transportation networks. PhD thesis, Technische Universitaet Muenchen.)

With the support of the network model and the concept of switch point, the overall data model of the multimodal transportation network can be completed by properly combining them. By giving necessary switch conditions, some mode graphs in the MMGS may form a subset containing the elements conditionally *pluggable* to each other. It should be noted that these input conditions are case dependent, which means they may differ from one concrete multimodal routing task to another. According to the definitions of switch condition, switch relation, and switch point, it is clear that a set of switch points can be retrieved with the given conditions. And every vertex pair within one switch point is just like a pair of plug and socket, which allows *plug-and-play* of the related mode graphs.

For a mode graph $G_1 \in G_M$, V_1' is the set of *plug vertices* in G_1 if $V_1' = \{v \mid v \in \sigma_{m_f = m_1} (\Gamma).v_f\}$, and the set of *socket vertices* if $V_1' = \{v \mid v \in \sigma_{m_f = m_1} (\Gamma).v_f\}$, where σ is the selection operation in relational algebra. A graph with plug–socket vertices is illustrated in Figure 4.5. It should be noted that the analogy between plug–socket and the vertex pair within a switch point implies that the connection from a plug to a socket vertex is directional.

For any ordered mode graph pair $(G_1, G_2) \in G_M \times G_M$, given a switch condition $\varphi_{1,2}$ whose subscript implies that it at least satisfies the criteria $(m_f = m_1) \wedge (m_t = m_2)$, the ordered mode graph pair (G_1, G_2) is said to be *one-way pluggable* iff $\Gamma_{1,2}' = \sigma_{1,2} (\Gamma)$ is not $\varnothing$. Furthermore, (G_1, G_2) is *double-way pluggable* iff none of $\Gamma_{1,2}'$ and $\Gamma_{2,1}'$ is $\varnothing$. Figure 4.6 illustrates three pluggable graphs where (G_1, G_2) and (G_3, G_1) are one-way pluggable, while (G_2, G_3) is double-way pluggable.

It is obvious that the motorized and pedestrian graphs pair is one-way pluggable under the condition of switch type being parking lot for cars, while the pedestrian and underground graphs pair are double-way pluggable under the condition of switch type being underground stations. With the support of the aforementioned definitions and discussions, the plug-and-play multimodal graph set can be defined.

Definition 4.6: Plug-and-Play Multimodal Graph Set (PnPMMGS)

Given a switch relation Γ, a MMGS G_M is a plug-and-play multimodal graph set (PnPMMGS) if and only if $\exists (G_1, G_2) \in G_M \times G_M$ is one-way or double-way pluggable. ∎

For a concrete multimodal path-finding job, let $M' \subseteq M$ be the set containing n modes that will be involved in the route calculation and $G_{M'} \subseteq G_M$ correspondingly.

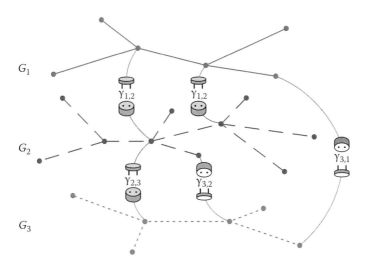

FIGURE 4.6 Example of one-way and double-way pluggable graphs. (After Liu, L., 2011, Data model and algorithms for multimodal route planning with transportation networks, PhD thesis, Technische Universitaet Muenchen.)

If an ordered graph pair $(G_1, G_2) \in G_{M'} \times G_{M'}$ is one-way pluggable for a given $\varphi_{1,2}$, $\varphi_{1,2}$, is recorded. Otherwise, nil is recorded if $\Gamma'_{1,2} = \varnothing$. All the $\varphi_{i,j}$ constitute the switch condition matrix (SCM), which is denoted as SCM = $[\varphi_{i,j}]_{n \times n}$.

The SCM can be utilized to connect the multiple discrete mode graphs (or some of them) under the given conditions, which may reduce design difficulty of the routing algorithm.

4.3   MULTIMODAL ROUTING ALGORITHMS

The design of the routing algorithms for computing multimodal optimal paths heavily couples with the underlying data model. Accordingly, different types of algorithms have been developed to find paths on the data models mentioned in the previous section.

Specifically, label-constrained (or regular language constrained) multimodal routing algorithms are designed to find optimal paths on a planar graph with multimodal labels on edges/vertices. The theory and application of these algorithms are introduced in Barrett et al. (2008), Delling, Pajor, et al. (2009), Delling, Sanders, et al. (2009), and Pajor (2009).

In a nutshell, the multimodal urban transportation network is modeled as a set of weighted, directed time-independent graphs (i.e., MMGS), and a set of time-independent or time-dependent switch points. When there exist pluggable graph pairs in the graph set, MMGS becomes PnPMMGS. For a concrete multimodal route-planning task with defined context and concrete input parameter values, the modes may either be manually input by the user or determined by the algorithm. Accordingly, the problems are formalized as two types of multimodal shortest path problems.

4.3.1 MULTIMODAL SHORTEST PATH PROBLEM WITH DETERMINISTIC MODE SEQUENCE

For the multimodal shortest-path problem with deterministic mode sequence, the inputs of the algorithm are:

- a mode set M;
- a nonnegative weighted, digraph set $G_M = \{G_i = \{V_i, E_i\}\}$ denoting MMGS;
- a set of cost functions $C_M = \{c_i : E_i \to R^+\}$;
- and a switch relation Γ.

Besides these four items, for each concrete MMSP calculation task, four extra components are given:

- a sequential list $M' = (m_1, m_2, \ldots, m_n), n \geq 1, m_k \in M$, indicating the modes to be involved in the path-finding process;
- a switch condition list $\Phi = (\varphi_1, \varphi_2, \ldots, \varphi_{n-1})$, where φ_k corresponds to the switch condition from m_k to m_{k+1};
- and a source vertex $s \in V_1$.

For generality, we solely consider the single-source multimodal shortest-paths problem here since the solution to the point-to-point case can be acquired by applying stop conditions in the iterations. Then the algorithm for single-source multimodal shortest paths problem can be generally described by the algorithmic framework (Algorithm I).

Algorithm I

Input: M', Φ, s
Step 1: Set k ← 1.
Step 2: Set $Q \leftarrow \varnothing$; Execute MultimodalInitialize(k, nil, s, Q) when k = 1, or MultimodalInitialize(k, φ_{k-1}, s, Q) when k > 1; IterativeRelax(k, Q); k ← k +1.
Step 3: Repeat step 2 until k > n.
The routine MultimodalInitialize works as following:

MULTIMODALINITIALIZE

Input: k, φ, s, Q
Step 1: Begin by setting $distance[k][u] \leftarrow \infty$ and $predecessor[k][u] \leftarrow$ nil for each vertex $u \in V_k$.
Step 2: If φ is nil, set $distance[k][s] \leftarrow 0$, $predecessor[k][s] \leftarrow s$ and add s to Q; otherwise go to step 3.
Step 3: If $distance[k][\gamma.v_t] > distance[k-1][\gamma.v_f] + \gamma.c$, set $distance[k][\gamma.v_t] \leftarrow distance[k-1][\gamma.v_f] + \gamma.c$, $predecessor[k][\gamma.v_t] \leftarrow \gamma.v_f$ and add $\gamma.v_t$ to Q for each tuple $\gamma \in \sigma_\varphi(\Gamma)$.

The IterativeRelax routine iteratively relaxes the edges in G_k and can be substituted by any existing single-source shortest path (SSSP) algorithm based on labeling method with the initialization phase removed. A generic IterativeRelax routine can be described as a three-step process as follows:

ITERATIVERELAX

Input: k, Q

Step 1: Remove a vertex u from the candidate list Q.

Step 2: If $distance[k][v] > distance[k][u] + c_k(u,v)$, set $distance[k][v]$ $\leftarrow distance[k][u] + c_k(u,v)$ and $predecessor[k][v] \leftarrow u$; add v to Q if $v \notin Q$ for each outgoing edge (u, v).

Step 3: If $Q \neq \varnothing$, then go to step 1.

Both label-setting and label-correcting methods can be applied into the algorithmic framework illustrated in Algorithm I. Their fundamental difference lies in the order according to which the distance labels are updated. Label-setting algorithms set one label as permanent in each step, whereas label-correcting algorithms consider all labels as temporary until the final step, when they all become permanent. Algorithm I has a computational complexity of $O\left(\sum_{k=1}^{n} g(|V_k|, |E_k|)\right)$, where $O\left(g(|V_k|, |E_k|)\right)$ is the computational complexity of IterativeRelax. ∎

Algorithm I finds shortest paths on the corresponding mode graph sequentially. The key action is the relay of the distance values on switch points during the initialization step in each round of path searching from the mode m_2 to m_n through which the MMSP can be finally found.

4.3.2 MULTIMODAL SHORTEST PATH PROBLEM WITH NONDETERMINISTIC MODE SEQUENCE

Instead of having the mode sequence as a part of the input, the optimal mode combination is to be automatically generated in this second type of the problem. Compared with the deterministic mode sequence, the basic data structures of M, G_M, C_M, and Γ are the same. Besides, the case-dependent part of input contains a subset $M' \subseteq M$ including n modes and a switch condition matrix (SCM).

This second type of MMSP problem is defined by introducing an auxiliary graph G whose adjacency matrix is expressed by the SCM.

Similar to the first type of MMSP problem, we consider the problem with single source in this type of problem. With regard to the inputs of the algorithm, the only difference from that of the first type of MMSP problem is that the switch conditions are expressed by the matrix SCM instead of a switch condition list Φ.

By applying a so-called SCM-Plug operation, which enables the connection of the involved mode graphs, the algorithm for a single-source MMSP problem with nondeterministic mode sequence can be depicted as a two-step process:

Algorithm II

Input: M', SCM, s
Step 1: $G, c \leftarrow$ SCM-Plug(M', SCM).
Step 2: Execute ShortestPathSearch on G with s.

The SCM-Plug is described as:

SCM-Plug

Input: M', SCM
Step 1: Initialize a new graph G consisting of an empty vertex set V and an empty edge set E, and its edge cost function is initialized to nil.
Step 2: Assign to V and E with vertex/edge sets union: $V \leftarrow \bigcup_{i \in M'} V_i$; $E \leftarrow \bigcup_{i \in M'} E_i$; the cost function on E is inherited from those on its constituent edge subsets.
Step 3: Let $(i, j) \in M' \times M'$ be an ordered mode pair.
Step 4: Execute the selection operation on switch relation table Γ by giving switch condition obtained from SCM, and get a result set of switch points Γ'.
Step 5: For each Switch Point γ in Γ', construct an edge e by assigning the switch-from, switch-to vertices and switching cost values to e as its starting, ending vertices, and edge cost, and append e to the edge set E.
Step 6: Repeat steps 4 and 5 for every mode pair $(i, j) \in M' \times M'$ if $\varphi_{i,j}$ in SCM is not nil.
Step 7: Return G consisting of V and E, and its edge cost function.

It turns out that the switch condition list Φ for Algorithm II is a special case of SCM$_\Phi$, which is expressed in Equation (4.2). In other words, the plug-and-play multimodal shortest-path problem can be solved by applying the SCM-Plug operation first and then a traditional SPA on the monomodal graph. The routing results by the two approaches are exactly the same as declared by Theorem 4.1.

$$
\text{SCM}_\Phi =
\begin{array}{c}
\\
m_1 \\
m_2 \\
\vdots \\
m_{n-1} \\
m_n
\end{array}
\begin{array}{cccccc}
m_1 & m_2 & m_3 & \cdots & m_{n-1} & m_n \\
\left[\begin{array}{cccccc}
\text{nil} & \varphi_1 & \text{nil} & \cdots & \text{nil} & \text{nil} \\
\text{nil} & \text{nil} & \varphi_2 & \cdots & \text{nil} & \text{nil} \\
\vdots & \vdots & \vdots & \ddots & \ddots & \vdots & \vdots \\
\text{nil} & \text{nil} & \text{nil} & \cdots & \text{nil} & \varphi_{n-1} \\
\text{nil} & \text{nil} & \text{nil} & \cdots & \text{nil} & \text{nil}
\end{array}\right]_{n \times n}
\end{array}
\tag{4.2}
$$

Lemma 4.1: Effectiveness of the Algorithm I

Given M, G_M, C_M, Γ, M', Φ, and $s \in V_1$ as the inputs to Algorithm I, we can have the shortest distance values from s on all the vertices $v \in V_{M'} = \bigcup V_k, k \in M'$ after the algorithm terminates.   ∎

Theorem 4.1: Equivalence of Algorithm I and Algorithm II

Given M, G_M, C_M, Γ, M', and its corresponding set M', Φ and its matrix expression SCM_Φ, $s \in V_1$, we have the following equivalence relation:

$$\text{Algorithm I}(M', \Phi, s) \Leftrightarrow \text{Algorithm II}(M', SCM_\Phi, s)$$

And after both algorithms terminate, we have the same shortest distance values from s on all the vertices $v \in V_M = \bigcup V_k, k \in M'$.

The advantage of Algorithm II lies in that the final optimal mode sequence can be given by the algorithm rather than input manually. This characteristic makes it suitable in the situation where a definite mode sequence is difficult to determine in advance, for example, the route planning task in a public transit network consisting of underground, suburban, tram, and bus lines altogether or some of them. In practice, Algorithm I and Algorithm II can be used in a hybrid fashion by treating the graph obtained by the SCM-Plug operation as one of the monomodal graphs in the sequential list of input for Algorithm I. In a prototype system, the problem of route planning for the combination of car driving and public transit is solved by applying the SCM-Plug operation on all the selected public transit networks and pedestrian network first, then constructing a (*car-driving, public transit*) mode list as well as feasible switch conditions and delivering them to Algorithm I.

4.4 MULTIMODAL ROUTE PLANNING IN REAL APPLICATIONS

The multimodal routing approach introduced in this chapter has already been applied in an online path-planning system. Case studies covering two German cities have been made in detail by Liu (2011) with the system. Figure 4.7 and Figure 4.8 show some typical routing results provided by the system. Different modes of path segments are indicated in different colors.

In the case illustrated by Figure 4.7, the user with a private electric car requests the fastest path between the origin and the destination. Given that the battery of the car can only support 7 km, the route planner suggests the user driving to a charging station first, then go to the destination by tram. In the case shown in Figure 4.8, the user indicates to the system that no private car is available on departure, while all the public transit means can be taken into account. Then the system provides a multimodal path consisting of tram lines, suburban lines, and pedestrian ways.

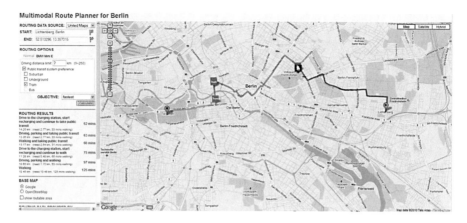

FIGURE 4.7   (See color insert.) Multimodal route from Lichtenberg (Siegfriedstraße 203) to a pedestrian junction with coordinates of longitude 52.513, latitude 13.357 in Tiergarten in Berlin. (After Liu, L., 2011, Data model and algorithms for multimodal route planning with transportation networks, PhD thesis, Technische Universitaet Muenchen.)

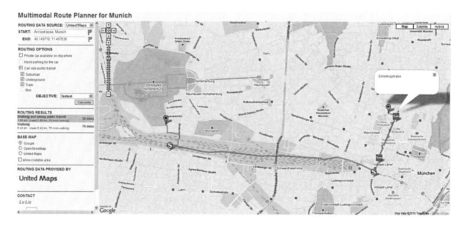

FIGURE 4.8   (See color insert.) Multimodal route from the crossing of Arcisstraße and Heßstraße near Technische Universität München to a pedestrian path junction with coordinates of longitude 48.150 and latitude 11.497 in Schloßpark Nymphenburg. (After Liu, L., 2011, Data model and algorithms for multimodal route planning with transportation networks, PhD thesis, Technische Universitaet Muenchen.)

4.5   SUMMARY

The increasing demand on intelligent multimodal navigation services and location-based services has stimulated our study on the automatic multimodal route planning system with the goal of finding reasonable and optimal paths in the context of multimodal transportation networks involving motorized roads, pedestrian ways, underground, suburban, and tram lines. To achieve the goal, great efforts have been made on how to properly express the multimodal transportation infrastructure including networks and intermodal facilities on one hand. On the other hand, the optimal path-finding approaches have been investigated in depth based on the multimodal network data model.

To sum up, the transportation networks can be modeled as a set of mode graphs according to their functional types, and the intermodal facilities, for example, parking lots and public transit stations, can be modeled using the concept of switch point. The discrete mode graphs can be plugged to each other by switch points in a flexible way in the process of path planning. The routing algorithms developed on the basis of the data model are rooted in the traditional shortest path algorithms. They can deal with the multimodal optimal path finding requests regardless of whether the mode sequence can be determined in advance.

At present, the vast application of SNS (social networking service), location-based services and their combination, that is, location-based social networking, are changing the way people use the Internet and mobile devices as well as the transportation facilities. The social–location–mobile (SoLoMo) applications are attracting more interests from academic and industrial fields. Following such a trend shaped by mobile computing, social computing, and location-based services, the socially enabled multimodal routing service will play an important role. In this sense, the approach introduced in this chapter may provide strong support to the location-based social networking services both in theory and practice.

REFERENCES

Ayed, H., Zineb, H., Carlos, G.-F., and Djamel, K., 2009. Hybrid algorithm for solving a multimodal transport problems using a transfer graph model. *Proceedings of the Second International Conference on Global Information Infrastructure Symposium.* Hammamet, Tunisia: IEEE Press.

Barrett, C., Bisset, K., Holzer, M., Konjevod, G., Marathe, M., and Wagner, D., 2008. Engineering label-constrained shortest-path algorithms. In *Algorithmic aspects in information and management,* R. Fleischer and J. Xu, eds., 27–37. Berlin/Heidelberg: Springer.

Barrett, C., Jacob, R., and Marathe, M., 2000. Formal-language-constrained path problems. *SIAM Journal on Computing,* 30 (3), 809–837.

Cormen, T. H., Leiserson, C.E., Rivest, R. L., and Stein, C., 2009. *Introduction to algorithms,* 3rd ed. Cambridge, MA: MIT Press.

David, E., and Michael, T. G., 2008. Studying (non-planar) road networks through an algorithmic lens. *Proceedings of the 16th ACM SIGSPATIAL international conference on Advances in Geographic Information Systems.* Irvine, California: ACM.

Delling, D., Pajor, T., and Wagner, D., 2009. Accelerating multi-modal route planning by access-nodes. In *ESA 2009,* A. Fiat and P. Sander, eds., 587–598. Heidelberg: Springer.

Delling, D., Sanders, P., Schultes, D., and Wagner, D., 2009. Engineering route planning algorithms. In *Algorithmics of large and complex networks,* J. Lerner, D. Wagner, and K. Zweig, eds., 117–139. Berlin: Springer-Verlag.

Diestel, R., 2006. *Graph theory.* New York: Springer-Verlag.

Fohl, P., Curtin, K. M., Goodchild, M. F., and Church, R. L., 1996. A non-planar, lane-based navigable data model for its. *Seventh International Symposium on Spatial Data Handling.* Delft, 7B.17–7B.29.

Goodchild, M. F., 1998. Geographic information systems and disaggregate transportation modeling. *Geographical Systems,* 5, 19–44.

Hoel, E. G., Heng, W.-L., and Honeycutt, D., 2005. High performance multimodal networks. In *Advances in spatial and temporal databases,* C. B. Medeiros, M. Egenhofer, and E. Bertino, 308–327. Berlin/Heidelberg: Springer.

Joyner, D., Nguyen, M. V., and Cohen, N., 2012. *Algorithmic graph theory,* Version 0.7-r1984. GNU-FDL. https://code.google.com/p/graphbook

Liu, L., 2011. Data model and algorithms for multimodal route planning with transportation networks. PhD thesis, Technische Universitaet Muenchen.

Liu, L., and Meng, L., 2009. Algorithms of multi-modal route planning based on the concept of switch point. *Photogrammetrie Fernerkundung Geoinformation,* 2009 (5), 431–444.

Pajor, T., 2009. Multi-modal route planning. Thesis, Universität Karlsruhe (TH).

Rodrigue, J.-P., Comtois, C., and Slack, B., 2009. *The geography of transport systems,* 2nd ed. New York: Routledge.

Section 2

New Trends

5 Location-Based Social Networks

Prashant Krishnamurthy and
Konstantinos Pelechrinis

CONTENTS

ABSTRACT

The emergence of smartphones with rich applications and with the ability of positioning such devices in real-time has made location-based social networks (LBSNs) increasingly popular. In this chapter, we discuss the basics of LBSNs with specific examples of commercial LBSN services offering different scopes and objectives. We also discuss how the data from LBSNs are being employed by the research community for understanding and addressing a variety of problems.

5.1 INTRODUCTION

The analysis and understanding of social networks and related social phenomenon has been ongoing for many decades now. The emergence of the Internet has created a new ability to examine, analyze, and understand social networks. The reasons for this include the ability to create new kinds of social links through the Internet, the ability to obtain the data that characterizes these links, and finally the ability to have sufficient numbers of social links to make meaningful analyses. For example, it is now possible to analyze the relationships and dynamics between the authors of Wikipedia articles at as fine granularity as possible in time, and at the same time be able to connect these dynamics to some context in terms of the messages exchanged, the content being edited, the sentiments associated with the communications, and so on (Yasseri et al. 2012). The availability of the technology and analytical tools to perform comprehensive and complicated analyses adds to the richness of the research in social networks. Such a possibility did not exist a decade or so back, and this has invigorated the work in social network analysis and its potential implications on society. The use of online social networks (OSNs) to (a) understand social effects and (b) exploit the understanding for applications such as business intelligence, marketing, social welfare, study of epidemics, and so on has exploded in recent years (Easley and Kleinberg 2010).

Online social networks typically assume that *links* or connections exist between individuals or members of the network without regard to physical distance or location (although there are exceptions). For example, a social graph of friends on Facebook simply assumes that there is a connection between two individuals on Facebook who are friends without considering where they are located. If Alice lives in Pittsburgh and has friends in Pittsburgh and San Francisco, they are both equivalent in terms of the OSN structure. Further, the lack of location information and lack of understanding of the influence of location on the social network (e.g., the extent of the role of geography and physical distance on developing social connections between individuals and development of a community), in some ways, limit the benefits of analyses on such OSNs. This is particularly important because traditionally social networks have been dependent on closeness in physical space. The probability that two individuals maintain connections decreases with increasing physical distance in traditional social settings (e.g., frequency of face-to-face contact described by Mok, Wellman, and Basu (2007). With the emergence of the Internet, the question arises whether this is still the case or whether social connections are independent of geographical distance. As we discuss in Section 5.3, it is still the case that connections (friendship links in an LBSN service) are *more likely* to exist at closer geographical distances and this conclusion is made possible through the study of social networks with embedded location information (Scellato, Noulas, and Lambiotte 2011).

The emergence of smart phones now equipped with the ability to position the device using the Global Positioning System (GPS) or other technologies (such as WiFi access points) has enabled the inclusion of location information with people, over time, over venues and points of interest, with images, with contextual text, and so on at a scale that was previously impossible before the advent of smartphones. With the inclusion of social links, there now exists a mechanism to analyze social networks with location information included. We broadly refer to such networks as *location-based social networks* (LBSNs) in this chapter. LBSNs provide location information, often with context as we explain later, and social information in terms of friendships and social links. However, they are also potential causes for privacy problems, since revealing location information and temporal information can allow identification of individuals, tracking of individuals, inferring an individual's income level or associations, and so forth as described in Section 5.3.

In this chapter, we provide an overview of LBSNs. In Section 5.2, we consider the basics of LBSNs and some example LBSNs. Section 5.3 provides some interesting analyses of LBSNs that have been recently performed. Section 5.4 provides a summary of the chapter.

5.2 BASICS OF LOCATION-BASED SOCIAL NETWORKING

In this section, we discuss the basic features of LBSNs. We first describe the types of location information that an LBSN may record, describe the concept of check-in, and discuss some types of LBSNs with real examples. We note here that the landscape of LBSN services is rapidly changing and some of the commercial services described in this chapter may change or even disappear over time.

We first briefly discuss two important aspects of studying social networks. Social network data have been used for understanding how communities are created. The data have also been employed to understand and solve problems in various domains such as business, marketing, and medicine. In the former, among the primary concerns that are considered include why individuals make connections with other individuals. Sometimes the connections are not with individuals but are rather with affiliations such as churches, restaurants, and organizations. It is believed that there are two ways in which social connections are created, namely, *homophily* and *focus constraint* (Easley and Kleinberg 2010; Expert et al. 2011). Homophily refers to the creation of social connections because of the similarity of individuals (e.g., two individuals both work in the same place, have similar incomes or interests, or social status). Focus constraint refers to the potential opportunity for social connections to be created (i.e., individuals must have an opportunity to interact). As previously mentioned, closeness in physical space affords such opportunities but the emergence of virtual interactions through the Internet could potentially reduce this need for physical colocation. Further, other aspects of social community creation are examined in the former. For instance, once a social connection gets created between *A* and *B* (dyadic connection) and a link gets created between *B* and *C*, a *triadic closure* is likely (i.e., *A* and *C* also get socially connected eventually). In the latter, the impact of social networks on potential application domains is examined. For instance, are there influential people in a social network that can spread information for marketing? Similarly, are there hubs that cause the spread of diseases in a community? Here the emphasis is on how the *structure* of a social network can be exploited for social or commercial benefit or to predict or explain phenomena that are heavily dependent on the social network structure.

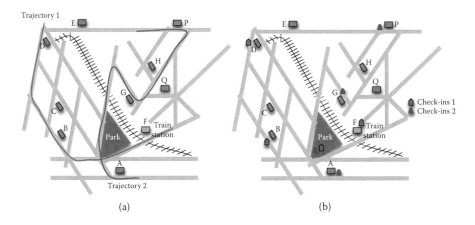

FIGURE 5.1 **(See color insert.)** Types of location information in LBSNs (a) trajectory based and (b) venue based.

5.2.1   Location Information in LBSNs

LBSNs include *location* information about the members of a social network at various points in time. The location information might be the latitude or longitude of a geographical position that the member was present at a given time, or it may be *semantic* or *contextual* information such as a venue or point of interest where the member was present at a given time.

Figure 5.1 shows examples of two types of location information that may be available in an LBSN. In Figure 5.1a, the continuous trajectory of a member of a social network as a function of time is maintained and often displayed on a map. In Figure 5.1b, the complete trajectory is not maintained, but *discrete* locations at *discrete* times when a member was present are available. Such locations have contextual information (e.g., train station) that is often missing in a trajectory-based LBSN. For example, in Figure 5.1a, the complete trajectories of two members of an LBSN are shown. Although the starting points, destinations, and routes are available, either trajectory does not clearly identify specific contextual information about points along the trajectory. In Figure 5.1b, the trajectories of the two members are not available, but instead, *check-ins* of the members at various venues are available. Check-ins correspond to identification of specific venues where a member was present (we return to check-ins later in this section). Both members of the LBSN have checked-in at a park, indicating that this is perhaps a common focus of interest for them. The check-ins of the member identified by the red color indicate that the member may be traveling by train or may have picked up a visitor at the train station.

Each type of location information has advantages and disadvantages. The former collects the positions of members of the LBSN at a much finer granularity that shows their movement but misses the contextual information. The latter includes rich contextual information but does not include every location that a member of the LBSN may have visited.

5.2.2   THE CONCEPT OF CHECK-INS

The social aspect of LBSNs works in a manner similar to many other OSNs. It is possible to see other members of OSNs that are close by (in location) by a member. The member can send requests to become a friend of such members. Offline communications can also be used by members to discover that they are using a particular LBSN service and later connect using the application provided by the service. Friendship graphs similar to other OSNs are possible with LBSN services.

The location aspect of most commercial and widely popular LBSN services uses the concept of "check-ins." Generally, the user or member of an LBSN service has to manually let the service know that he or she is at a particular location at a particular time. The LBSN service usually provides an application that runs on the mobile device that the member uses. The application accesses the location information provided by the device using GPS or WiFi positioning, and presents the user with a list of venues or locales that are close by. The member manually selects the correct venue and may decide to check-in to the LBSN service. In other words, the member is letting the service know that he or she is there at that venue at that time. The accuracy and precision of positioning are the reasons why a venue is not automatically selected by the application. If a venue is not in the presented list, the member can manually enter the name and details of the venue that he is visiting. This also helps the LBSN service populate its database with new venues. The details of such venues may include category, address, and phone numbers.

We note here that check-ins are voluntary. That is, a member of an LBSN service need not to check-in at a location that he or she has visited for any reason. For instance the member may not have had time to check-in. The member may have decided not to check-in to avoid revealing this information. The member may have simply forgotten to check-in. Many LBSN services also allow *off-grid* check-ins, by which a member can check-in to the LBSN service but does not make this information public, generally, or even to his or her friends in the social network. It is also possible to send or forward check-ins in some LBSN services to other social network sites such as Twitter or Facebook to alert a wider circle of friends.

Typically, there is no verification of the exact location of the member or the venue he or she is visiting when the member checks-in to the LBSN service. Many early LBSNs allowed a user to send short messages, e-mails, or use the Web to check-in to the service. Such e-mails or text messages could be sent from anywhere, not necessarily the venue that was reported. This can lead to misuse of the service as described later. Recently, some LBSN services that are tailored for business use automatic detection of the venue when the member visits, using technologies that are available on the mobile devices and with additional infrastructure deployed in venues. For example, the service called Shopkick (Frommer 2010) uses an inaudible sound that is transmitted by a special box installed in the store. This sound is detected by the application running on the mobile phone that then checks-in with the service over a cellular data or WiFi network.

Check-ins are now used by LBSN services to offer virtual rewards (such as points or badges) and actual organizations to encourage visitors to physical locations, as discussed later.

5.2.3　Example LBSN Services

In this section, we will briefly describe some commercial LBSN services and what we believe are their primary objectives. As we will see later, the objective of an LBSN may influence the usage and behavior of its members.

5.2.3.1　Location Sharing Services

The primary objective of the early LBSN services was to allow members of a social network to share their locations with friends. The earliest LBSN service was Dodgeball, which was founded in 2000. Dodgeball allowed members to check-in using short text messages. Dodgeball was acquired by Google, which has since discontinued the service and instead has created its own service called Google Latitude (www.google.com/latitude). Although Google latitude allows users with a Google account to check-in using mobile applications on the iPhone and Android phones, check-ins are not possible on computers or mobile phones with other operating systems. However, members can still share their locations using a computer or other networked devices using a Web browser. Brightkite was another LBSN that was primarily oriented for location sharing. It was started in 2007 and has since been discontinued. Glympse (www.glympse.com) is an LBSN service that allows users with smartphones to be tracked as they move and restrict this sharing for a specific period of time (including single-use situations). The shared location with Glympse can be viewed on a mobile phone or a computer with a fine granularity on who can see the shared location.

5.2.3.2　Location Guides

Gowalla was a very popular LBSN that morphed into a service that supported curated city and location guides using information provided by its members. The primary objective of Gowalla was for the members to explore interesting locations called "spots" and check-in at such locations, using categories such as nature walks, pubs, and landmarks. Gowalla was able to provide place recommendations for its members and anyone accessing the service with information, tips, photos, and other information about places, locales, and venues in a city. Gowalla was started in 2007, acquired by Facebook in 2011, and has subsequently been shut down.

5.2.3.3　Business-Oriented LBSNs

So far we have mentioned several LBSN services that were either acquired or discontinued as they did not have a viable business model. Checking-in at commercial venues provides an obvious opportunity to exploit location information and interests of humans for enhancing commerce at such venues. This provides LBSN services an avenue for generating revenues and enhances sustainability. This approach has been taken up by several LBSN services, including the most popular one to date, namely, Foursquare. Shopkick and Yelp are other most popular business-oriented LBSN services to date.

Foursquare claims to have over 2 billion check-ins as of this writing, with more than 20 million members and 750,000 businesses being part of its ecosystem. Foursquare aims to provide both personal and commercial benefits. On the one hand, it allows members to share their locations, tips, and likes with friends allowing the service to recommend places and ideas for activity to its members. On the other hand, it allows businesses to benefit through information about potential customers, offer promotions, and snag and maintain customer interest in a product or venue. For example, in the last Foursquare day, which was April 16, 2012, McDonald's launched a promotion campaign that increased traffic to its stores by a claimed 33% on the day and on the following days as well (Foursquare 2012c). Foursquare is enabling brands and nonprofits to gain from the concept of promotions with check-ins. Several case studies by Fousquare show how organizations that are as disparate as TV channels (History channel and MTV), magazines (*Fast Company*), airlines (Lufthansa), and nonprofits (New York Public Library) can benefit by capturing the attention of members that are *physically* present in locations of interest and may also socially spread the word through their friends.

Yelp (www.yelp.com) started out as a business review site and has now expanded to become an LBSN service. Yelp claims 71 million unique visitors in the first quarter of 2012, and its business model includes selling advertisements to local businesses while allowing check-ins, reviews, and so on by its members.

5.2.3.4 Gaming-Oriented LBSNs

Many recent LBSN services are gaming oriented. Services such as Foursquare implement gaming in a limited manner by offering badges and mayorships based on check-ins. For example, the member who has the most number of check-ins at a venue is declared the mayor of that venue and gets a "virtual" badge for this accomplishment. Similarly, checking in at diverse venues and for special events allows members to collect and display badges. The idea behind more sophisticated gaming-oriented LBSN services is to exploit the geographical locations of members for fun activity. Some of the gaming-oriented LBSN services are Scvngr (www.scvngr.com) and Torchbear (www.torchbear.com).

Scvngr portrays itself as a location-based mobile gaming platform that can be used for gaming, and also for sharing locations and a platform for businesses to advertise. In addition to check-ins that earn them points, members of Scvngr have to tackle challenges at specific locations, which increases the engagement of its members. Scvngr cites a case study where a jewelry store in Philadelphia launched a competitive hunt for a hidden diamond ring by 300 couples using either short messages or the Scvngr platform on mobile phones to deliver clues (Robbins Diamond Jewelers and SCVNGR 2008). The buzz around this event increased the visits to the store's Website by 150% along with free spots on the regional news networks.

5.3 RESEARCH ON LBSNS

In this section, we will discuss some works that have been conducted in recent years on LBSNs. In particular, we will briefly discuss the works that have looked at how and why people use LBSN services, investigated the creation of friendship links using data from LBSNs, examined understanding of sociospatial aspects of neighborhoods and regions using LBSNs, and finally investigated privacy issues and malicious behavior in LBSN services.

5.3.1 Usage of LBSN Services

Some recent research works have focused on why and how people use LBSN services. We summarize some of such services and their results next.

Lindqvist et al. (2012) use a marketing research-based methodology to understand the usage of Foursquare. Three sets of data were used for this research. The first set comprised of interviews with six early adopters, a second qualitative survey consisted of 18 participants who applied through ads placed on Craigslist, and the third quantitative survey was conducted with 219 participants (158 from the United States, 46 from Europe) whose participation was requested through flyers, Twitter, and Facebook. The study showed a huge variation in behavior of people using Fourquare. For example, in the second survey, surveyed individuals had only one friend and other surveyed individuals had 48 friends. In the third survey, there were individuals with 0 friends and individuals with 2250 friends. There were individuals with 0 mayorships and others who had collected 141 mayorships in the third survey. People who were surveyed stated that they used Foursquare as something to do when they were bored or they used it to collect badges. The study seemed to indicate that the quest for badges decreased over time. Some people were conscious about how they would be perceived by friends when they checked in at certain types of venues such as fast food places and avoided checking in where there was potential for negative perceptions. There was also a bimodal distribution of people who would (or would not) check-in from work or homes.

The study by Cramer, Rost, and Holmquist (2011) used 20 in-depth interviews and 47 survey responses to arrive at some conclusions similar to those in Lindqvist et al. (2012). This study identified the utilitarian aspect of check-ins (e.g., to facilitate meetings) and the self-representation aspect of check-ins (i.e., I am a person who visits venues of a certain type). Again, a wide variation in the sample was observed (some interviewees had 0 friends while others had 92 friends on Foursquare). People were likely to think about spamming their friends with check-in updates. Thus, the audience of the check-in information played a role as to what types of check-ins were reported. Certain check-ins were sometimes used as "personal bookmarks" or for "life-logging" for a user to remember where he or she had been. An interesting result of this study was the possibility of users *creating* venues with names that referred to activities rather than a real name. For instance, the study reports that venues do need not to be real places and are user generated. Some users of Foursquare used it to generate fantasy venues. The gaming aspect of check-ins was observed to be strong (e.g., some users tried to capture the mayorship of a venue from other users). The conclusion from this study is that the *voluntary nature* of check-ins allows a large degree of freedom for users to express themselves as they want rather than allow the LBSN service to track them all the time.

The *type* of an LBSN service appears to play a role on how users may employ the service based on the results in Pelechrinis and Krishnamurthy (2012a). This work uses analyses of actual check-in data rather than interviews and surveys to infer differences across LBSN services. As mentioned in Section 5.2, LBSN services have different scopes. The primary purpose of Gowalla, for example, evolved into creating city guides, whereas Scvngr has focused on gaming. Pelechrinis and Krishnamurthy

(2012a) analyzed data from two commercial LBSNs—Gowalla and Brightkite—to examine the temporal evolution of usage patterns to see what the data reveals. Users of two social networks that were examined increase their level of activity as they use the LBSN service. However, users of the two services exhibit different behaviors over time as described next. The data sets used were made available by Cho, Myers, and Leskovec (2011) and are as follows. The Gowalla data set consists of 6,442,892 public check-in data performed by 196,591 Gowalla users in 647,923 distinct places, during the period between February 2009 and October 2010. Every check-in log includes a tuple in the form <User ID, Time, Latitude, Longitude, Venue ID>. Gowalla users also participate in a friendship network with reciprocal relations, which consists of 950,327 links. The Brightkite data set consists of 4,491,143 public check-in data performed by 58,228 Brightkite users in 772,966 distinct places, during the period between April 2008 and October 2010. The check-in information is in exactly the same aforementioned format. Brightkite users also participate in a friendship network, which consists of 214,078 links that are originally asymmetric.

The analysis of the data sets considers among other things (1) the inter-check-in time, that is, the time that has elapsed between two consecutive check-in events; (2) the number of unique venues visited by a member of each LBSN service; and (3) the entropy of a member, which captures his or her *diversity* with regard to the places visited. The analysis considers these metrics as a function of check-in counts of members of the LBSN service. The check-in count is simply a temporal measure (it simply counts how many check-ins a user has). The distribution of the inter-check-in time indicates that as the check-in count increases, the inter-check-in time reduces. That is, a user checks in more often as usage of the service increases (he or she is more comfortable checking in frequently).

The significant difference in the behavior of members of Gowalla and Brightkite occurs when the number of unique check-ins is analyzed over time. As the check-in count is increased, the number of unique venues also increases linearly as shown in Figure 5.2 in both cases. However, the slope of the least-square linear fit in the data is very different for the two data sets. In particular, for Gowalla the slope is equal to 0.65, and for Brightkite the slope is much smaller, 0.085. Essentially, this means that every check-in a Gowalla (Brightkite) user is performing has a probability of 0.65 (0.085) of being at a previously unseen venue in his or her history log of locations. Clearly, these are two very different kinds of behaviors of users across the data sets. Since the check-ins of a user do not reveal all of the actual places that they have been and they capture only the places that users are willing to share in the network, an important factor that can affect the sharing attitudes of people is related to the objective and the nature of the underlying network, that is, its main application and purpose. Gowalla evolved to become a city guide application. People that visit a city for the first time could make use of the check-ins of Gowalla users (and possibly textual comments accompanying them) and explore locales in this new environment. Hence, Gowalla users may be tempted (perhaps even encouraged) to check-in at new spots in order to provide a more comprehensive guide of their city, which can explain the large slope of the linear curve. On the contrary, Brightkite did not have a similar objective and it was mainly a social-driven application. It aimed at connecting people in the physical world through location sharing. Hence, Brightkite users have

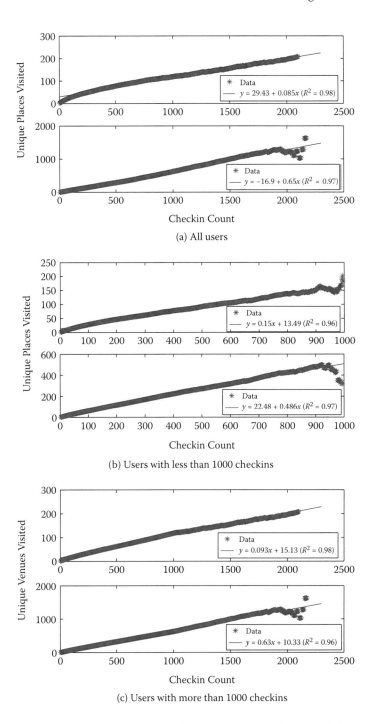

FIGURE 5.2   (See color insert.) Number of unique venues as a function of check-in count for (top) Brightkite users and (bottom) Gowalla users.

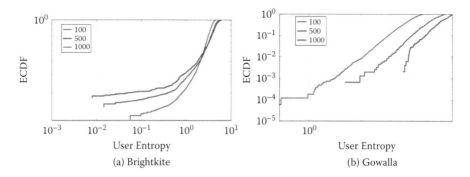

FIGURE 5.3 **(See color insert.)** The distribution of the user entropy for different check-in counts.

no such incentive causing them to be more skeptical when sharing their presence, which can account for the much smaller slope of the corresponding linear fit.

Pelechrinis and Krishnamurthy (2012a) also consider the "entropy" of a user of an LBSN service, which captures the diversity of a user with regard to the places he has visited. It not only considers the number of distinct locations visited by him, but it also takes into account the frequency of these visits. Let us assume that L_u is a set containing all the locations shared by user u. If $P_l(u)$, $l \in L_u$, is the fraction of check-ins of user u that happened in location l, then the entropy e_u of user u is defined as

$$e_u = -\sum_{l \in L_u} P_l(u) \cdot \log\big(P_l(u)\big)$$

From this equation, one can notice that when a member of an LBSN service visits many venues in fairly equal proportions, his entropy will be large. On the contrary, when most of a member's activity is restricted to a few locales only, his entropy will be low. In other words, a (non)diverse user with respect to the venues he visits will exhibit (low) high entropy.

Using the two data sets, Pelechrinis and Krishnamurthy (2012a) have calculated the average user entropy as a function of the check-in order. The results are presented in Figure 5.3. Brightkite users exhibit much lower (average) entropy as compared to the Gowalla users. Recall that the latter have a much larger number of unique venues, which means that they "distribute" their activity in more places, exhibiting higher diversity and thus entropy. Further, the entropy of a Brightkite user stabilizes fairly quickly, after only a few check-ins. On the contrary, the entropy of a Gowalla user slowly increases as the check-in count increases. This reinforces the conclusion that the type or scope of an LBSN service will likely influence its usage behavior.

5.3.2 Social Aspects

We next consider some works that have focused on the social aspects of LBSNs, namely, the properties of social connections in LBSNs. We focus this section on

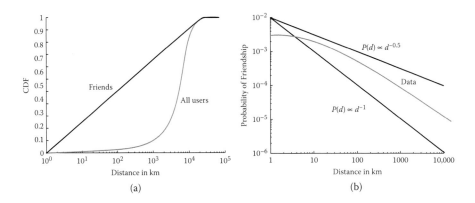

FIGURE 5.4   (See color insert.) (a) CDF of physical distance between all users and physical distance between friends and (b) probability of friendship as a function of distance.

some results from Scellato, Noulas, and Lambiotte (2011) and Pelechrinis and Krishnamurthy (2012b), but other works that consider the social aspects have appeared in recent years.

As previously discussed, there are two reasons why social connections may develop: common interests and opportunity for interaction. Sociological studies indicate that friendships are likely to persist when people are closer in physical space as this affords the opportunity to interact and also reduces the effort to maintain friendship connections. The question that has been recently posed is whether online interactions reduce the necessity to be physically close and whether distance no longer plays a role in maintaining social connections. Physical distance still appears to play a preferential role in maintaining connections (Scellato, Noulas, and Lambiotte 2011).

The work by Scellato, Noulas, and Lambiotte (2011) looks at data from three LBSN services—Foursquare, Brightkite, and Gowalla—to analyze the spatial properties of social relationships. The Foursquare data are mined from check-ins that have been forwarded or pushed to Twitter, the Brightkite data are from September 2009, and the Gowalla data from August 2010. All three data sets mostly exhibit similar behavior. Figure 5.4 shows a summary of the findings in this work (the plots are not from real data but only illustrate the trends reported by Scellato, Noulas, and Lambiotte, 2011). The physical distance between friends is much smaller than that of any two users picked at random as shown in Figure 5.4(a). For instance, the probability that the physical distance between two friends is at most 100 km is 0.4, while it is less than 0.1 for two users in general. This implies that friends are more likely at smaller physical distances. Figure 5.4b shows the probability of two members of an LBSN service being friends as a function of physical distance. The actual data lie between $P(d) = d^{-1}$ and $P(d) = d^{-0.5}$ with a flat part between 0 and 10 km or so. This indicates that long-range friendships are quite likely indicating a fairly large amount of heterogeneity in friendships.

Pelechrinis and Krishnamurthy (2012b) model an LBSN as an affiliation network rather than a friendship network and analyze the data from a commercial network using this model. The idea in this work is to look at the *opportunity for interaction* in a different way to determine whether people that visit the same places have a higher probability of being friends. In other words, the authors find there are signs of *homophily* with regard to the spatial behavior of the users. They find that friends exhibit in general much larger *similarity* (explained briefly next) with regard to the number of common venues visited as compared to nonfriends. Considering only the number of common venues between two members of an LBSN service is not very helpful for strongly tying the two components of the network. For instance, two members of an LBSN may both have been at an airport, at a large supermarket store, and at a very popular restaurant in town. However, these are venues that would be likely for any individual in that geographical area. Instead, it is the *diversity* of these common venues with regard to people visiting them that is more informative. For example, if two individuals both visit a rare book store that is unlikely to be visited by almost everyone, it is likely that these two individuals have a friendship connection. This result is also supported by their evaluations and results from simple, unsupervised social link classifiers.

The similarity of two members of an LBSN service is defined by Pelechrinis and Krishnamurthy (2012b) as the ratio of the number of venues that have been visited by both members to the total number of venues visited by either of them. That is, if A has visited venues p, q, r, and s, and B has visited venues p, s, and x, the similarity between them is 2/5 since there are two common venues p and s, and all together five venues p, q, r, s, and x. A threshold of 120 miles is used to group members as being "nearby" or "distant." Figure 5.5a presents the distribution of similarity for three classes of pairs: nearby pairs of friends, distant pairs of friends, and nearby pairs of nonfriends. Clearly, friends that reside in geographic proximity to each other have the highest similarity scores. Figure 5.5b shows the similarity values for nearby friends as a function of distance between home locations. As we can see, distance does not appear to have any effect on the similarity for these users. A slight decrease of the (average) coefficient can be observed, but it is not significant. However, distance appears to be critical for friends that live far apart (as one might have expected). As we see in Figure 5.5c, after some distance (approximately 2500 miles) the similarity values are drastically reduced. An LBSN member B will have fewer opportunities to "follow" the trails of his or her friend A if he or she lives far away.

5.3.3 Geographical and Temporal Activity

The availability of time-stamped location information of people on a large scale, often annotated by context (such as venues, categories, and other semantic information), provides an opportunity for understanding geospatial activity of human beings on a scale that was previously not possible. The LBSN service Foursquare itself has analyzed check-in data to look at *similar neighborhoods* across different cities (e.g., what may be considered equivalent to Pittsburgh's strip district in say Seattle, Washington) (Foursquare 2012a). We briefly present the results of work reported in Cheng et al. (2011) and Noulas et al. (2011a, 2011b).

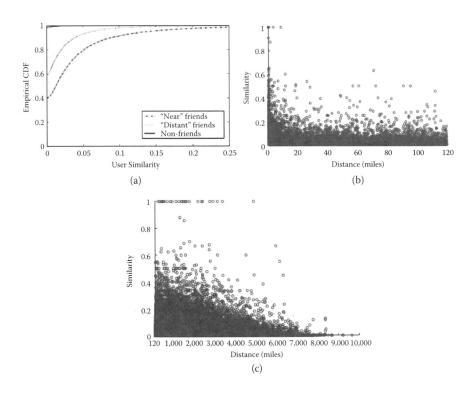

FIGURE 5.5 **(See color insert.)** (a) CDF of user similarity, (b) similarity of nearby friends, and (c) similarity of distant friends.

Cheng et al. (2011) used Twitter feeds with geotags instead of data directly from specific LBSN services and included about 22 million check-ins. About 54% of the data was from Foursquare check-ins, but over 1200 applications were reportedly used. The study assumed that the "home" of a member was the place with the most number of check-ins. The study created a word cloud to see where most of the check-ins were located. The word cloud reveals that check-ins are most common in general coffee shops, restaurants, and a specific coffee shop (e.g., Starbucks). Also, the frequency of check-ins showed three distinct peaks or "heart-beats" as shown in Figure 5.6 (this figure illustrates the trends reported in Cheng et al., 2011, and is not from real data) and there were some variations across cities. For instance, Amsterdam exhibits a peak at an earlier time while New York City had the most check-ins at night indicating the variation of activity in these cities. This study also looked at the distance between consecutive check-ins by people and this was found to follow a power law. The study also reported strong weekly periodic return probabilities for locations, that is, members tend to check-in at the same places periodically, especially on a weekly basis.

The work by Noulas et al. (2011a) looks at the activity of 700,000 Foursquare users over a 100-day period. This work concurred with the heart-beats reported in Cheng et al. (2011) through the demonstration of peaks in the number of check-ins during weekdays, but a relatively flat check-in number over the daytime on weekends. This work further classified the check-ins by category. Check-ins were more common in

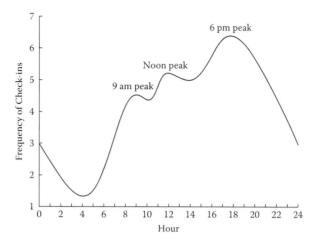

FIGURE 5.6 Illustration of three major heart-beats of check-ins reported in Cheng et al. (2011).

corporate offices between 8:00 a.m. and 10:00 a.m. while they were more common in homes in the evenings. Activity and place transitions by users of LBSN services were also evaluated in this work.

Like Foursquare's work (2012a), the work by Noulas et al. (2011b) clusters geographical areas in cities based on the number and categories of check-ins using a Foursquare dataset as in Noulas et al. (2011a). The geographical area of a city is divided into square blocks and each block is represented by the category and types and numbers of check-ins of venues located in that square block. This allowed the researchers to categorize neighborhoods in cities like New York and London based on the activity. This work also suggested the ability to tailor recommendations or facilitate urban planning.

5.3.4 Privacy

Privacy issues that arise when people report their locations and times publicly have been discussed in many mainstream media articles and research papers. Check-ins can enable pinpointing the area of activity and perhaps even the residence of an individual, and through related data (e.g., where does A shop and how frequently?) it may be possible to infer income levels and other types of private information. The work by Li and Chen (2010) uses data from Brightkite to investigate how people perceive the privacy problem. The way this is measured is by computing the fraction of off-grid (protected) location updates by a member of Brightkite's LBSN service, which provides the number of public location updates and the total number of updates. Figure 5.7 illustrates the trends reported in Li and Chen (2010). Again, the plots are illustrative of the trends and are not from actual data. Clearly, older people and those that have more activity appear to be more conscious of privacy issues.

The work by Tena and Raivio (2011) proposes the use of a location broker with policies to facilitate privacy, while sharing appropriate location information as necessary.

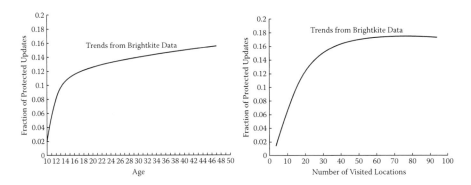

FIGURE 5.7   Illustration of trends reported in Li and Chen (2010) regarding privacy perceptions by members of Brightkite.

5.3.5   MALICIOUS BEHAVIOR

It is quite possible for malicious behavior to emerge with regard to fake check-ins, especially as the rewards for checking in with LBSN services become attractive. For example, a member of an LBSN can falsely check-in to a location to snatch the mayorship of a venue from someone else. This seems to be a harmless activity when only mayorships are considered. However, some venues offer promotions based on the number of check-ins and there could be economic harm to such venues because of malicious behavior (He, Liu, and Ren 2011; Zhang et al. 2012). The use of challenges that users can solve only by actually being present at a location and geofencing (as in Shopkick) has been suggested as an option for addressing this problem. In the former case, a member of an LBSN has to be physically present in a venue to enter a code or take a picture or identify a specific item in a menu to check-in. In the latter case, the mobile device cannot be outside the venue if it is to hear the audio signal emitted by the special box located inside a given venue. Also, the use of WiFi received signal-strength-based location proofs for fake check-in detection has been proposed and evaluated through simulations in (Zhang et al. 2012). The idea in this case is that the received signal-strength with its temporal variability provides a fingerprint that may be used to determine whether a user is actually inside a venue.

5.3.6   EMERGING USES OF LBSNs

LBSN data is being used in innovative ways as already described. In addition, there are other uses being considered. Scellato, Noulas, and Mascolo (2011) use location information to improve *friend recommendations.* They focus on the temporal evolution of the social graph and they utilize a combination of information drawn from both the social and location component to improve friend recommendations. Similarly, *location recommendations* can make use of LBSN data as may also local and business searches. The use of LBSN for targeted marketing and promotions is ongoing through many LBSN services. Recently, Foursquare members have started to also contribute to the improvement of Open Street Maps creating a huge spike in the number of contributors to the open source mapping service (Foursquare 2012b).

5.4 SUMMARY

LBSN services are becoming very popular and fairly pervasive with the emergence of smartphones worldwide. Data on the most popular LBSN service, Foursquare's Web site, indicate that over 50% of its members are outside the United States with check-ins from every country including North Korea! People are somewhat shy of revealing their locations, but the benefits of LBSN services, namely, sharing locations with friends and family, fun activities like collecting and displaying virtual rewards, real monetary incentives such as coupons and discounts, are encouraging an increasing number of people to check-in to LBSN services.

In this chapter, we provided an overview of LBSNs. We discussed the basics of LBSNs such as check-ins, and also considered specific examples of commercial LBSN services with different scopes and objectives. We also discussed how the data from LBSNs are being employed by the research community for examining a variety of problems.

The inclusion of location data, semantically enhanced with tips, categories, and other data with social networking information, has created a rich set of data that could be exploited for solving or understanding a variety of diverse problems including the spread of epidemics, movement of people across geographical areas in time, categorization of neighborhoods in cities and regions to discover similarities and differences, and for commercial activity such as advertising and promotions by local businesses. Research in this area has only recently burgeoned and it appears that there are potential discoveries yet to be made. The commercial success of LBSN services is likely to induce malicious behavior and security problems that will also need to be addressed in the coming years.

REFERENCES

Cheng, J., et al. 2011. Exploring Millions of Footprints in Location Sharing Services. *Proceedings of the Fifth International AAAI Conference on Weblogs and Social Media*, 81–88.

Cho, E., S. A. Myers, and J. Leskovec. 2011. Friendship and Mobility: User Movement in Location-Based Social Networks. *Proceedings of ACM KDD*, 279–311.

Cramer, H., M. Rost, and L. E. Holmquist. 2011. Performing a Check-In: Emerging Practices, Norms and "Conflicts" in Location-Sharing Using Foursquare. *Proceedings of ACM MobileHCI*, 57–66.

Easley, D., and J. Kleinberg. 2010. *Networks, Crowds, and Markets*. Cambridge University Press.

Expert, P., et al. 2011. Uncovering Space-Independent Communities in Spatial Networks. *PNAS*, vol. 108, no. 19, pp. 7663–7668.

Foursquare. 2012a. A Hackday Project: What Neighborhood is the "East Village" of San Francisco? http://engineering.foursquare.com/2012/03/08/a-hackday-project-what-neighborhood-is-the-'east-village'-of-san-francisco/.

Foursquare. 2012b. Making a Better Map: Four Months of @OpenStreetMap with @MapBox & @foursquare. http://blog.foursquare.com/2012/07/10/making-a-better-map-four-months-of-openstreetmap-with-mapbox-foursquare/.

Foursquare. 2012c. McDonald's Case Study: McDonald's Foursquare Day 4/16. https://foursquare.com/business/merchants/casestudies/mcdonalds.

Frommer, D. 2010. Here's Shopkick's Special Sauce: A Box in Every Store That Verifies You're Really There. http://www.businessinsider.com/heres-shopkicks-special-sauce-a-box-in-every-store-that-verifies-youre-really-there-2010-8.

He, W., X. Liu, and M. Ren. 2011. Location Cheating: A Security Challenge to Location-based Social Network Services. *2011 31st International Conference on Distributed Computing Systems (ICDCS)*, 740–749.

Li, N., and G. Chen. 2010. Sharing Location in Online Social Networks. *IEEE Network*, vol. 24, no. 5, 20–25.

Lindqvist, J., et al. 2012. I'm the Mayor of My House: Examining Why People Use Foursquare: A Social-Driven Location Sharing Application. *Proceedings of ACM CHI*, 2409–2418.

Mok, D., B. Wellman, and R. Basu. 2007. Did Distance Matter before the Internet? Interpersonal Contact and Support in the 1970s. *Social Networks*, vol. 29, pp. 430–461.

Noulas, A., S. Scellato, and C. Mascolo et al. 2011. An Empirical Study of Geographic User Activity Patterns in Foursquare. Fifth International AAAI Conference on Weblogs and Social Media.

Noulas, A., S. Scellato, and C. Mascolo, et al. 2011. Exploiting Semantic Annotations for Clustering Geographic Areas and Users in Location-Based Social Networks. Fifth International AAAI Conference on Weblogs and Social Media.

Pelechrinis, K., and P. Krishnamurthy. 2012. Location-Based Social Network Users through a Lense: Examining Temporal User Patterns. Accepted in AAAI Fall Symposium on Social Networks and Contagion Spreading.

Pelechrinis, K., and P. Krishnamurthy. 2012. Location Affiliation Networks: Inferring Social Information via Location History. Accepted in ECML-PKDD.

Robbins Diamonds Jewelers and SCVNGR. 2008. Robbins Diamonds Dash Case Study. http://www.scvngr.com/pdfs/Robbins_Diamonds_Dash_Case_Study.pdf.

Scellato, S., A. Noulas, and R. Lambiotte, et al. 2011. Socio-Spatial Properties of Online Location-Based Social Networks. *Proceedings of Fifth International AAAI Conference on Weblogs and Social Media*, 329–336.

Scellato, S., A. Noulas, and C. Mascolo. 2011. Exploiting Place Features in Link Prediction on Location-Based Social Networks. *Proceedings of ACM KDD*, 1046–1054.

Tena, A. V., and Y. Raivio. 2011. Privacy Challenges of Open APIs: Case Location Based Services. *Ninth IEEE Annual International Conference on Privacy, Security and Trust*, 213–220.

Yasseri, T., et al. 2012. Dynamics of Conflicts in Wikipedia. PLoS ONE 7(6): e38869. doi:10.1371/journal.pone.0038869.

Zhang, K., W. Jeng, F. Fofie, K. Pelechrinis, and P. Krishnamurthy. 2012. Towards Reliable Spatial Information in LBSNs. Accepted in ACM LBSN 2012.

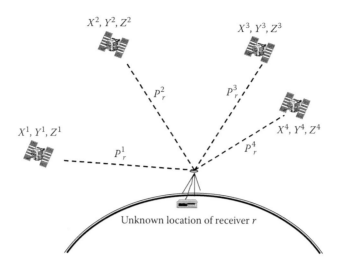

FIGURE 1.6 Determination of position in space by ranging to multiple satellites.

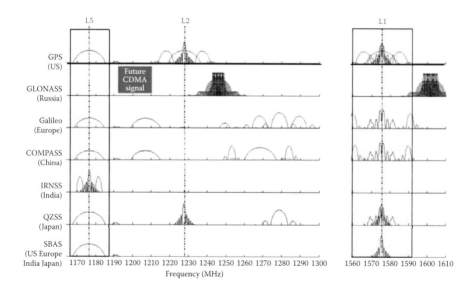

FIGURE 1.9 Proposed Signal Overlay for all available GNSS, RNSS and SBAS (Turner, 2010). With Permission.

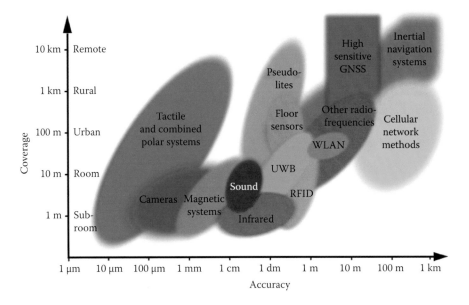

FIGURE 1.10 Relationship between coverage and accuracy for a range of indoor positioning technologies. (From Mautz, R., 2012, Indoor Positioning Technologies, Habilitation thesis, ETH Zurich. With permission.)

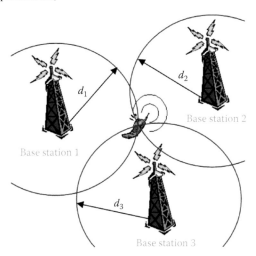

FIGURE 1.12 Triangulation of the user's position based on the distance measurements to three base stations (cell towers). The latitude and longitude of the user are obtained as the intersection of three circles centered at the towers, with radii of d_1, d_2, and d_3.

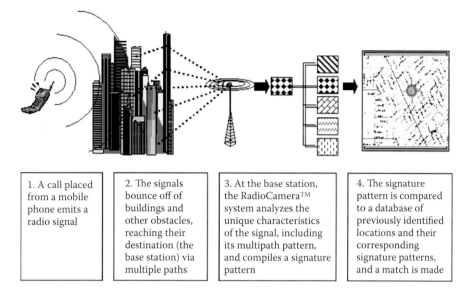

| 1. A call placed from a mobile phone emits a radio signal | 2. The signals bounce off of buildings and other obstacles, reaching their destination (the base station) via multiple paths | 3. At the base station, the RadioCamera™ system analyzes the unique characteristics of the signal, including its multipath pattern, and compiles a signature pattern | 4. The signature pattern is compared to a database of previously identified locations and their corresponding signature patterns, and a match is made |

FIGURE 1.13 Location pattern matching.

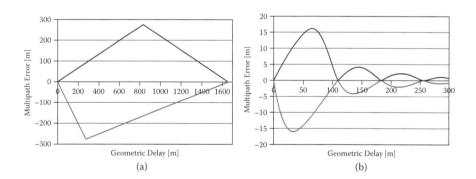

FIGURE 1.14 Multipath error envelopes for (a) GSM and (b) UMTS. (From Hein, G., 2001, On the Integration of Satellite Navigation and UMTS, CASAN-1 International Congress, Munich, Germany.)

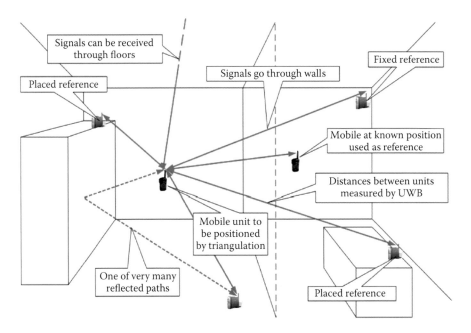

FIGURE 1.15 Ultra-Wideband positioning System (Thales, 2004).

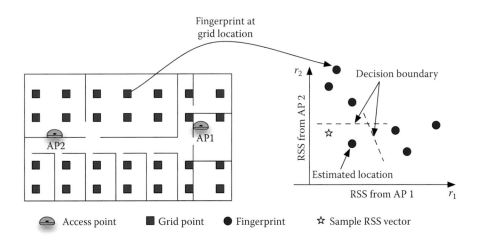

FIGURE 2.2 WiFi RSS-based location fingerprinting and Euclidean distance for estimating location.

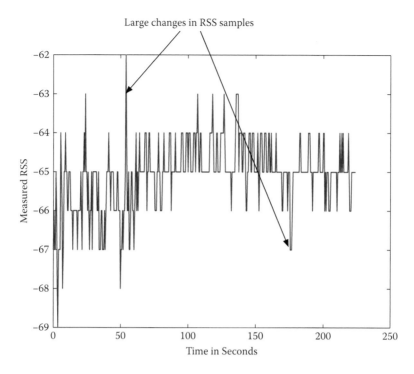

FIGURE 2.3 Time series of RSS samples (in dBm) measured on a laptop.

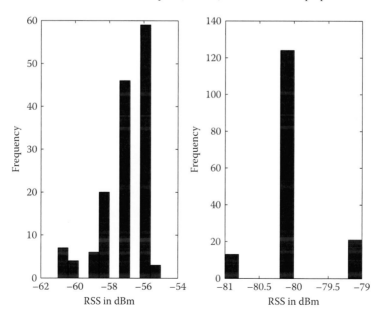

FIGURE 2.4 Histograms of RSS samples.

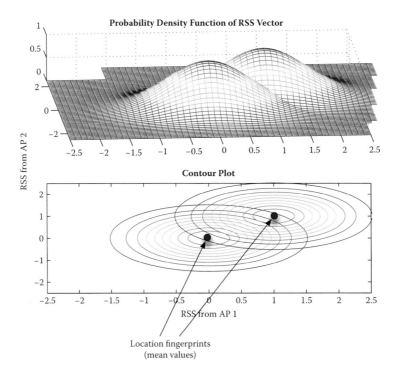

FIGURE 2.5 Gaussian RSS distributions in 2D with two location fingerprints.

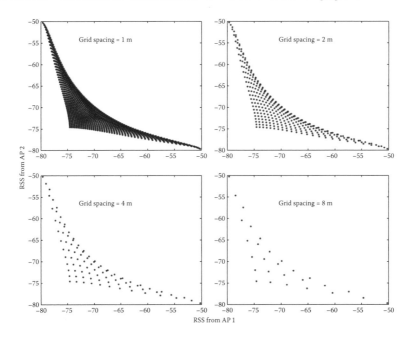

FIGURE 2.7 Location fingerprints in a square area as a function of grid spacing.

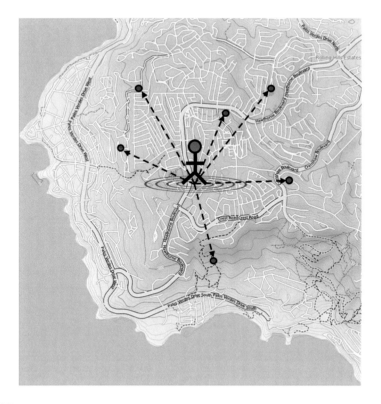

FIGURE 3.1   Geographically contextualizing user location information in an LBS.

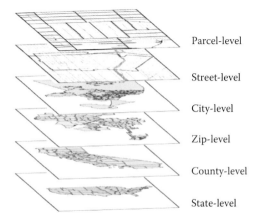

Parcel-level

Street-level

City-level

Zip-level

County-level

State-level

FIGURE 3.3   Geocoding reference data layers.

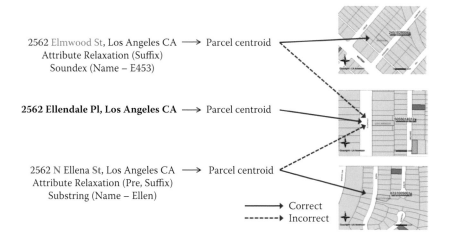

2562 Elmwood St, Los Angeles CA ⟶ Parcel centroid
Attribute Relaxation (Suffix)
Soundex (Name – E453)

2562 Ellendale Pl, Los Angeles CA ⟶ Parcel centroid

2562 N Ellena St, Los Angeles CA ⟶ Parcel centroid
Attribute Relaxation (Pre, Suffix)
Substring (Name – Ellen)

⟶ Correct
-----▶ Incorrect

FIGURE 3.6   Example geocoder mismatches.

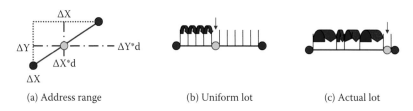

(a) Address range   (b) Uniform lot   (c) Actual lot

FIGURE 3.7   Linear interpolation methods.

FIGURE 3.13   3D indoor geocoding.

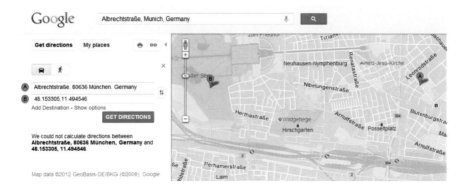

FIGURE 4.2   No direct path provided by Google Maps.

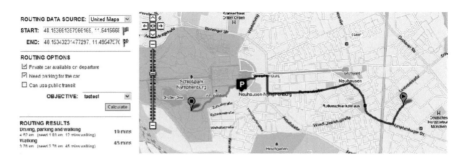

FIGURE 4.3   A feasible double-modal route provided by multimodal route planner. (After Liu, L., 2011, Data model and algorithms for multimodal route planning with transportation networks, PhD thesis, Technische Universitaet Muenchen.)

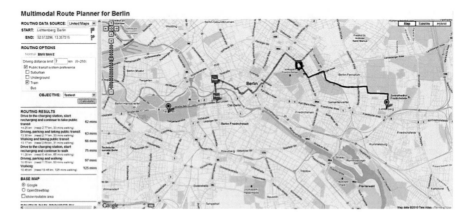

FIGURE 4.7   Multimodal route from Lichtenberg (Siegfriedstraße 203) to a pedestrian junction with coordinates of longitude 52.513, latitude 13.357 in Tiergarten in Berlin. (After Liu, L., 2011, Data model and algorithms for multimodal route planning with transportation networks, PhD thesis, Technische Universitaet Muenchen.)

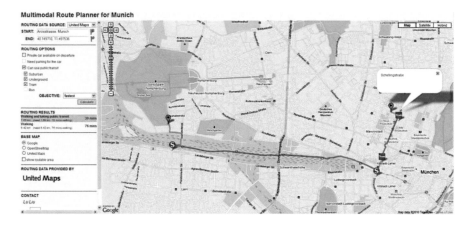

FIGURE 4.8 Multimodal route from the crossing of Arcisstraße and Heßstraße near Technische Universität München to a pedestrian path junction with coordinates of longitude 48.150 and latitude 11.497 in Schloßpark Nymphenburg. (After Liu, L., 2011, Data model and algorithms for multimodal route planning with transportation networks, PhD thesis, Technische Universitaet Muenchen.)

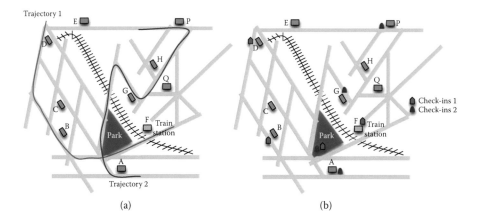

FIGURE 5.1 Types of location information in LBSNs (a) trajectory based and (b) venue based.

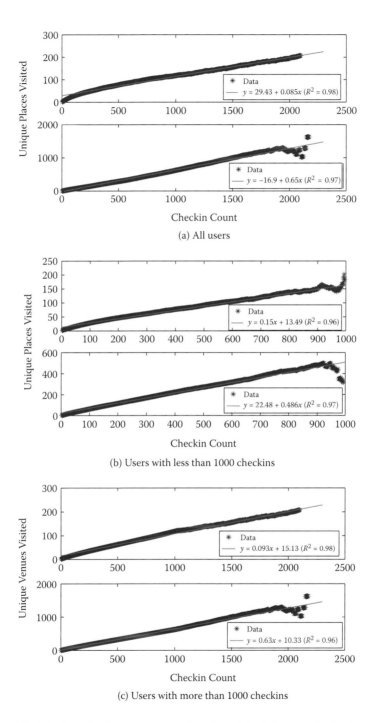

FIGURE 5.2 Number of unique venues as a function of check-in count for (top) Brightkite users and (bottom) Gowalla users.

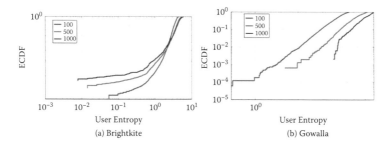

FIGURE 5.3 The distribution of the user entropy for different check-in counts.

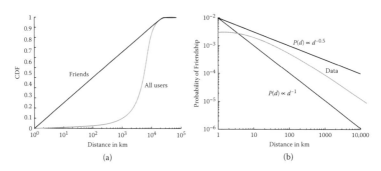

FIGURE 5.4 (a) CDF of physical distance between all users and physical distance between friends and (b) probability of friendship as a function of distance.

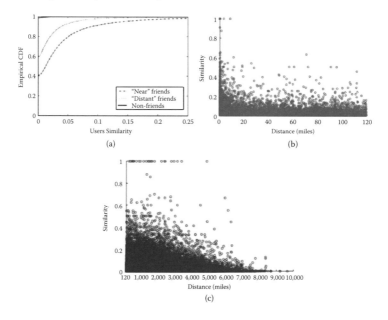

FIGURE 5.5 (a) CDF of user similarity, (b) similarity of nearby friends, and (c) similarity of distant friends.

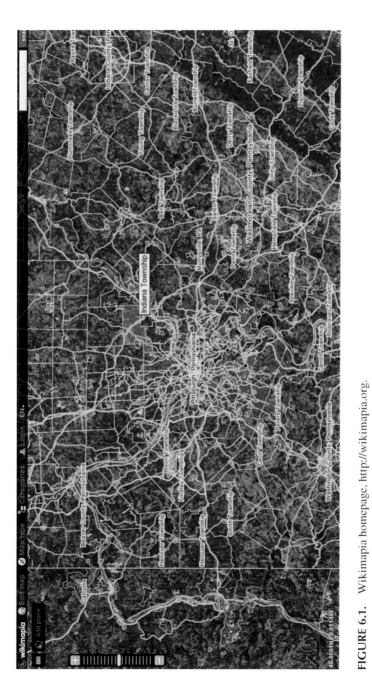

FIGURE 6.1. Wikimapia homepage, http://wikimapia.org.

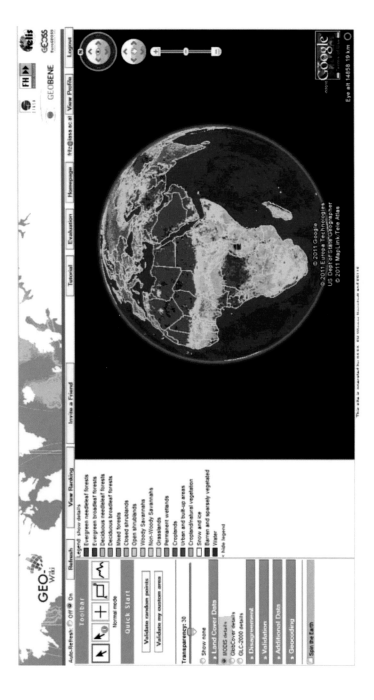

FIGURE 6.2 Geo-Wiki Project homepage. (From Geo-Wiki.org instructions page, http://www.geo-wiki.org/instructions.php.)

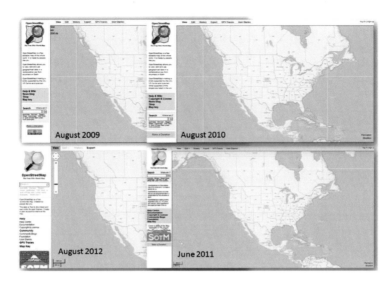

FIGURE 6.7 OSM interface overtime, clockwise from top left, August 2009, August 2010, June 2011, and August 2012, © OpenStreetMap contributors, CC BY-SA. (From 2009, 2010, and 2011 images from the Wayback Machine at the Internet Archive August 2012; August 2012 image from http://www.openstreetmap.org/.)

FIGURE 6.8 OSM Map of Weybridge. This map was created by OpenStreetMap Map. © OpenStreetMap contributors, CC BY-SA. (Retrieved from Wikipedia, http://en.wikipedia.org/wiki/File:KT13_Area.4.png.)

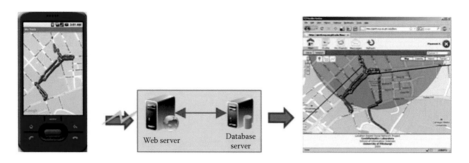

FIGURE 7.1 Screenshots of GPS traces recording and sharing.

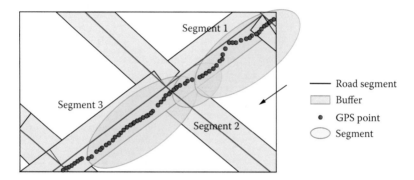

FIGURE 7.3 Example of GPS point segmentation.

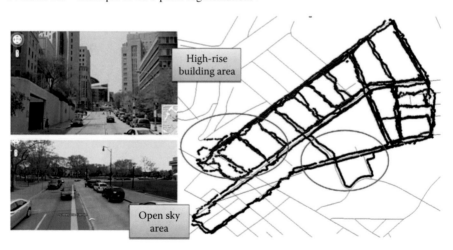

FIGURE 7.8 Ten GPS traces.

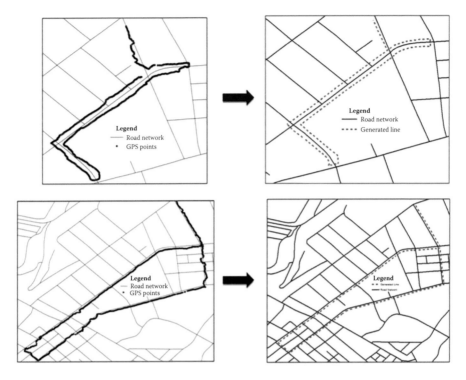

FIGURE 7.9 Examples of generated sidewalks/crosswalks using GPS traces.

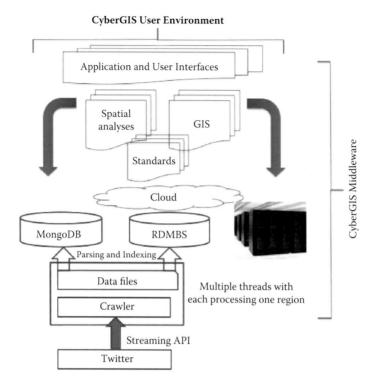

CyberGIS User Environment
Application and User Interfaces
Spatial analyses
GIS
Standards
Cloud
MongoDB
RDMBS
Parsing and Indexing
Data files
Crawler
Multiple threads with each processing one region
Streaming API
Twitter
CyberGIS Middleware

FIGURE 8.1 CyberGIS architecture.

Tweets Spider
Manual Mode
Auto Mode
settings:
From 12/01/2011
To 05/31/2012
X-cells: 10
Y-cells: 10
Location: 10
Click Mode:
Hover Mode:
United States
Mexico

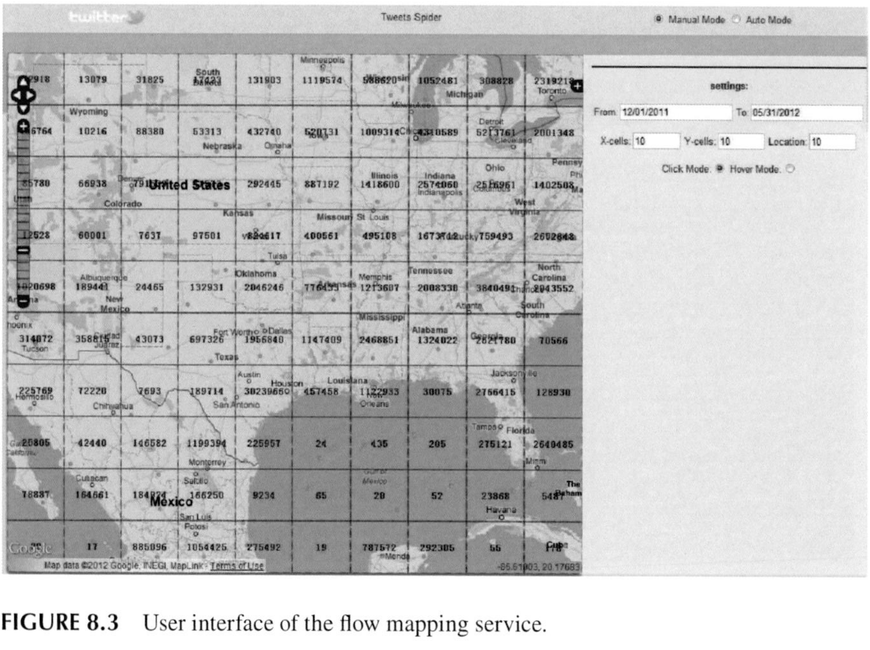

FIGURE 8.3 User interface of the flow mapping service.

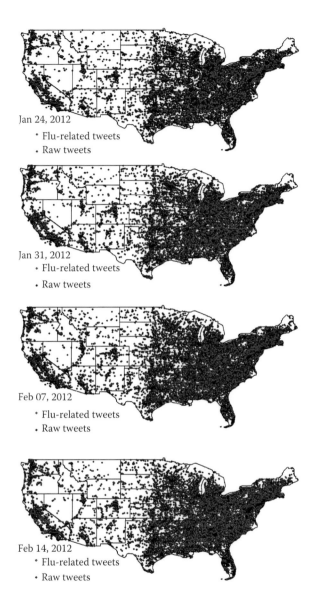

FIGURE 8.4 Maps of raw tweets and flu-related tweets of four different sample days in the conterminous United States.

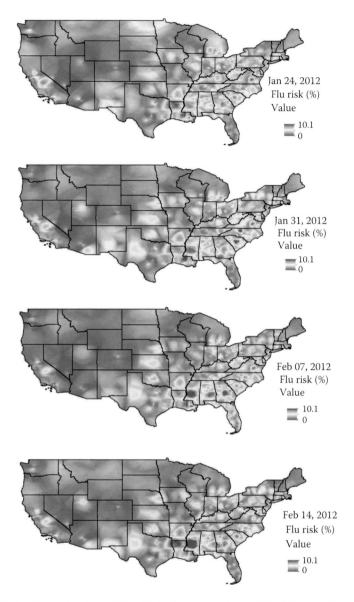

FIGURE 8.5 Maps of estimated flu-risk in the conterminous United States of four different sample days.

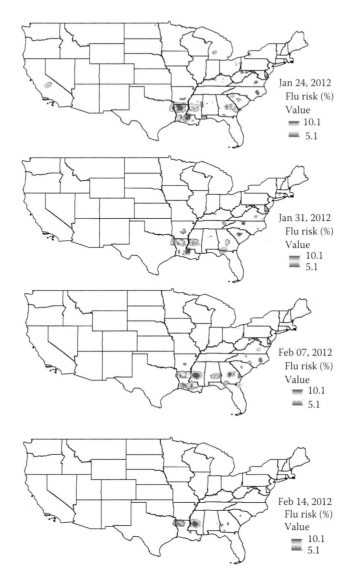

FIGURE 8.6 Maps of significantly high flu-risk areas in the conterminous United States of four different sample days: estimated flu-risk >5.1% and *p*-value <0.005.

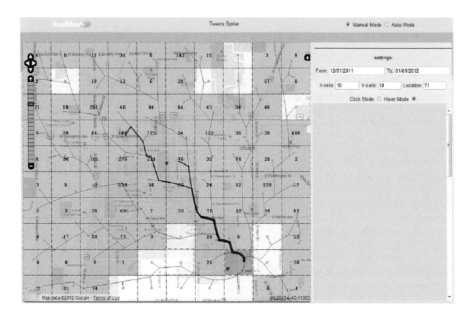

FIGURE 8.7 Sample flow mapping result I: Movement patterns of Orchard Downs residents in the area of the University of Illinois from Dec. 1st 2011 to Jan. 1st 2012.

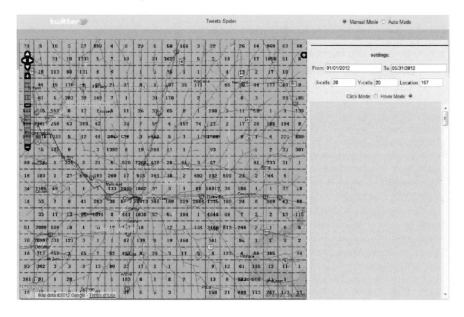

FIGURE 8.8 Sample flow mapping result II: Movement patterns of Urbana residents in the area of Urbana–Champaign–Bloomington, Illinois from Jan. 1st 2012 to May 31st 2012.

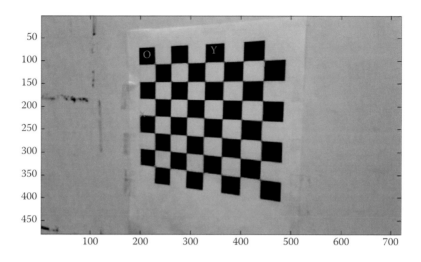

FIGURE 9.9   A checkerboard to calibrate camera.

FIGURE 9.10   A sequence of images extracted from a video.

FIGURE 9.11   SIFT features of a streetview image.

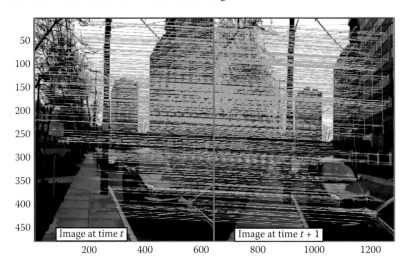

FIGURE 9.12   Matched SIFT feature points.

FIGURE 9.13 Feature points in Image 1 versus Epipolar lines in Image 2.

FIGURE 9.14 Positioning results by using visual odometry, top view (left) and street view (right).

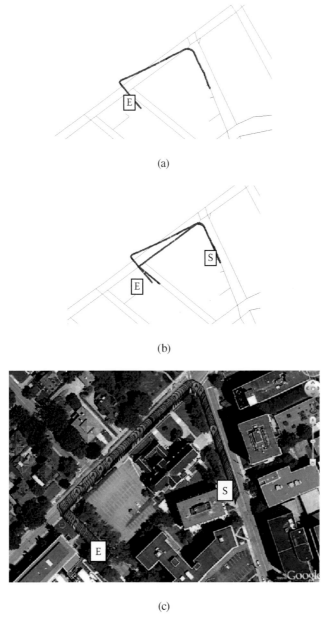

(a)

(b)

(c)

FIGURE 9.15 Position estimations and map matching results. (a) Monocular visual odometry results in one route before map matching. (b) Estimated locations on one route before map matching and after map matching. (c) Map matching results overlaid on Google Maps.

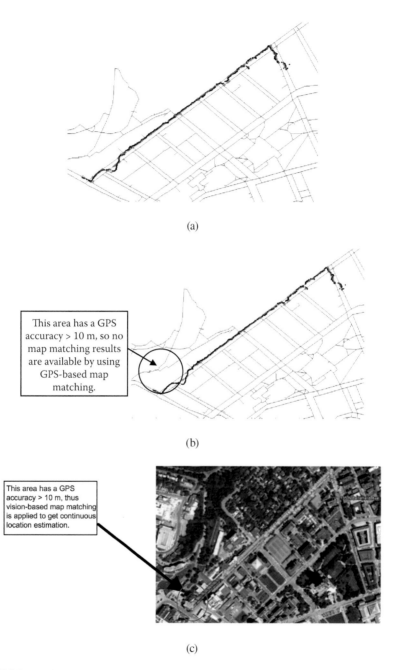

(a)

This area has a GPS accuracy > 10 m, so no map matching results are available by using GPS-based map matching.

(b)

This area has a GPS accuracy > 10 m, thus vision-based map matching is applied to get continuous location estimation.

(c)

FIGURE 9.17 Comparison of GPS-based map matching results with multisensor map matching results. (a) GPS-based map matching results, as compared with raw GPS data overlaid on the sidewalk map. (b) GPS-based map matching results in GPS accuracy ≤10m, compared with raw GPS data overlaid on the sidewalk map. (c) Multisensor map matching results overlaid on Google Maps.

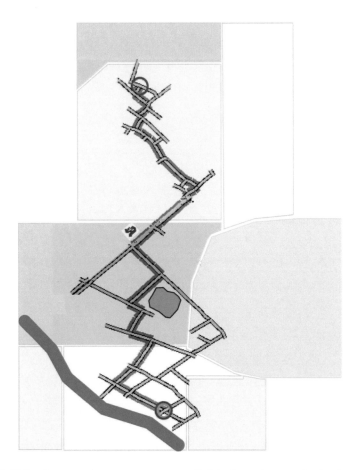

FIGURE 11.2 An example route-aware map. Origin is at the top, destination is at the bottom (denoted by the circles). The main route is shown in bold; alternative routes help to recover from potential wayfinding errors, as do increased details around origin and destination, and the inclusion of landmark and region information. (Modified from Schmid, Falko, Denise Peters, and Kai-Florian Richter, 2008, You are not lost—You are somewhere here, in *You-Are-Here-Maps: Creating a Sense of Place through Map-Like Representations*, edited by A. Klippel and S. Hirtle, Workshop at International Conference Spatial Cognition 2008.)

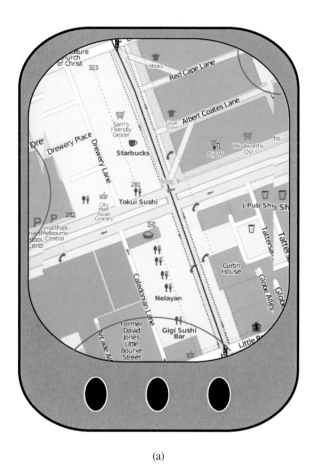

(a)

FIGURE 11.3 Three different approaches to overcome the keyhole problem: (a) Halo, (b) Wedge, and (c) YAH^x maps. (Data from OpenStreetMap.)

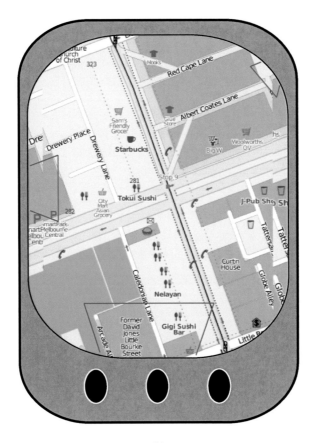

(b)

FIGURE 11.3 (continued)

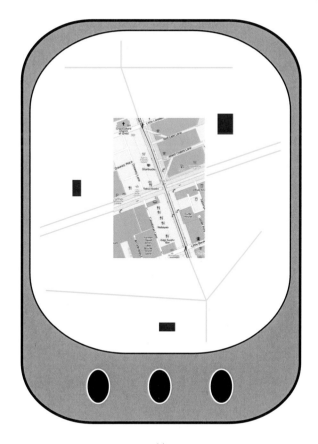

(c)

FIGURE 11.3 (continued)

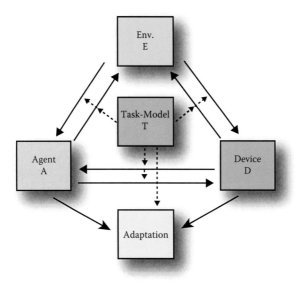

(a)

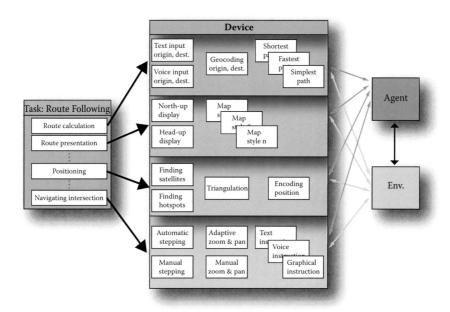

(b)

FIGURE 11.4 (a) Task-based context model. (b) Modular LBS architecture. (Modified from Richter, Kai-Florian, Drew Dara-Abrams, and Martin Raubal, 2010, Navigating and learning with location based services: A user-centric design, in *7th International Symposium on LBS and Telecartography*, edited by G. Gartner and Y. Li, 261–276.)

6 Geo-Crowdsourcing

Jessica G. Benner and Hassan A. Karimi

CONTENTS

ABSTRACT

A wave of new collaboration features and mapping tools has given rise to geo-crowdsourcing services. In this chapter, we review geo-crowdsourcing as a new means for the creation of online maps by groups of distributed users and discuss OpenStreetMap as a case study along with some challenges and issues for this new mapping trend.

6.1 INTRODUCTION

Until very recently, geospatial professionals in state organizations, national militaries, and civilian National Mapping and Cadastre Agencies were the main creators of maps (Heipke 2010) and custodians of geospatial data collections. Today, new applications for crowdsourcing geospatial data enable the public to participate in mapping and manage the shared content with an open license for use and community-based repositories called commons (Van den Berg et al. 2011). These new geo-crowdsourcing applications are vehicles for quick and local acquisition of geospatial data (Chilton 2009) as well as sample data sources for location-based services (LBS). In a geo-sensed world, data streams from everyday users of LBSs. Today's geo-crowdsourcing applications represent a test bed for managing and utilizing these large data sets in creative ways in the future. Given their importance for future LBSs and mapping, this chapter reviews geo-crowdsourcing and offers a detailed description of a successful and popular system, as of this writing in 2012, OpenStreetMap (OSM) (www.openstreetmap.org).

Google Earth (www.google.com/earth/index.html) and OSM are two examples of early sites that support geo-crowdsourcing. After several years of use, two terms to describe these applications emerged in research literature: *crowdsourcing* and *volunteered geographic information* (VGI). The term crowdsourcing was coined in 2006 (Yuen et al. 2011), one year after the release of Google Earth and two years after the creation of OSM. Since 2006, this term has been widely applied in discussions of collaborative Web-based geospatial applications in the literature.

Hudson-Smith et al. (2009) describe "mapping for the masses" and define crowdsourcing as "essential ways in which large groups of users come together to create data and add value by sharing." Fritz et al. (2009) list various terms applicable to the creation of geospatial user-created content but use the term crowdsourcing in the title of their publication describing the Geo-Wiki project. Heipke (2010) prefers "the term crowdsourcing to describe data acquisition by large and diverse groups of people [… largely untrained…] using web technology." Finally, Dodge and Kitchin (2011) describe crowdsourcing as "the collective generation of media, ideas and data undertaken voluntarily by many people" which is "premised on mass participation, with distributed voluntary effort, and a degree of coordination."

In an effort to describe the growing phenomenon of geospatial data collection and sharing, Goodchild (2007a) coined the term VGI as "a special case of the more general Web phenomenon of user generated content." Since then, researchers and practitioners have used the term VGI in discussions of the societal implications of the Geospatial Web (Elwood 2009); introducing the concept of WikiGIS (Roche et al. 2012); best practices for VGI commons (Van den Berg 2011); and numerous evaluations of OSM (Schmitz et al. 2008; Girres and Touya 2010; Haklay 2010; Mooney and Corcoran 2012b; Neis et al. 2012).

In this chapter, we adopt the term geo-crowdsourcing to refer to Web applications in which large, distributed groups of volunteers collect and share VGI. We choose OSM as our case study because of its size and its wide set of services for managing and using geo-crowdsourced data. What is more, the OSM project has been described as a prime (Neis et al. 2012), striking (Haklay 2010), complex and promising (Zielstra and Zipf 2010), most significant (Girres and Touya, 2010), and most famous (Mooney and Corcoran 2012a) example of the collection of VGI (i.e., geo-crowdsourcing).

6.2   GEO-CROWDSOURCING

The capacity for collaborative data creation, under the Web 2.0 paradigm, supports technologies that are user-driven allowing large groups of users to contribute content normally collected by experts. Technologies of this nature, commonly called crowdsourcing systems, have grown in magnitude due to widespread use of the Internet and WWW (Yuen et al. 2011). These collaborative sites enable communication and participation essential to supporting people, separated by distance, to collaborate. The Geospatial Web represents an important catalyst for geo-crowdsourcing by supporting the sharing of geospatial content and maps over the Internet (Haklay et al. 2008). Today, Web mapping 2.0 can be viewed as a combination of public mapping sites (e.g., view only) and Web map servers (e.g., viewing and editing) (Haklay et al. 2008). Goodchild (2007a) lists the technologies enabling VGI as the participatory Web, georeferencing, GPS, geotags, cameras, and broadband communication. Heipke (2012) echoes a similar listing, divided into two categories, basic technologies and data sources. Heipke lists georeferencing and broadband communication as the basic technologies, and GPS tracks and orthophotos as the main data sources that support the crowdsourcing of geospatial data.

Building on the foundation of Web 2.0 and geospatial Web technologies previously discussed, two types of geo-crowdsourcing sites have emerged: location-based social networks (LBSNs) and geowikis. Each of these systems inherits features from their more "traditional" parents. LBSNs are more socially oriented, geared toward location sharing (Barkhuus et al. 2008). Geowikis' focus on the collaborative editing of (Priedhorsky and Terveen 2008) and traceability of changes (Roche et al. 2012) to geospatial features is in line with the collaborative editing common in traditional wiki sites like Wikipedia. Sui (2008) uses the term *wikification* of GIS to describe fundamental GIS components and functions extended to the Web and "performed in the wiki spirit." Since LBSNs are covered in another chapter of this book and focus on location sharing, the remainder of this chapter will focus on geowiki projects in which users tend to engage more fully in geo-crowdsourcing tasks.

With an understanding of the catalysts that ushered geo-crowdsourcing sites into existence, we can now turn to their uses. In addition to the collection of geospatial data, geo-crowdsourcing sites can contribute to the organization and naming of geospatial data by serving as digital gazetteers. This use of geo-crowdsourcing content could lead to improved sets of descriptors that semantically enrich the VGI collected by these sites and offer a mechanism to update the terminology used in traditional gazetteers. A second use of VGI is informing users about local activities and "life at a local level" (Goodchild 2007a). This value of local information is the foundation for the field of neogeography. Many definitions of neogeography have come forward from its original definition by Di-Ann Eisnor as a "socially network linked mapping platform" (Haklay et al. 2008) to another definition linking neogeography to the "everywhereness" of spatial information in our daily lives (Elwood 2009) and finally a "narrow definition" discussed in Warf and Sui (2010). This narrow view of neogeography is a "process whereby varied groups of people use an eclectic set of online geospatial tools to describe and document aspects of their lives, society, or environment in terms that are meaningful to them." This last definition highlights the impact of mapping local data on the usefulness of the data. Now we provide a few example applications for illustration.

6.2.1 EXAMPLE APPLICATIONS

Wikimapia (http://wikimapia.org) is a geowiki that provides users with a means to "describe the whole world" as their tagline reads. On its guidelines page, Wikimapia lists its aim to "create and maintain a free, complete, multilingual, up-to-date map of the whole world." The interface (Figure 6.1) consists of an interactive map and wiki

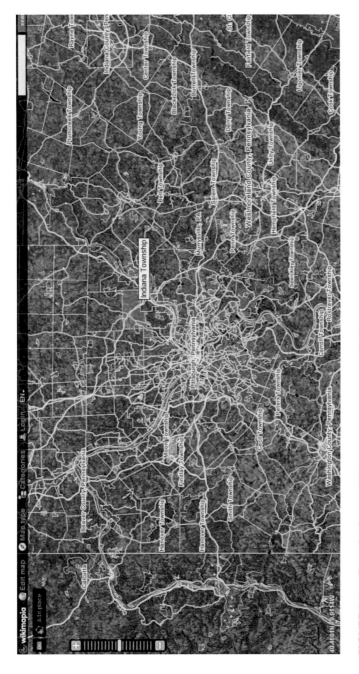

FIGURE 6.1. (See color insert.) Wikimapia homepage. http://wikimapia.org.

tools that support and keep track of editing. Users of Wikimapia can add point, line, and area features to the interactive map and categorize each feature using a set of categories provided by the system. Wikimapia encourages a "neutral point of view" when editing and offers a set of guidelines for users (Wikimapia 2012b). Finally, following the model of the commons, Wikimapia shares the content of its maps with developers through a free API (Wikimapia 2012a).

The Geo-Wiki Project is a geowiki where users can help validate global land cover maps (Fritz et al. 2009). The interface (Figure 6.2) consists of an interactive map supported by a Google Earth browser plug-in and several editing tools. It has been noted that the geo-crowdsourcing approach of the Geo-Wiki project is "inexpensive and allows Internet users from any region of the world to get involved" (McCallum et al. 2010). In the validation process, volunteers review hotspot maps of disagreement between different land cover and compare target locations using satellite images in Google Earth and their personal knowledge to verify the correct land cover (Fritz et al. 2009). Finally, the Geo-Wiki project offers a downloadable data set of its validation results in the native Google Earth KMZ format (Geo-Wiki Project 2012).

These two examples represent different uses of crowds in geo-crowdsourcing. Wikimapia uses the crowd to create maps of data around the world while the Geo-Wiki project uses the crowd to both create new and validate existing data. Now that we have introduced what geo-crowdsourcing is and provided a few examples, we will provide a detailed case study of OSM.

6.3   OPENSTREETMAP

OpenStreetMap (OSM) is a geowiki and platform for geo-crowdsourcing. The OSM geowiki is an interactive map with editing and export tools as well as the ability to view a history of edits for any location on the map. The OSM platform consists of dozens of mapping projects that utilize the OSM API to access OSM data and share their content back to the project, and a suite of geoprocessing and wiki-based tools for developers of these projects. In this section, we begin with a history of OSM collected from Web pages in the Internet Archive, other sites related to OSM, and research literature. Then we describe the current landscape of OSM.

6.3.1   HISTORICAL VIEW OF OSM

6.3.1.1   Creation of OSM and the Interface

Steve Coast, a physics student at the University College of London, created OSM in 2004 because maps in the United Kingdom were not freely available and he thought that the strict control of geodata limited geospatial applications (Gyford 2004). The first version included a GPS receiver, a laptop with long battery life, an OpenSQL database, and trace and editing software owned by Coast (Gyford 2004). On August 16, 2004, the first public OSM Web site was deployed (OpenStreetMap August 2004). This first version of OSM was simple and did not include any map-editing features. One month after the release, the first post occurred on the listserv (OpenStreetMap

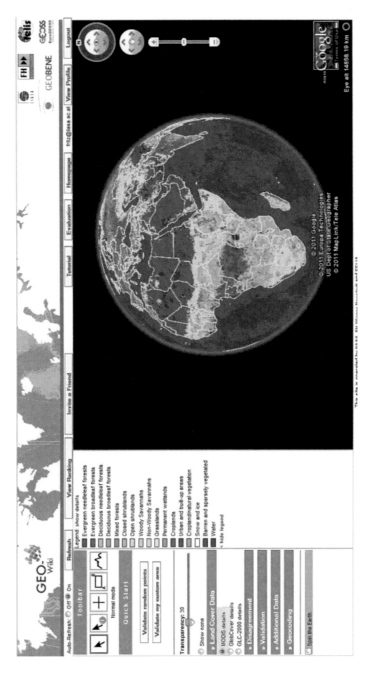

FIGURE 6.2 (See color insert.) Geo-Wiki Project homepage. (From Geo-Wiki.org instructions page, http://www.geo-wiki.org/instructions.php.)

OpenStreetMap

Links:

mailing list
svn access

Welcome to OpenStreetMap.org

This is an effort to produce free (CC licensed) streetmaps of the world. The idea is to run around with a gps and annotate the data using the applet below in a wiki-type way. I use a linux laptop with an insane battery life, broken keyboard and screen with a gps and wifi card in it. I cycle around and use a zaurus pda to talk wifi to the laptop to add data like street names. There is a **long** way to go in terms of getting data and writing code, so feel free to help. More info is in the blog below the applet, so scroll down. Right now what you see is a flight I took to london from berlin, lots of streets in london and cambridge. Also some train tracks to london, cambridge and leeds. The applet loads a set amount of data, and this data changes when you change view (zoom, move) so zooming in will show more data. Oh, you need a half-decent JVM with swing, sorry.

Error. Click for details

FIGURE 6.3   Early version of OSM interface using OpenMap, © OpenStreetMap contributors, CC BY-SA. (From Wayback Machine at the Internet Archive August 2012.)

September 2004), and the OSM Web site was rewritten to use OpenMap (BBN Technologies 2012) and to preload a "set amount of data" (OpenStreetMap September 2004). This new interface, shown in Figure 6.3, is the first OSM interface that incorporates the features of a geowiki.

In early 2005, OSM continued to evolve at a rapid pace with the consolidation of API, applet, and wiki onto a new server (OpenStreetMap February 2005) and a new front page (Figure 6.4) and editing interface (OpenStreetMap March 2005). Under this architecture, a user visited the front page and entered a set of desired coordinates to view the map.

Continuing their rapid evolution, a new interface merging the front page and map page was released nine months later in December 2005 (Figure 6.5). Note the similarities between this interface from late 2005 and the main map page of OSM as it looks as of this writing in 2012 (Figure 6.7).

OpenStreetMap

...The Free Wiki World Map

51.526447 -0.14746371   Grab map!
● Lat/Lon ○ Street/Place ○ Post/Zipcode

FIGURE 6.4   New OSM front page with familiar logo. © OpenStreetMap contributors, CC BY-SA. (From Wayback Machine at the Internet Archive August 2012.)

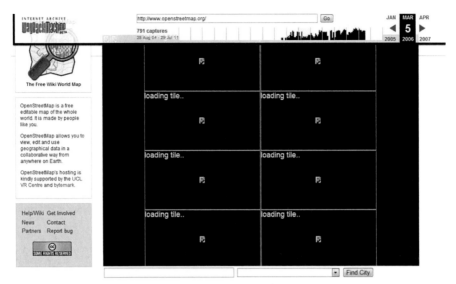

We're trialing adverts to support the project. Login and they go away.

FIGURE 6.5 Merged OSM interface. © OpenStreetMap contributors, CC BY-SA. (From Wayback Machine at the Internet Archive August 2012.)

In January 2006, software called Java OpenStreetMap Editor (JOSM) that allowed users to edit OSM data both online and offline (OSM List Contributor January 2006) was released. At this time, OSM participation increased and through the support of a company called bytemark, OSM was able to quickly decentralize their system and split the wiki, mailing list, and other features to a virtual machine (OpenStreetMap February 2006). Building on the new merged interface, OSM released a new interactive page for reviewing and searching shared GPS traces in March 2006 (OpenStreetMap March 2006). In October 2006, version 1.4 of JOSM called "JOSM Birthday Release" debuted (OpenStreetMap October 2006). This new version improved developers' ability to extend the features of JOSM through plug-ins and optimized JOSM performance. An example of a JOSM plug-in is shown in Figure 6.6 (White 2006). In November 2006, OSM began implementing a technology called a Slippy map, still used today, to render the main map page and incorporate project layers into the map (OpenStreetMap November 2006).

After 2006, the rapid development of the OSM interface slowed. According to the OSM Wiki History page, in May 2007, OSM released the Rails Port (OSM server code) with the newest version of the API and upgraded their map editor. A capture from the Internet Archive on May 12, 2007, shows a new tab for collecting users' diaries and one on August 9, 2007, shows a new search feature. Much of the developments since 2007 have not focused on the OSM interface. Figure 6.7 shows the OSM interface for the last four years: August 2009, August 2010, June 2011, and August 2012. It is easy to see that the main features of the interface have been somewhat rearranged (e.g., the search bar and user diaries tab) but have remained largely the same since 2009.

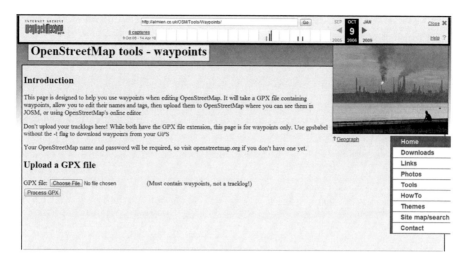

FIGURE 6.6 Waypoints plug-in for JOSM editor. © OpenStreetMap contributors, CC BY-SA. (From Wayback Machine at the Internet Archive August 2012.)

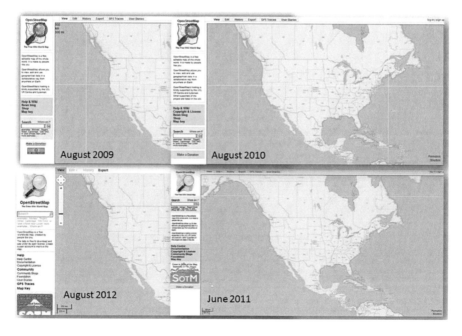

FIGURE 6.7 **(See color insert.)** OSM interface overtime, clockwise from top left, August 2009, August 2010, June 2011, and August 2012, © OpenStreetMap contributors, CC BY-SA. (From 2009, 2010, and 2011 images from the Wayback Machine at the Internet Archive August 2012; August 2012 image from http://www.openstreetmap.org/.)

6.3.1.2   Popularity and Users

In his initial development of OSM at the Euro Foo Camp in 2004, Coast mentions that he envisions multiple users contributing and editing data to OSM (Gyford 2004). According to the OSM wiki entry for the Map Limehouse event, the first mapping party was held in the town of Limehouse in July 2005. Mapping parties are events in which several users of OSM gather together to map the same physical location. Several months after the successful mapping party in Limehouse, a group of registered users began using OSM to map their hometown of Birmingham (OpenStreetMap November 2005). In January 2006, OSM boasted 1000 registered users as well as the fact that participation was beginning to exceed the capacity of OSM's editing software (OpenStreetMap January 2006). In May 2006, a mapping workshop was held in the Isle of Wight. The workshop, written about in the *Guardian* (Mathieson 2006), included 30 to 40 participants who gathered with the goal of producing a released map of the island in one weekend. After the success of the Isle of Wight mapping party, a series of mapping parties were held in 2006 (OpenStreetMap November 2006).

Near the time of the Isle of Wight workshop, discussions began toward the creation of the OpenStreetMap Foundation (OpenStreetMap May 2006). According to the OSM Wiki History page, the OSM Foundation was established in August 2006 and remains an established not-for-profit organization that acts as the OSM legal entity, the custodian of the computer servers and services for hosting OSM, engages in fundraising for the platform, and organizes the yearly State of the Map conference (Open Street Map Foundation 2012). By February 2009, the main map of OSM was being updated more than once per week. Indeed the uptake in events like mapping parties, the establishment of a foundation to manage the OSM project, and the increase in map updates are evidence for the growing popularity of OSM over time.

Now we will look at the increase in the number of users from the 1000th registered user in December 2005 to the present. Table 6.1 shows a subset of the history milestones for registered users from the OSM Wiki History page and the most recent number of registered OSM users available in the OSM Wiki Stats page. Sixteen months after the release of OSM, there were 1000 registered users with an average

TABLE 6.1

Milestones for Number of OSM Users since 2004

Date	Number of Users	Months since Last Milestone	Average Increase per Month since Last Milestone	Magnitude of Increase in Users per Month
December 2005	1000	17	59 users per month	—
August 2007	10,000	20	450 users per month	7.6 times
March 2009	100,000	19	4,737 users per month	10.5 times
November 2011	500,000	32	12,500 users per month	2.6 times
August 2012	718,000	10	21,800 users per month	1.7 times

Note: Number of users data is from the OSM Wiki History page and OSM Wiki Stats page.

of 59 users registering each month. At 3 years, OSM reaches 10,000 users with an average of 450 users registering each month. At roughly 4.5 years, OSM has 100,000 users and an average of 4,737 users registering each month. More recently, just over 6 years into the OSM project, there are 500,000 registered users and an average of 12,500 users who registered for the project each month. Today, we see that OSM has 718,000 registered users and an average of 21,800 users registering each month. These numbers are astonishing; however, when we compare the magnitude of increase in users per month for each milestone, we see an increase in magnitude from August 2007 to March 2009 and a subsequent decrease in magnitude from March 2009 to November 2011 and November 2011 to August 2012.

We have seen the large increase in registered users of OSM over time, but are they actively participating in the project or are they merely curious? Mooney and Corcoran (2012a) investigated the activity of contributors to OSM London. Using the entire history of edits for the city of London from April 2005 to October 2011, they found 2795 unique users over the period. Selecting the top 20 users for their analysis, they find that these top 20 users contributed over 60% and edited over 50% of all the ways and 12% of all nodes in London. Interestingly, they find 72% of the users have contributed 20 or less edits to OSM London. Expanding this first work, Mooney and Corcoran (2012b) used the same London data set to study users contributing 200 or more edits to ways with five or more edits (n = 33,230 objects) in the OSM London. They find that a low number of users collaboratively edit features and that most editing occurs over time by a small set of users. So, while the number of registered users of OSM increases each month, the work of Mooney and Corcoran indicates that the main contributions continue to come from small sets of devoted mappers.

6.3.1.3 Data

OSM data refers to the content that is shared by users. The original OSM map elements (basic components of OSM) were nodes, segments, and ways. In 2007, under API v. 0.5, segments were no longer supported and they were replaced by relations. Today, there are three elements: nodes (point in space), ways (linear feature or area), and relations (optional elements that define relations between other elements). In December 2004, the first street was added to OSM (OpenStreetMap December 2004). In January 2006, OSM users completed the first set of motorways in the United Kingdom, which was 3000 km (OpenStreetMap January 2006). By March 2006, OSM introduces map features (fields to describe map elements), OSM data is successfully loaded into a Garmin GPS device, and an OSM map is added to the Weybridge Wikipedia article (OpenStreetMap March 2006). In April 2006, the first data dump occurred and developers created a tool, Osmarender, to output OSM data to SVG and make maps with street names and place names (OpenStreetMap April 2006). Figure 6.8 shows an SVG Web map using XSLT for Weybridge.

In June 2006, the 20 millionth GPS point was uploaded and OSM began support for a common API between Google, Yahoo!, and Microsoft, called Mapstraction (OpenStreetMap June 2006). This common API provided increased data interoperability for OSM to import larger data sets from data providers. Over the next 2 years, OSM acquired aerial imagery from Yahoo!, automotive navigation data for the Netherlands, and TIGER line files for the United States. Bing followed suit in

FIGURE 6.8 (See color insert.) OSM Map of Weybridge. This map was created by OpenStreetMap Map. © OpenStreetMap contributors, CC BY-SA. (Retrieved from Wikipedia, http://en.wikipedia.org/wiki/File:KT13_Area.4.png.)

late 2010. Significantly, according to the OSM Wiki Ordinance Survey Opendata Page, in April 2010, almost 6 years after the creation of OSM, the Ordinance Survey agrees to share data with OSM under the Opendata license.

Table 6.2 displays the number of GPS points, nodes, segments, ways, and relations for the OSM database on two dates, January 2007 and May 2012. The number of GPS points is 50 times larger, there are 200 times more edited nodes, and 400 times more ways in 2012. Looking at the average number of GPS points, nodes, and ways added per month, we get a glimpse of the mapping activity occurring in OSM over this period of 64 months. One important note is the addition of several large data sets from 2007 to 2010 including the TIGER line files for the United States that are included in these totals.

With any geospatial data, measures of data quality are fundamental. Several studies investigate the quality of OSM data. We only discuss a few to show the trend of results. Haklay (2010) compared OSM London to the proprietary Ordinance Survey data set and found that OSM data are fairly accurate (within 6 meters on

TABLE 6.2

Size of OSM in Number of Items in January 2007 and May 2012

World Stats	January 2007	May 2012	Average per Month
GPS points (raw)	48.8 million	2.8 billion	43 million
Nodes (edited)	7.2 million	1.45 billion	22 million
Segments	7.4 million	—	
Ways	332,000	135.1 million	2 million
Relations	—	1.4 million	

Note: 2007 and 2012 data are from the OSM Wiki Database Statistics and OSM Wiki Stats pages.

average) when compared to the Ordinance Survey data. Further findings include an 80% overlap of data for motorway objects and coverage of 29% of England over 4 years of data collection by OSM with 24% of these missing attributes. Finally, Haklay (2010) discovers a lack of mapping in poorer cities in England that indicates stronger participation in more affluent areas.

Girres and Touya (2010) analyze OSM France data. In this work, they compare OSM France to the BD TOPO reference data set. They find an average positional difference of 6.65 meters and small geometric differences between polygon features (lakes). The attributes for lakes in the OSM France were less robust than the reference data set; however, features with more contributors tended to have better attributes. For completeness, they find that smaller objects are more likely unmapped and that the density of contributors in an area has an effect on the completeness. Finally, due to a lack of integrity constraints in OSM, they find many logical inconsistencies especially along political boundaries featuring rivers. The paper closes with a discussion of the importance of specifications to ensure data quality.

Two investigations are reported for Germany. Zielstra and Zipf (2010) compare the OSM Germany data set with a proprietary navigation data set TeleAtlas. They find that OSM Germany is more complete in German cities and decreases with distance from the city center. They also note the OSM street network includes more pedestrian paths but has inadequate support for car navigation applications. Zielstra and Zipf conclude that although OSM offers a large amount of data, it is not an adequate alternative to TeleAtlas due to its lack of coverage in rural areas. Neis et al. (2012) focus on the expansion of the street network and route network for car navigation in OSM Germany. Comparing OSM Germany to a TomTom (TeleAtlas) data set, they find OSM Germany covers 27% more data for the total street network and routes for pedestrians but still lacks 9% of the data available in TomTom for car navigation. They offer a projection that the OSM Germany data set will be a geometrically complete alternative to the TomTom proprietary data set by the end of 2012; however, their findings regarding attribute quality and turn restrictions will keep OSM Germany from being a full alternative to TomTom for now.

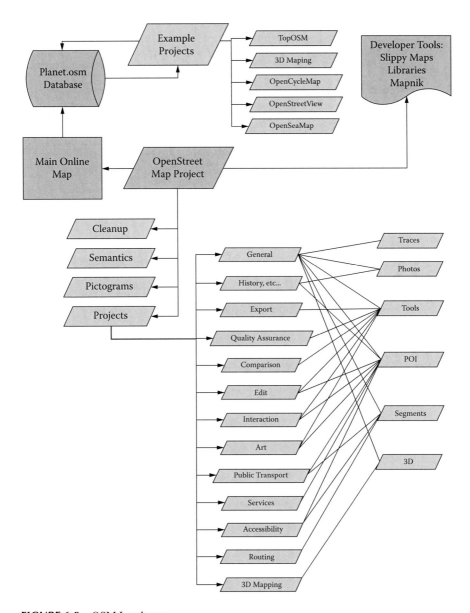

FIGURE 6.9   OSM Landscape.

6.3.2   VIEW OF THE CURRENT OSM LANDSCAPE

The OSM landscape includes the main map, the wiki, planet.osm database, a set of OSM Project tasks, a suite of developer tools, and a large collection of projects tuned to specific kinds of data. Figure 6.9 depicts this landscape and the relationships between its components. The geowiki is represented as Main Map. Main Map is interactive and includes various editing and other tools. The OSM Wiki contains pages of text-based content such as statistics of use, links to various projects, and

TABLE 6.3
Size of OSM in Bytes

2012	Size in GB (Compressed)	Size in GB (Uncompressed)
Full History	25	500
Current Database	20	250
Europe	11	~180
Asia	2.5	~42.5
Africa	0.336	~5.7
Australia	0.337	~5.7
Germany	1.7	~28.9
Italy	0.69	~11
U.S.	~5.24	~89
California (U.S.)	0.62	~10.5
Pennsylvania (U.S.)	0.087	~1.5

Source: Estimate of planet.osm database is from Mooney and Corcoran (2012a), the current database (HEAnet 2012), and all other estimates (Geofabrik 2012).

descriptions of map features, to name a few. The OSM Projects including Cleanup, Semantics, Pictograms, and Projects represent efforts by the OSM community to control and manage data collection, naming, visualization, and uses, respectively. The Cleanup project helps to ensure that the collected data meet a certain level of geospatial accuracy and quality. The Semantics project is devoted to the quality and interoperability of the naming and tagging of features collected in Main Map and outside projects that share content with the platform. The Pictograms project aims to control the symbols that represent various phenomena that users collect and to guarantee that symbols are appropriate and distinct. The Projects project monitors numerous outside projects that use OSM data and share data with the planet.osm database.

The planet.osm database stores the totality of OSM data and editing history. All the data stored in the planet.osm are freely available to the public. This data set includes the data for the entire earth and is quite large (Table 6.3). Because of the size of the planet.osm database, third-party services offer downloadable data sets for smaller areas such as continents, countries, and select cities. The example projects listed in the top of Figure 6.9 show various types of data shared with the OSM platform. These include topographic data, 3D data, bicycle paths/trajectories, images, and seafaring data. This view illustrates the enormity of the OSM landscape and comprehensive nature of the data collection at OSM.

The bottom of Figure 6.9 lists the categories of outside projects collaborating with OSM. These 13 categories of projects reflect both data collection projects and editing services that outside collaborators have shared with the OSM infrastructure. The column to the right of the projects lists the potential content types that each project category can include. For example, the General category includes projects collecting trace data, images, points of interest (POIs), segments, and 3D data as well as mapping or editing tools; whereas the Export category only includes projects focused on developing tools for exporting OSM data.

Today, OSM is the largest geo-crowdsourcing platform in the world. The OSM landscape includes dozens of projects that use OSM data as a basis for collecting thematic data for their specific goals and who subsequently share the collected data with OSM. This relationship between OSM and its projects has led to the large master database, planet.osm. Table 6.3 shows statistics for the size of planet.osm, the current OSM database, and databases of varying sizes (i.e., continent, country, and city). The values in the table are from Geofabrik (2012), an index of mirror/openstreetmap.org (HEAnet 2012), and estimates in Mooney and Corcoran (2012a). The estimates (with a tilde, ~) are based on a conversion factor of 1024 MB in 1 GB and a compression ratio of 17:1 in which uncompressed data is 17 times larger than the compressed data. The full history database is planet.osm. The current database holds the most recent version of all the features shared with OSM. The data sets for continents, countries, and cities contain all the features created for that respective geographic extent. As shown in Table 6.3, the size of the current database is quite large. Consequently, data processing of OSM data for large geographic areas requires high-performance computing, hardware and software, platforms. It is interesting to note the differences between the mapping activities in the continent of Africa versus the country of Italy. The entire continent of Africa has less data in OSM than the much smaller country of Italy.

6.4 CHALLENGES AND ISSUES

This section discusses challenges and issues relevant to geo-crowdsourcing.

6.4.1 CHALLENGES

A challenge in any crowdsourcing system is the motivation of volunteers. Goodchild (2007a) lists user motivation as a challenge for VGI. Heipke (2010) lists crowds of users who contribute geo-crowdsourced data, adapted from a EuroSDR workshop. They include map lovers, casual mappers, experts, media mappers, passive mappers, open mappers, and mechanical turks. This exhaustive list highlights the wide range of users and their potential motivations to contribute to geo-crowdsourcing. Fritz et al. (2009) propose the use of competitive games and social networks to spread the word as potential solutions to the challenge of motivating the crowd.

A second challenge for geo-crowdsourcing is quality control. Yuen et al. (2011) call for crowdsourcing systems that are designed to include measures to ensure good output from the users. In the case of geo-crowdsourcing, good input is often measured by using aspects of geospatial data quality. Quality control is an important challenge in any mapping project, and when many distributed users are collaborating to collect data it is critical that quality control measures are embedded in the tools available to users. Girres and Touya (2010) discuss many data quality issues and encourage the development of data specifications in their analysis of OSM France. Other authors (Fritz et al. 2009; Van der Berg et al. 2011) assert that the implementation of tools that allow volunteers to monitor content, similar to Wikipedia, will aid in quality control for geo-crowdsourcing.

Hudson-Smith et al. (2009) describe the third challenge relevant to geo-crowdsourcing, the introduction of "intelligence into the ways users can create, share and apply data about themselves to themselves." The creation of new forms of information or new knowledge from data sets collected through geo-crowdsourcing is an ongoing issue related to the Semantic Web. Gruber (2008) describes the current state of the social web as "collec*ted* knowledge systems" built on "collec*ted* intelligence." In his view, there is "no emergence of truly new levels of understanding" in existing crowdsourcing systems (Gruber 2008). This criticism of the basic challenge for Web 2.0 has implications for the utilization of geo-crowdsourced data by future LBSs. Finally, Roche et al. (2012) mention collective intelligence and open networks as areas where WikiGIS sites can lead to "geo-referenced collective intelligence development." In the future, it will be critical to use the collected data for unique purposes to generate new knowledge rather than solely focus on its collection.

6.4.2 ISSUES

In an editorial on the wikification of GIS, Sui (2008) lists privacy, liability, and equity as issues for VGI. Issues of privacy are well known and are not detailed here. The issue of liability strongly ties to issues of authority. Goodchild (2007a) uses the term "asserted geographic information" to describe content that is asserted to be correct by the user who volunteers it but does not embody any strict standards commonly viewed as authoritative in the past. Dodge and Kitchin (2011) also comment on issues of authority noting that geo-crowdsourcing (or crowdsourced cartography as they call it) changes the roles of authorship, "normative ontology" used in cartography, and the traditional "fixity, objectivity and authority" of cartographic representations. The issue of equity is an important issue regarding who benefits from geo-crowdsourcing, the crowd or industry. Today, Google, TeleAtlas, and NAVTEQ, and other companies specializing in geospatial services have added features that capitalize on the crowd's participation to enhance their services (Chilton 2009).

Both Goodchild (2007a) and Heipke (2010) have raised the issues of the impact of the digital divide and language support on the success of geo-crowdsourcing. Goodchild notes that the majority of people living in developing countries lack "access to the Internet and broadband access in particular." Heipke discusses the digital divide as a "fundamental limitation for mapping the whole world" due to a lack of supporting technology in many parts of the world. The effect of the digital divide is present in Table 6.3. The lack of mapping activity in Africa is likely a direct result of the digital divide. Finally, both Heipke and Goodchild highlight limitations of language support because most systems can only be used by people with knowledge of English and the Latin alphabet.

6.5 SUMMARY

In this chapter, we discussed on geo-crowdsourcing as the creation of online maps by groups of distributed users. We discussed Web 2.0 and the Geospatial Web as catalysts for the emergence of geo-crowdsourcing projects and discussed a few examples of geowikis that are in wide use today. The rest of the chapter provided details of OSM, the largest collaborative mapping platform in existence, and thus a representative of existing geo-crowdsourcing sites. Finally, we closed the chapter with a discussion of issues and challenges for the future of geo-crowdsourcing.

REFERENCES

Barkuus, Louise, Barry Brown, Marek Bell, Malcolm Hall, Sherwood Scott, and Matthew Chalmers. 2008. "From Awareness to Repartee: Sharing Location within Social Groups." In CHI 2008 Proceedings: I am here. Where are you? Florence, Italy.

BBN Technologies. 2012. "Openmap Open Systems Mapping Technology." http://openmap.bbn.com/.

Chilton, Steve. 2009. "Crowdsourcing Is Radically Changing the Geodata Landscape: Case Study of Openstreetmap." In 24th International Cartographic Conference, November 15–12.

Dodge, Martin, and Rob Kitchin. 2011. "Mapping Experience: Crowd sourced Cartography." (August 28, 2011). Available: http://papers.ssrn.com/S013/papers.cfm?abstract_id=1921340.

Elwood, Sarah. 2009. "Geographic Information Science: Emerging Research on the Societal Implications of the Geospatial Web." *Progress in Human Geography* 1–9.

Fritz, Steffen, Ian McCallum, Christian Schill, Cristoph Perger, Roland Grillmayer, Frédéric Achard, Florian Kraxner, and Michael Obersteiner. 2009. "Geo-Wiki.Org: The User of Crowdsourcing to Improve Global Land Cover." *Remote Sensing* 1(3): 345–354.

Geofabrik. 2012. "Download Openstreetmaps Extracts." Geofabrik, http://download.geofabrik.de/osm/.

Geo-Wiki Project. 2012. "Download Data Webpage." http://www.geo-wiki.org/login.php?menu = results.

Girres, Jean-François, and Guillaume Touya. 2010. "Quality Assessment of the French Openstreetmap Dataset." *Transactions in GIS* 14(4): 435–459.

Goodchild, Michael F. 2007a. "Citizens as Sensors: The World of Volunteered Geography." *GeoJournal* 69: 211–221.

Goodchild, Michael F. 2007b. "Citizens as Sensors: Web 2.0 and the Volunteering of Geographic Information." *GeoFocus* (Editorial) 7: 8–10.

Google Earth. 2012. http://www.google.com/earth/index.html,.

Gruber, Tom. 2008. "Collective Knowledge Systems: Where the Social Web Meets the Semantic Web." *Web Semantics: Science, Services and Agents on the World Wide Web* 6(1): 4–13.

Gyford, Phil. 2004. "Euro Foo Camp: Steve Coast—Opentextbook & Openstreetmap." In Blog of Phil Gyford. Retrieved from the Internet Archive: http://web.archive.org/web/20051104193142/http://www.gyford.com/phil/notes/2004/08/21/euro_foo_camp_stev.php.

Haklay, Mordechai. 2010. "How Good Is Volunteered Geographical Information? A Comparative Study of Openstreetmap and Ordinance Survey Datasets." *Environment and Planning B: Planning and Design* 37, 682–703.

Haklay, Muki, Alex Singleton, and Chris Parker. 2008. "Web Mapping 2.0: The Neogeography of the Geoweb." *Geography Compass* 2(6): 2011–2039.

HEAnet. 2012. "Index of/Mirrors/Openstreetmap.Org." Ireland's National Education and Research Network, http://ftp.heanet.ie/mirrors/openstreetmap.org/.

Heipke, Christian. 2010. "Crowdsourcing Geospatial Data." *ISPRS Journal of Photogrammetry and Remote Sensing* 65: 550–557.

Hudson-Smith, Andrew, Michael Batty, Andrew Crooks, and Richard Milton. 2009. "Mapping for the Masses: Accessing Web 2.0 through Crowdsourcing." *Social Science Computer Review* 27(4): 524–538.

Internet Archive. 2012. "Wayback Machine, Internet Archive, San Francisco, Ca." http://archive.org/.

Mathieson, Steve. 2006, May 11. "A Sidestep in the Right Direction: An Innovative Exercise in Electronic Mapping Aims to Bypass the Block on Access to Data Subject to Crown Copyright." In *The Guardian*. Retrieved from the Internet Archive: http://web.archive.org/web/20061231062343/http://technology.guardian.co.uk/weekly/story/0,,1771598,00.html.

McCallum, Ian, Steffen Fritz, Christian Schill, Cristoph Perger, Frédéric Achard, Roland Grillmayer, Barabara Koch, Florian Kraxner, Michael Obersteiner, and Markus Quinten. 2010. "Earth Information Systems/Capacity Building—Geo-Wiki.Org: Harnessing the Power of Volunteers, the Internet and Google Earth to Collect and Validate Global Spatial Information." *earthzine*.

Mooney, Peter, and Padraig Corcoran. 2012a. "Who Are the Contributors to Openstreetmap and What Do They Do?" Paper presented at the GISRUK 2012, Preprint.

Mooney, Peter, and Padraig Corcoran. 2012b. "How Social Is Openstreetmap?" Paper presented at the AGILE 2012,.

Neis, Pascal, Dennis Zielstra, and Alexander Zipf. 2012. "The Street Network Evolution of Crowdsourced Maps: Openstreetmap in Germany 2007–2011." *Future Internet* 4: 1–21.

OpenStreetMap. 2004, August. "Archive for August 2004." Opengeodata blog. Retrieved from the Internet Archive: http://web.archive.org/web/20070109223605/http://www.opengeodata.org/?m = 200408.

OpenStreetMap. 2004, September. "Archive for September 2004." Opengeodata blog. Retrieved from the Internet Archive: http://web.archive.org/web/20070109223652/http://www.opengeodata.org/?m = 200409.

OpenStreetMap. 2004, December. "Archive for December 2004." Opengeodata blog. Retrieved from the Internet Archive: http://web.archive.org/web/20070109223613/http://www.opengeodata.org/?m = 200412.

OpenStreetMap. 2005, February. "Archive for February 2005." Opengeodata blog. Retrieved from the Internet Archive: http://web.archive.org/web/20070109223659/http://www.opengeodata.org/?m = 200502.

OpenStreetMap. 2005, March. "Archive for March 2005." Opengeodata blog. Retrieved from the Internet Archive: http://web.archive.org/web/20070109223709/http://www.opengeodata.org/?m = 200503.

OpenStreetMap. 2005, November. "Archive for November 2005." Opengeodata blog. Retrieved from the Internet Archive: http://web.archive.org/web/20070109223759/http://www.opengeodata.org/?m = 200511.

OpenStreetMap. 2006, January. "Archive for January 2006." Opengeodata blog. Retrieved from the Internet Archive: http://web.archive.org/web/20070109223810/http://www.opengeodata.org/?m = 200601.

OpenStreetMap. 2006, February. "Archive for February 2006." Opengeodata blog. Retrieved from the Internet Archive: http://web.archive.org/web/20070109223821/http://www.opengeodata.org/?m = 200602.

OpenStreetMap. 2006, April. "Archives for April 2006." Opengeodata blog. Retrieved from the Internet Archive: http://web.archive.org/web/20070109223833/http://www.opengeodata.org/?m = 200604.

OpenStreetMap. 2006, June. "Archives for June 2006." Opengeodata blog. Retrieved from the Internet Archive: http://web.archive.org/web/20070109223454/http://www.opengeodata.org/?m = 200606.

OpenStreetMap. 2006, October. "Archives for October 2006." Opengeodata blog. Retrieved from the Internet Archive: http://web.archive.org/web/20070109223916/http://www. opengeodata.org/?m = 200610.

OpenStreetMap. 2006, November. "Archives for November 2006." Opengeodata blog. Retrieved from the Internet Archive: http://web.archive.org/web/20070109223926/http://www. opengeodata.org/?m = 200611.

OpenStreetMap. 2012. http://www.openstreetmap.org/.

OpenStreetMap Foundation. 2012. "Homepage." http://www.osmfoundation.org/wiki/Main_Page.

OpenStreetMap Wiki. 2012a. "History." http://wiki.openstreetmap.org/wiki/History.

OpenStreetMap Wiki. 2012b. "Ordinance Survey Opendata Page." http://wiki.openstreetmap. org/wiki/Ordnance_Survey_Opendata.

OpenStreetMap Wiki. 2012c. "Stats." http://wiki.openstreetmap.org/wiki/Stats.

OSM List Contributor. 2006, January. "[Openstreetmap] Josm 1.0 Released." OSM Lists. http://lists.openstreetmap.org/pipermail/talk/2006-January/002163.html.

Priedhorsky, Reid, and Loren Terveen. 2008. "The Computational Geowiki: What, Why, How." In Computer Supported Collaborative Work, CSCW'08. San Diego, California: ACM.

Roche, Stephane, Boris Mericskay, Wided Batita, Matthieu Bach, and Mathieu Rondeau. 2012. "WikiGIS Basic Concepts: Web 2.0 for Geospatial Collaboration." *Future Internet* 4: 265–284.

Schmitz, Sebastian, Alexander Zipf, and Pascal Neis. 2008. "New Applications Based on Collaborative Geodata—The Case of Routing." XXVIII INCA International Congress on Collaborative Mapping and SpaceTechnology, Gandhinagar, Gujarat, India.

Sui, Daniel. 2008. "The Wikification of GIS and Its Consequences: Or Angelina Jolie's New Tattoo and the Future of GIS." Computers, *Environment and Urban Systems* 32: 1–5.

van den Berg, Heinrich, Serena Coetzee, and Anthony Cooper. 2011. "Analysing Commons to Improve the Design of Volunteered Geographic Information Repositories." http://africa-geodownloads.info/094_vandenberg_coetzee_cooper.pdf.

Warf, Barney, and Daniel Sui. 2010. "From GIS to Neogeography: Ontological Implications and Theories of Truth." *Annals of GIS* 16(4): 197–209.

White, Oliver. 2006. "Openstreetmap Tools—Waypoints." Almién.co.uk: http://web.archive.org/ web/20070111131146/http://almien.co.uk/OSM/Tools/Waypoints/.

Wikimapia. 2012a. "Wikimapia Api." http://wikimapia.org/api/.

Wikimapia. 2012b. "Wikimapia Guidelines." http://wikimapia.org/user/tools/guidelines/.

Wikimapia. 2012b. "Wikimapia Homepage." http://wikimapia.org.

Yuen, Man-Ching, Irwin King, and Kwong-Sak Leung. 2011. "A Survey of Crowdsourcing Systems." In *IEEE International Conference on Privacy, Security, Risk and Trust and IEEE International Conference on Social Computing*, 766–771.

Zielstra, Dennis, and Alexander Zipf. 2010. "A Comparative Study of Proprietary Geodata and Volunteered Geographic Information for Germany." 13th AGILE International Conference on Geographic Information Science, Guimaraes, Portugal.

7 Pedestrian Path Generation through GPS Traces

Piyawan Kasemsuppakorn and Hassan A. Karimi

CONTENTS

ABSTRACT

Advanced mobile devices equipped with positioning technologies have shown great promise in location-enabled applications, especially navigation systems. A spatial database is one necessary component of a navigation system as it provides the base data to perform navigation functions. Currently, pedestrian networks are not widely available for many areas, in part due to the relatively lower demand compared to road networks for car navigation and other applications. This lack of available pedestrian networks calls for the development of efficient and cost-effective techniques appropriate for collecting and constructing pedestrian networks. This chapter discusses a pedestrian network data model and an algorithm for automatically constructing pedestrian networks using GPS traces collected through collaborative mapping. Experimental results indicate that the proposed algorithm has a high potential to generate sidewalks and crosswalks from collected GPS traces.

7.1 INTRODUCTION

With the advanced capabilities of mobile devices and the success of car navigation systems, interest in pedestrian-centric navigation systems is on the rise. The backbone of a navigation system is its geospatial database that represents a network (e.g., a network of roads) used for computing optimal routes between pairs of locations and as a visual reference to orient users to the environment, among other functions. Road network databases of many countries in North America, Europe, and Asia, due to the popularity of car navigation systems, are publicly available. In contrast, pedestrian network databases that include pedestrian path segments, such as sidewalks, crosswalks, footpaths, and trails, are not publicly available in many countries; in some countries they are provided through commercial mapping companies at significant cost. This is the reason why current navigation systems that provide navigation assistance to pedestrians are based on road networks suitable for car navigation (Gaisbauer and Frank 2008). Road networks are not appropriate for assisting pedestrians with their navigation needs (Hampe and Elias 2003; Elias 2007; Scharl and Tochtermann 2007) since road networks do not adequately represent a model of the pedestrian navigation environment and do not cover all pedestrian paths (Walter, Kada, and Chen 2006; Elias 2007; Holone, Misund, and Holmstedt 2007; Gaisbauer and Frank 2008). Substitution of road networks for pedestrian networks may cause navigation guidance for pedestrians to be ineffective, especially for people with disabilities who require specialized guidance for mobility. Pedestrian navigation systems are not the only application where pedestrian network databases are needed. Other applications include transportation planning for measuring accessibility and physical activity studies to measure the walkability index.

Conventionally, map data are collected by cartographers and geographers, among others, by using advanced specialized technologies, and are disseminated by national mapping agencies and commercial mapping companies. This approach is expensive and is subject to strict copyright laws in most countries. A recent approach to collect map data is collaborative mapping, which potentially could turn everyone into a mapmaker (Gillavry 2006) and allows for a cost-effective means of collecting a large amount of GPS traces through the cooperative use of mobile devices, positioning technologies, and Web 2.0. The main objective of collaborative mapping is to create freely licensed geographic information. For example, OpenStreetMap (OSM) is a well-known collaborative mapping project that provides a set of tools for registered users to contribute their own Global Positioning System (GPS) traces as well as the ability to manually edit geographic information such as road maps or points of interest (OSM 2010). By GPS traces we mean the series of raw GPS data received by GPS receivers. It has been reported that thousands of volunteers from around the world have contributed data and collaboratively manually edited the world map (Haklay and Weber 2008) based on collected GPS traces. Although OSM is more scalable than the conventional approach, it still relies on human intervention for creating geographic information.

In this chapter, we explore the feasibility of automatically constructing a pedestrian network based on the data collected through collaborative mapping. The contributions of this chapter are pedestrian paths definition and analysis, and an algorithm for automatically generating sidewalks and crosswalks (two major pedestrian path types) using GPS traces. The proposed algorithm involves two major steps: collection and construction. GPS traces containing a series of GPS points are collected by volunteers as they travel along pedestrian paths. In the construction step, a series of collected GPS points are first grouped into segments, which are then used to determine the geometries of pedestrian paths with the assistance of road segments. The challenges of this approach include accounting for GPS quality, determining the correct side of the road segment that users actually travel, and assuring the geometrical accuracy of constructed pedestrian paths.

The structure of the chapter is as follows. Section 7.2 provides related work on applications requiring a pedestrian network, existing techniques for generating pedestrian networks, and background on techniques used in this chapter. Section 7.3 discusses pedestrian network databases. Section 7.4 discusses pedestrian network data collection, and Section 7.5 provides details of the algorithm for generating sidewalks and crosswalks. Section 7.6 explains the evaluation parameters and the experimental results. Conclusions and future research are explained in Section 7.7.

7.2 RELATED WORK

Two main applications whose operations can greatly benefit from pedestrian networks are pedestrian navigation services and physical activity research. Pedestrian navigation services can use pedestrian networks to calculate routes for pedestrians including people with special needs. Physical activity research can benefit from pedestrian networks by utilizing information such as connectivity, which has significant associations with physical activity analysis (Humpel, Owen, and Leslie 2002). Examples of research projects using routing services for wheelchair-bound and visually impaired individuals include Magus (Beale et al. 2006), U-Access (Sobek and Miller 2006), ODILA (Mayerhofer, Pressel, and Wieser 2008), Ourway (Holone, Misund, and Holmstedt 2007), and those by Karimanzira, Otto, and Wernstedt (2006), and Kasemsuppakorn and Karimi (2009b). Overall these research projects reported that the required pedestrian networks are not available and road networks are not suitable for accurately assisting pedestrians. For these reasons these studies manually constructed their own pedestrian networks to suit their purposes. In the absence of pedestrian network databases, most pedestrian auditing and accessibility/connectivity studies substitute pedestrian networks with road networks on the assumption that all streets have adjacent sidewalks, while completely ignoring other pedestrian walkways that are not adjacent to streets (Handy et al. 2002; Achuthan, Titheridge, and Mackett 2007). Chin et al. (2008) studied the differences between road and pedestrian networks and how these differences influenced the walkability index. The pedestrian network in this study included parks and walkways, and was manually digitized from aerial photos. The result showed that using a pedestrian network offers a more realistic means of measuring the level of connectivity than a road network.

Currently, there is a void in the literature about the challenges, methods, and best practices for constructing pedestrian networks. However, there exist techniques that have been employed for construction of tailored pedestrian networks. The first technique, road network proxy, uses portions of road networks selected from road attributes such as speed or road type (Achuthan, Titheridge, and Mackett 2007). For example, all roads with a posted speed limit of less than or equal to 35 mph are selected and all roads that prohibit pedestrian traffic (e.g., freeway or highway) could be excluded (MTC 2006). Although road networks are widely available and well defined, they do not reflect the movement characteristics of pedestrians. The second technique, image digitization, generates a pedestrian network by using Geospatial Information System (GIS) tools to manually digitize satellite or aerial images and validate the results by incorporating field survey or knowledge from local people. Due to the widespread availability of high-resolution raster images and easy-to-use GIS tools, this technique has been employed in many research studies related to pedestrian navigation services (Beale et al. 2006; Holone, Misund, and Holmstedt 2007; Chin et al. 2008; Kasemsuppakorn and Karimi 2009a). Today, there are many sources for high-resolution satellite images including the U.S. Geological Survey (USGS), which provides 0.305 m pixel resolution images with natural color ortho-images (USGS 2009) covering many urban areas. The advantage of the image digitization technique is its ability to generate custom pedestrian networks for specific applications with elements such as walking pathways and trails. However, this technique is generally suited for a small area as it requires field survey in order to complete and validate the data collection process. The third technique is based on image processing to extract pedestrian paths from raster images. Walter, Kada, and Chen (2006) presented an algorithm of the shortest path calculation for pedestrians based on raster maps. The proposed algorithm creates an undirected graph representing pedestrian paths from raster images in two steps. The first step (preprocessing) generates a binary raster map, where "1" represents a pedestrian path and "0" represents an obstacle. The algorithm requires human input to select segments on the map that are considered walkable areas. The second step (skeleton) generates pedestrian paths from a binary raster map and employs mathematical morphology operations. The limitation of this algorithm is that it requires manual input from humans to indicate pixels of walkable areas.

Clearly, none of the aforementioned techniques is suitable for automatically constructing pedestrian networks. This chapter explores techniques for extracting pedestrian networks from road networks and collected GPS traces. Interest in location-based social networking (LBSN) (Karimi et al. 2009; Fusco, Michael, and Michael 2010) and the rapid adoption of mobile devices by a wide variety of users are facilitating the collection of GPS traces. Through LBSN people can broadcast their current location, search for nearby friends, and share their opinions or activities with friends using blogs, photos, or music with spatial coordinates and time (Fusco, Michael, and Michael 2010). Members of LBSNs can be both contributors, providing digital content, and consumers, using the content provided by other members. Any user-generated content related to spatial data is called volunteered geographic information (VGI) (see Goodchild 2007). With the availability of GPS-enabled mobile phones, people can now collect GPS traces of where they are and where they have been in an unobtrusive and continuous manner. VGI has facilitated collaborative geospatial content and crowdsourcing as a means of collecting real-world GPS traces that can be used for map generation. Potential contributors of real-world GPS traces include volunteers and general mobile social network members. Volunteers can collect GPS traces for the creation of a base map, for example, see the OSM project (Haklay and Weber2008), while members of LBSNs could provide data such as health and leisure, where GPS traces are a by-product of users' activities.

Considering that some pedestrian path segments exist alongside roads, a road network could be used as a starting point for constructing a pedestrian network. Data sources for road networks are provided by government agencies (e.g., U.S. Census Bureau's Topographically Integrated Geographic Encoding and Reference [TIGER]), nonprofit organizations (e.g., Pennsylvania Spatial Data Access [PASDA]), and commercial mapping companies (e.g., NAVTEQ). Line buffering, a common GIS operation that creates a zone of a specified distance around a line (Worboys and Duckham 2004), is suitable to estimate the location of pedestrian paths using the road network as a reference. The size of the buffer distance may vary depending on the road segment's characteristics, such as number of lanes and direction. Line buffering, a potential approach for construction of pedestrian networks, is described and experimented in this chapter.

7.3 PEDESTRIAN NETWORK

A pedestrian is "any person who is afoot or who is using a wheelchair or a means of conveyance propelled by human power other than a bicycle" (Washington State Legislature 2003). We call any pathway that is designed for pedestrians as a pedestrian path segment, whose purpose is to improve pedestrian safety, reduce potential accidents, and promote mobility and accessibility. We distinguish between seven different types of pedestrian path segments: sidewalk, pedestrian walkway or footpath, accessible entrance, crosswalk, pedestrian bridge, pedestrian tunnel, and trail (Kasemsuppakorn and Karimi 2009a). A sidewalk is the most general structure that is designed for pedestrian traffic alongside a road. A pedestrian walkway, or footpath, is a path that is not along the side of a road such as a walkway between buildings or a footpath to a plaza. A building entrance is a part of a pedestrian walkway, but it is specifically at the entrance of the building. The following three types are related to crossing a street. Crosswalk is a facility that is marked on a road indicating the part of the road which pedestrians should use for crossing the road. The grade-separated crossings are built structures that assist pedestrians to cross areas unnavigable by a simple crosswalk because the terrain is too dangerous or inconvenient to travel. A pedestrian bridge is constructed over a roadway, whereas a pedestrian tunnel is an underground passageway. A trail is mostly designed for recreational activities such as running trails or natural trails.

The vector data model (Lo and Yeung 2006) is suitable for representing pedestrian networks due to its ability to represent complex spatial objects using basic graphical elements. It allows explicit topological representation between objects and the computation of many tasks such as routing. The pedestrian path types defined earlier can be represented by two basic graphical elements: point and line. A point, defined by a pair of coordinates, identifies a topological junction between two or more lines, or the location of objects such as an accessible entrance of a building. A point should also contain the elevation value (z-coordinate) to calculate slope, which is important for pedestrians, especially wheelchair users. A line, described by end points and a list of shape points, represents pedestrian paths for all types. A network generally refers to a type of mathematical graph that represents relationships between nodes and lines using connectivity (Kothuri, Godfrind, and Beinat 2007). Connectivity information is known as topological information that contains nodes, links, and their relationship.

7.4 PEDESTRIAN NETWORK DATA COLLECTION

Pedestrian network data collection is the first step of constructing pedestrian networks through the collaborative mapping approach. This step involves collection of GPS traces, which may require a large number of community members to travel in a particular area with GPS-enabled mobile phones or GPS devices and make their collected GPS traces available online for other interested members. In this chapter, we employ a framework called Social Navigation Network (SoNavNet) for collecting GPS traces of a pedestrian network. SoNavNet has been developed in the Geoinformatics Laboratory of the School of Information Sciences at the University of Pittsburgh for recommendations on points of interests (POIs) and routes, locating, tracking (i.e., real-time trajectory recording), and sharing navigation-related information (Karimi et al. 2009). SoNavNet's architecture supports both Web and mobile application development on a social network that is available to everyone with access to the World Wide Web. SoNavNet is based on a three-tier architecture: client, middle-tier server, and database server. The client could be either mobile devices or Web interfaces. The middle-tier server or Web server provides the core services of the application and is able to request additional services from external service providers such as a map service provider. The database server is responsible for data storage and management of both spatial and non-spatial data. In its current version, the system can be used by its members to determine their current locations or to update locations at fixed time intervals (e.g., one second) via GPS-enabled mobile phones (e.g., T-Mobile G1) or GPS receivers. Once a trip is completed, the member can display the GPS points on Google maps and annotate the data by posting messages or notes. Then the data is ready to be made available online after it is uploaded to the server where privacy settings can control whether the data is shared with others or kept private. Although the data collection is labor intensive and a number of participants is needed, the data collection process is simple and inexpensive. Sample screenshots of the mobile client and the web client for recording and sharing GPS traces are shown in Figure 7.1.

7.5 PEDESTRIAN NETWORK GENERATION ALGORITHM

This section discusses our proposed algorithm to generate pedestrian networks from collected GPS traces. We focused on sidewalks and crosswalks in this algorithm because they constitute the bulk of pedestrian path segments and share a common characteristic, that is, they are generally along the road segments. For this, a road network is used as a starting point to generate sidewalks and crosswalks based on the

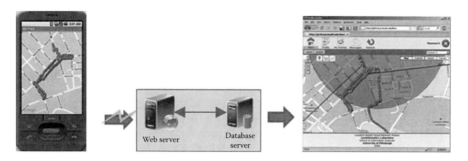

FIGURE 7.1   (See color insert.) Screenshots of GPS traces recording and sharing.

line buffering operation. Figure 7.2 shows the flowchart for the proposed algorithm that consists of three main steps: (1) preprocessing, (2) GPS point segmentation, and (3) line generation.

The objective of the preprocessing step is to eliminate outliers caused by GPS errors and the GPS time-to-first-fix problem (TTFF). Understanding quality of GPS traces is essential in generating pedestrian networks. This is due to GPS accuracy, which is susceptible to the multipath problem where signal from satellites is reflected by objects such as buildings, dense vegetation, tunnels, and large vehicles. Pedestrian path segments are closer to buildings than roads are, and buildings are one main source of interference with GPS signals in urban environments; GPS accuracy may be degraded while walking along pedestrian paths next to high-rise buildings. Moreover, because GPS data are constantly shifting, data recorded along the same path at different times may yield different accuracies. From the available information received from a GPS receiver, horizontal dilution of precision (HDOP), number of satellites used, and walking speed can be used for noise filtering. The filtering, the step takes GPS traces as input and after eliminating noises produces filtered GPS traces. Also eliminated are estimated positions at the beginning of the trip because GPS receivers typically use stored positions when they are first powered up.

The objective of the GPS point segmentation step is to cluster GPS points into segments. The inputs to this step are filtered GPS points and the road network in a given area. In this step, only road segments within the boundary of the filtered GPS traces are selected in order to reduce the number of required road segments. Then the selected road segments are buffered using road attributes such as a road lane category and direction of travel as the buffer size. Road lane category is determined by number of lanes in each direction and direction of travel identifies legal travel directions of a road segment. Generally, there are two types of direction of travel: one-way and two-way. The buffer size between roads and sidewalks is estimated by using the standard for minimum road width and sidewalk width defined by the American Association of State Highway and Transportation Officials (AASHTO) and the Federal Highway Administration (FHWA) and the Institute of Transportation Engineers (ITE). AASHTO specifies a minimum lane width of 4.267 m (including 0.61 m for inner shoulder width) and a minimum outside shoulder width of 3.048 m to provide refuge for disabled vehicles and bicyclists (AASHTO 2005). FHWA and ITE recommend a minimum width of 1.829 m for a sidewalk or walkway,

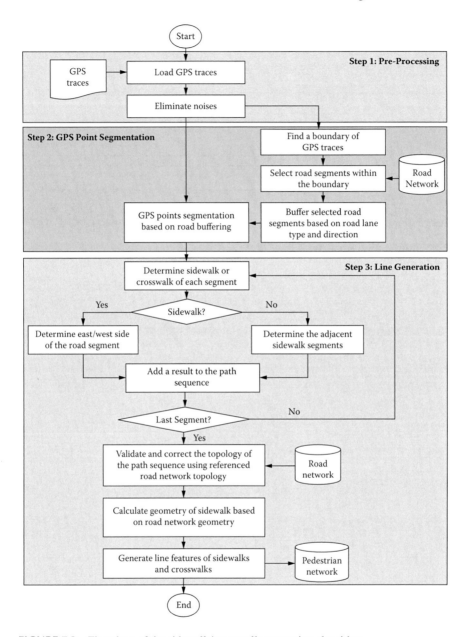

FIGURE 7.2 Flowchart of the sidewalk/crosswalk generation algorithm.

which allows two people to pass alongside comfortably (Pedestrian and Bicycle Information Center 2009). Based on these standards, the minimum buffer size along each side of road centerlines, by summation of road width, shoulder width, and sidewalk width for each lane category and direction of travel of a road segment, is estimated. Examples of minimum buffer sizes for both two-way and one-way direction are given in Table 7.1. After selecting road segments, the filtered GPS points are

TABLE 7.1

Examples of Minimum Buffer Sizes for Each Direction of Travel and Number of Lanes

Direction of Travel	Figure	Minimum Buffer Size

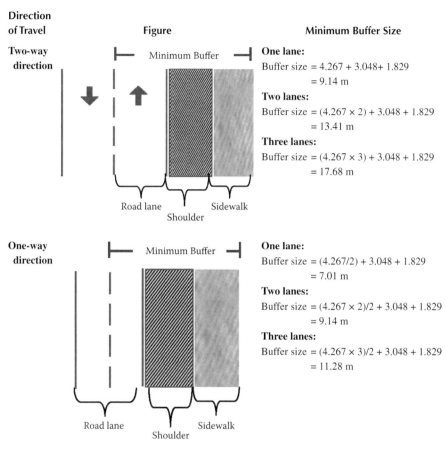

Two-way direction

One lane:
Buffer size = 4.267 + 3.048+ 1.829
= 9.14 m

Two lanes:
Buffer size = (4.267 × 2) + 3.048 + 1.829
= 13.41 m

Three lanes:
Buffer size = (4.267 × 3) + 3.048 + 1.829
= 17.68 m

One-way direction

One lane:
Buffer size = (4.267/2) + 3.048 + 1.829
= 7.01 m

Two lanes:
Buffer size = (4.267 × 2)/2 + 3.048 + 1.829
= 9.14 m

Three lanes:
Buffer size = (4.267 × 3)/2 + 3.048 + 1.829
= 11.28 m

clustered using the geometry relationship "within." GPS points are grouped into the same segment if their coordinates are within the same buffer geometry. Since selected road segments are buffered independently of each other, it might have overlapping regions between buffers. Therefore, a GPS point might belong to one or more buffers. The output from this step is a sequence of segments that contain referenced road information and associated GPS points. An example of GPS point segmentation is illustrated in Figure 7.3.

The objectives of the line generation step are to determine the feature type of each segment, to determine whether the segment is a sidewalk or crosswalk, to generate the geometry of each segment, and to ensure the connectivity between the generated segments. The input to this step, derived from the GPS point segmentation step, is the sequence of segments that contain related road information and associated GPS

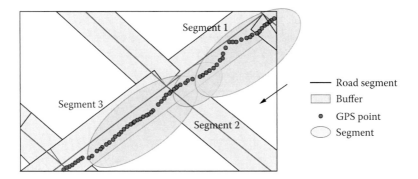

FIGURE 7.3   **(See color insert.)** Example of GPS point segmentation.

points. As shown in Figure 7.2, the first task in the line generation step is to determine whether the segment crosses the road and the side of the road on which the segment is. This requires that we first determine the geographic relationship between the road segment and associated GPS points using bearing and geographical distance. The bearing between successive GPS points, the bearing is calculated by using the great circle navigation formula (Williams 2009) and the geographical distance is calculated by using the haversine formula (Sinnot 1984). A segment is classified as a sidewalk when the average bearing between successive points is closely parallel to the bearing of a particular road segment. Moreover, the total distance between successive points in a segment should be close to the length of a road segment to be considered as a sidewalk segment; otherwise, these GPS points are considered as noise. A segment is identified as a crosswalk when the average bearing between successive points is nearly perpendicular to the bearing of the road segment. To determine the east or west side of a road segment, which is necessary for calculating the geometry of a sidewalk, the linear regression model is employed and the pseudocode of the side determination is illustrated in Figure 7.4.

In practice, the coordinates of two end points of each road segment in the road network are used. One point (x_1, y_1) is considered as the start and the other (x_2, y_2) as the end point. There are four possible cases of road segment alignment. The first is vertical line, which uses the longitude of a GPS point to determine the side. A GPS point lies to the west when the longitude of a GPS point is less than the longitude of either the start or the end point; otherwise it lies to the east. For the other three cases, the slope and intercept of each road segment are calculated to generate a line equation, using the linear regression model. The latitude of each coordinate indicates the heading of each road segment. In the second case, the heading of a road segment points to north ($m > 0$) if the latitude of the start point is less than the latitude of the end point; otherwise ($m < 0$), and it points to south (the third case). The side of a GPS point lies to the west of a road when the heading of a road segment points to north and the latitude of a GPS point (y_3) is greater than the calculated latitude from the line equation. If the heading of a road segment points to south, a GPS point lies to the east of a road when the latitude of a GPS point (y_3) is greater than the calculated latitude from the line equation. The last case, horizontal line, uses the latitude of a GPS point to determine the side of the segment. A GPS point lies to the west when

<u>**Side determination**</u>
Get the coordinate (x_1, y_1) of the beginning point of a road segment
Get the coordinate (x_2, y_2) of the end point of a road segment
Get the coordinate (x_3, y_3) of a GPS point
Case road segment of
       Vertical line $((x_2 - x_1) = 0)$:
           If $(x_3 < x_2)$ then a GPS point lies to the west of the road
           Else if $(x_3 > x_2)$ then a GPS point lies to the east of the road
           Else a GPS point lies on the road
           End if
       Heading North $(m > 0)$:
           If $(y_3 > m*x_3 + b)$ then a GPS point lies to the west of the road
           Else if $(y_3 < m*x_3 + b)$ then a GPS point lies to the east of the road
           Else a GPS point lies on the road
           End if
       Heading South $(m < 0)$
           If $(y_3 < m*x_3 + b)$ then a GPS point lies to the west of the road
           Else if $(y_3 > m*x_3 + b)$ then a GPS point lies to the east of the road
           Else a GPS point lies on the road
           End if
       Horizontal line $(m = 0)$
           If $(y_3 > y_2)$ then a GPS point lies to the west of the road
           Else if $(y_3 < y_2)$ then a GPS point lies to the east of the road
           Else a GPS point lies on the road
           End if
End Case

FIGURE 7.4   Pseudocode of side determination.

the latitude of a GPS point is greater than the latitude of either the start or end point; otherwise it lies to the east.

To illustrate the process of determining side of a road segment, two examples are given. The first example is a straight road segment that is composed of two points, as shown in Figure 7.5a, and the second example is a curved line that is composed of seven points, as shown in Figure 7.5b. A GPS point within the circle in the first example is determined to be on the west side of the road segment because the slope of the line is greater than zero and the calculated latitude is less than the latitude of this GPS point. On the other hand, a GPS point within the circle in the second

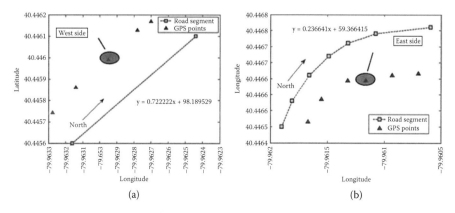

FIGURE 7.5   Examples of establishing sides of a road segment.

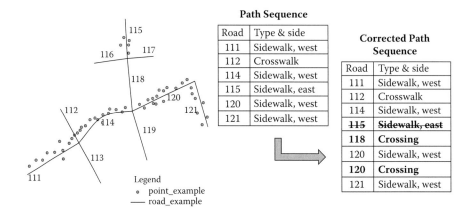

FIGURE 7.6   Example of topology validation.

example is determined to be on the east side of the road segment. Using only one GPS point along a road segment is not sufficient to determine the actual side of the road segment. To reduce biases in determining the side of a road segment to which every GPS point belongs, the probability of being east or west is calculated. The final result is determined by a majority of GPS points.

Once types and sides of all segments are determined and stored in the path sequence, the topology of the road network is used to validate the connectivity of the segments. This is necessary and must be processed before calculating the geometries of line features because of the uncertainty of GPS data that is susceptible to the multipath problem and the systematic drift. Therefore, GPS points may generate erroneous paths that are not possible to travel. For instance, an erroneous path can occur when sequentially generated segments abruptly jitter back and forth along both sides of a road, or when there are skipped segments that create gaps in a path. An example of validating the connection in the path sequence is shown in Figure 7.6.

After validating the connectivity of the segments, the geometry of each sidewalk segment is generated. Since the geometry of a road segment is used as a starting point, the beginning point, the ending point, and the shape points (if any) of each road segment are extracted from the road networks. The latitude and longitude of a point representing the geometry of a sidewalk is subsequently calculated using the great circle navigation formula (Williams 2009), as illustrated in Equations 7.1 and 7.2, respectively.

$$new\ lat = asin\left(sin\left(lat\right) * cos\left(dist\right) + cos\left(lat\right) * sin\left(dist\right) * cos\left(radial\right)\right) \qquad (7.1)$$

$$new\ lng = lng + atan2\left(sin\left(radial\right) * sin\left(dist\right) + cos\left(lat\right), cos\left(dist\right)\right.$$
$$\left. - sin\left(lat\right) * sin\left(new\ lat\right)\right) \qquad (7.2)$$

Three inputs are required in the geometry calculation: starting point, radial, and gap distance. A starting point, composed of latitude (*lat*) and longitude (*lng*), can be

TABLE 7.2

Examples of Gap Distance

Direction of Travel	Figure	Gap distance

Two-way direction	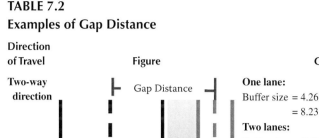 Road Sidewalk	**One lane:**
		Buffer size = 4.267 + 3.048 + (1.829/2)
		    = 8.23 m
		Two lanes:
		Buffer size = (4.267 × 2) + 3.048 + (1.829/2)
		    = 12.50 m
		Three lanes:
		Buffer size = (4.267 × 3) + 3.048 + (1.829/2)
		    = 16.76 m

One-way direction	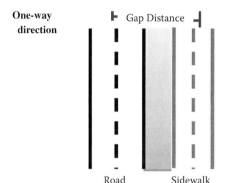 Road Sidewalk	**One lane:**
		Buffer size = (4.267/2) + 3.048+ (1.829/2)
		    = 6.1 m
		Two lanes:
		Buffer size = (4.267 × 2)/2 + 3.048 + (1.829/2)
		    = 8.23 m
		Three lanes:
		Buffer size = (4.267 × 3)/2 + 3.048 + (1.829/2)
		    = 10.36 m

one of the points at either end of the road segment or a shape point along the path of a road segment. A radial is the direction from a starting point, expressed as the angle measured from north in a clockwise direction. East (90°) and West (270°) are degrees used for calculating geometry of shape points on the east and west sides, respectively. North East (45°) and South East (135°) are used to calculate the geometry of the intersection points (a connection point between sidewalk and crosswalk) on the east side of the road. On the other hand, North West (315°) and South West (225°) are employed for the intersection points (a connection point between sidewalk and crosswalk) on the west side of the road. The gap distance (*dist*), the perpendicular distance between the road centerline and the sidewalk centerline, is estimated based on lane category, direction of travel, standard road width, and sidewalk width. Table 7.2 shows examples of gap distance (m) of both direction of travel for each number of lane.

It is true that the measurements on real roads and sidewalks may not follow the theoretical values suggested by statutory authorities, as we employed to estimate

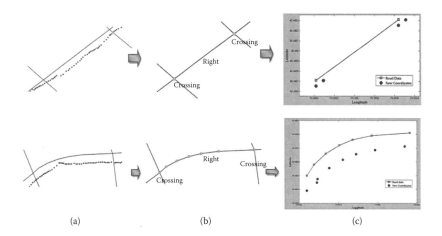

FIGURE 7.7   Example of geometry calculation.

gap distance. In practice, the high-resolution satellite images provided by Google Earth and its measurement tools can be used to measure the real road and sidewalk conditions, instead of field tests. Thirty samples of two-way roads with one lane for each direction (group 1) and 30 samples of one-way roads with one lane (group 2) were randomly selected, and the gap distances between the road centerlines and the sidewalk centerlines were measured. The average and standard deviation of gap distance of group 1 is 7.50 m and 1 m, respectively. For group 2, the average gap distance is 5.72 m and the standard deviation is 0.8 m. Comparing with the estimated gap distance presented in Table 7.2, the gap difference between the estimated value suggested by authorities and the actual value for group 1 is 0.73 m and for group 2 is 0.38 m, which is a small difference. Since there is no information on real road and sidewalk condition, using the theoretical values to estimate the actual value is possible and is experimented in this work.

Figure 7.7 shows a summary of the steps from collected GPS points (Figure 7.7a), to path type and side determination (Figure 7.7b), to sidewalk geometry calculation (Figure 7.7c). The sidewalk segment is generated by topology between shape points. To ensure topological correctness, the end node of each segment is used as the beginning node of the next segment, since the sequence of segments is determined based on the sequence of GPS points and the topology of the road network.

7.6   EVALUATION

Four metrics were used to evaluate the algorithm: (1) accuracy to distinguishing between sidewalks and crosswalks, (2) success rate of sidewalk/crosswalk determination, (3) accuracy to determine side of the road for sidewalk segments, and (4) geometrical accuracy for generated sidewalks and crosswalks. Two experiments were performed to evaluate the algorithm. In the first experiment the first three metrics were measured and the fourth metric was measured in the second experiment. Collected GPS traces through SoNavNet, NAVTEQ road network, and Geotools 2.7-M1 were used in these experiments.

Two parameters required in the algorithm, buffer size and gap distance between road and sidewalk, were explained in Table 7.1 and 7.2, respectively. For the experiments, due to GPS errors, the constant value of 5 m was added to the minimum buffer size suggested in Table 7.1 to cover sufficient GPS points. The value of 5 was found suitable after experimenting with values between 3 to 9; the largest value of 9 was considered because of the accuracy of the GPS Standard Positioning System (SPS), which is less than 9 m (Inside GNSS 2008).

7.6.1 TEST DATA

Ten members of the Geoinformatics Laboratory participated in collecting GPS traces for testing the SoNavNet's prototype and completed a questionnaire expressing their feedback on the data collection experience. The testing environment was confined within the University of Pittsburgh's main campus, which includes both high-rise buildings and open-sky environments. Ten GPS traces (with fixed interval of 1 s) along sidewalks and crosswalks were collected and treated as separate inputs to the algorithm and experimented one at a time. The new generated pedestrian path segment from each GPS trace was then incrementally stored in the pedestrian network. Number of GPS points, number of sidewalk segments, crosswalk segments, and total length of each actual walking path were collected to validate the performance of the algorithm. Figure 7.8 shows a map of the campus and GPS traces, and Table 7.3 shows the characteristics of the 10 GPS traces used in the experiments.

As mentioned earlier, HDOP can be used to eliminate noises from GPS errors. Lower HDOP indicates better GPS positional precision. The average HDOP value of the 10 traces was 1.11 and the highest HDOP value was 7.22. GPS observations with the HDOP value greater than 2 were detected as noises and were eliminated in the experiments. Moreover, GPS observations with no speed and number of satellites used were also eliminated.

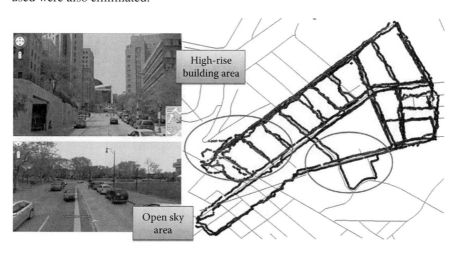

FIGURE 7.8 (**See color insert.**) Ten GPS traces.

TABLE 7.3

GPS Traces Used in Experiments

Trip Number	Number of GPS points	Actual Walking Paths			
		Number of segments	Number of sidewalks	Number of crosswalks	Total Length (m)
1	992	32	24	8	1,486.90
2	325	22	13	9	1,234.00
3	282	20	14	6	1,244.50
4	1,264	73	44	29	3,019.10
5	1,247	68	45	23	3,159.60
6	767	37	28	9	2,395.50
7	1,127	42	30	12	2,784.50
8	1,311	49	37	12	3,121.70
9	1,307	56	41	15	3,756.40
10	881	28	20	8	1,704.10
Total	**9,503**	**427**	**296**	**131**	**23,906.30**

7.6.2 EXPERIMENT 1

The purposes of this experiment were to measure the accuracy of the algorithm in classifying path type, sidewalk and crosswalk, and to examine the accuracy of the algorithm in identifying the correct side of the road for each sidewalk segment. Inputs to this experiment included 10 GPS traces and the road network of the testing area. The result from the algorithm was the segments of pedestrian path type along with road segment number and the side of the road. This result was compared against actual pedestrian paths traveled (identified from data collectors after trips). The number of actual sidewalks and crosswalks traveled and those generated by the algorithm in each trip were reported. For comparison, the numbers of correctly identified sidewalks (S_c), crosswalks (C_c), and side of sidewalks (SS_c) were calculated. Three evaluation parameters to validate the performance of the algorithm are accuracy of path type identification (A_p), success rate of sidewalk/crosswalk determination (SR), and accuracy of side identification (A_s). A_p, SR, and A_s values range from 0 to 1, with 1 being highest. A_p, SR, and A_s are calculated as follows:

$$A_p = (S_c + C_c)/\text{Number of generated sidewalk and crosswalk segments} \quad (7.3)$$

$$SR = (S_c + C_c)/\text{Actual number of travelled sidewalks and crosswalks} \quad (7.4)$$

$$A_s = SS_c/\text{Number of generated sidewalk segments} \quad (7.5)$$

A_p was calculated by dividing the number of correctly identified sidewalks and crosswalks by the number of generated segments (sidewalk and crosswalk segments) by the algorithm. SR was calculated by dividing the number of correctly identified sidewalks and crosswalks by the actual number of traveled segments for both sidewalks and crosswalks. A_s was calculated by dividing the number of correctly

TABLE 7.4

Results of Three Evaluation Metrics from Experiment 1

Trip Number	Number of Actual Traveled Sidewalks	Number of Actual Traveled Crosswalks	Number of Generated Sidewalks	Number of Generated Crosswalks	S_c	C_c	SS_c	A_p	SR	A_s
Trip 1	24	8	19	10	18	7	18	0.862	0.781	0.947
Trip 2	13	9	11	9	10	9	10	0.95	0.864	0.909
Trip 3	14	6	11	8	11	1	4	0.632	0.600	0.364
Trip 4	44	29	41	28	39	18	30	0.826	0.781	0.732
Trip 5	45	23	40	23	39	23	39	0.984	0.912	0.975
Trip 6	28	9	25	8	22	8	21	0.909	0.811	0.840
Trip 7	30	12	30	10	28	7	20	0.875	0.833	0.667
Trip 8	37	12	35	15	31	14	30	0.9	0.918	0.857
Trip 9	41	15	34	13	31	12	30	0.915	0.768	0.882
Trip 10	20	8	21	9	19	5	19	0.8	0.857	0.905
					Average value			0.865	0.812	0.808

identified sides by the number of generated sidewalk segments by the algorithm. Table 7.4 shows the result.

The average accuracy of path type identification, success rate of sidewalk/crosswalk determination, and accuracy of side identification are 0.865, 0.812, and 0.808, respectively. The results from most trips are near optimum (greater than 85%), which means that the algorithm is able to determine pedestrian path type and side of the road segments from collected walking GPS traces. However, the algorithm performed poorly around high-rise buildings and narrow road areas as it identified wrong walking sides (e.g., Trip 3 and Trip 7) and generated nonexistent sidewalk/crosswalk segments.

7.6.3 EXPERIMENT 2

The purpose of this experiment was to measure the geometrical accuracy for generated sidewalks and crosswalks. Inputs to this experiment included the 10 GPS traces, result from Experiment 1 (sequence of identified segments), and road segments. Examples of generated sidewalks and crosswalks using GPS traces are shown in Figure 7.9. To measure the performance of the algorithm and the estimated gap distance between road and sidewalk, the external evaluation methodology (Wiedemann 2003) of automatic road extraction in image processing was employed. The methodology compares the result of the algorithm in each trip with a reference network, a manually digitized vector map of the pedestrian paths traveled. The two networks were matched by using the buffering approach where the buffer distance was set to 1.829 m as recommended by ITE as a minimum width for a sidewalk or walkway. A buffer polygon around the reference network was first created by using the proximity toolset in ArcGIS. Then the reference network and the result

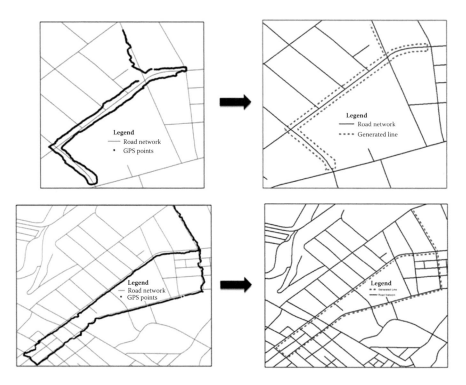

FIGURE 7.9 **(See color insert.)** Examples of generated sidewalks/crosswalks using GPS traces.

by the algorithm were matched. Each segment of the generated paths is considered "matched" if its geometry is within the buffer polygon geometry of the reference network; otherwise, it is considered "unmatched." Next, the geometrical accuracy is calculated by dividing the length of matched segments by the total length of generated segments. The range value of the geometrical accuracy is between 0 and 1, with 1 being the optimal value.

In this experiment, the result from Experiment 1, which is the sequence of sidewalk and crosswalk segment of all 10 GPS traces, was used to generate line features. Then the performance of the algorithm was evaluated using the methodology as described earlier. The geometrical accuracy of each GPS trace of Experiment 2 is shown in Table 7.5. The percentage of geometrical accuracy from eight trips is greater than 75%, which means that the estimated gap distance between road network and sidewalk network in the algorithm is mostly correct, and it is proved that the geometry of road segments can be used as the starting point for the geometry calculation of sidewalk segments. The generated line of Trip 3 has low geometrical accuracy because of the errors from side identification of Experiment 1. For Trip 1, pedestrian path type and side identification are mostly correct, but the geometrical accuracy of the generated line is low because some sidewalks are not exactly parallel to roads and the road segments may be geometrically inaccurate.

TABLE 7.5

Results of Geometrical Accuracy from Experiment 2

Trip	Length of Generated Paths (m)	Length of Matched (m)	Geometrical Accuracy
Trip 1	1,435.06	707.72	0.493
Trip 2	1,048.69	907.12	0.865
Trip 3	1,003.23	398.178	0.397
Trip 4	3,135.63	2,379.13	0.759
Trip 5	3,043.37	2,434.80	0.800
Trip 6	2,116.05	1,589.67	0.751
Trip 7	2,405.47	1,977.10	0.822
Trip 8	2,870.69	2,331.88	0.812
Trip 9	3,182.90	2,786.60	0.875
Trip 10	1,471.33	1,161.97	0.790
		Average Value	0.736

7.7   CONCLUSIONS AND FUTURE RESEARCH

A pedestrian network is an essential resource in a variety of applications, especially in pedestrian navigation systems and urban planning projects. In the absence of approaches for automatically generating pedestrian networks, alternative approaches are deemed necessary. This chapter discussed seven pedestrian path types with their relations to a pedestrian network. Since each of these pedestrian path types has different characteristics, in this chapter, an algorithm for generating sidewalks and crosswalks was investigated using the GPS traces and road networks. Collaborative mapping through LBSNs to collect GPS traces was discussed. Line buffering operation and chosen parameters (e.g., gap distance) were investigated. Two experiments using ten GPS traces were conducted to evaluate the algorithm performance. The result of the first experiment showed that the algorithm is able to distinguish between sidewalk and crosswalk types, and that it can identify correct sides. However, the algorithm suffers from the inaccuracy of GPS points in some areas (e.g., highrise buildings). The result from the second experiment showed that the average geometrical accuracy of the generated line from 10 traces using the algorithm is about 74%. In future research, we plan to improve the accuracy of the algorithm in classifying path type and geometrical accuracy of generated lines and to enhance the algorithm to generate other pedestrian path types such as footpath and trail.

REFERENCES

Achuthan, Kamalasudhan, Helena Titheridge, and Roger Mackett. 2007. Measuring pedestrian accessibility. Paper presented at Proceedings of the Geographical Information Science Research UK (GISRUK) conference.

American Association of State Highway and Transportation Officials (AASHTO). 2005. A policy on design standards.

Beale, Linda, Kenneth Field, David Briggs, Phil Picton, and Hugh Matthews. 2006. Mapping for wheelchair users: Route navigation in urban spaces. *The Cartographic Journal* 43:68–81.

Chin, Gary K. W., Kimberly P. Van Niel, Billie Giles-Corti, and Mathew Knuiman. 2008. Accessibility and connectivity in physically activity studies: The impact of missing pedestrian data. *Journal of Preventive Medicine* 46:41–45.

Elias, Birgit. 2007. Pedestrian navigation: Creating a tailored geodatabase for routing. In *4th Workshop on Positioning, Navigation and Communication (WPNC'07)*. Hannover, Germany: IEEE.

Fusco, Sarah Jean, Katina Michael, and M. G. Michael. 2010. Using a social informatics framework to study the effects of location-based social networking on relationships between people: A review of literature. *IEEE International Symposium on Technology and Society (ISTAS10)*, pp. 157–171.

Gaisbauer, Christian, and Andrew U. Frank. 2008. Wayfinding model for pedestrian navigation. In *11th International Conference on Geographic Information Science,* University of Girona, Spain.

Gillavry, Edward Mac. 2006. Collaborative mapping and GIS: An alternative geographic information framework. In *Collaborative Geographic Information Systems*, edited by Shivanand Balram and Suzana Dragicevic, pp. 103–120. Hershey, PA: Idea Group Publishing.

Goodchild, Michael F. 2007. Citizens as sensors: The world of volunteered geography. *GeoJournal* 69:211–221.

Haklay, Mordechai, and Patrick Weber. 2008. OpenStreetMap: User-generated Street Maps. *IEEE Pervasive Computing* 7(4):12–18.

Hampe, Mark, and Birgit Elias. 2003. Integrating topographic information and landmarks for mobile navigation. Available from http://www.ikg.uni-hannover.de/fileadmin/ikg/staff/publications/konferenzebeitraege_abstract_review/hampeelias_telecarto2003.pdf.

Handy, Susan L., Marion G. Boarnet, Reid Ewing, and Richard E. Killingworth. 2002. How the built environment affects physical activity: Views from urban planning. *American Journal of Preventive Medicine* 23(2):64–73.

Holone, Harald, Gunnar Misund, and Hakon Holmstedt. 2007. Users Are Doing It for Themselves: Pedestrian Navigation with User Generated Content. In *Internation Conference On Next Generation Mobile Applications, Services and Technologies (NGMAST'07)*. Cardiff, UK: IEEE.

Humpel, Nancy, Neville Owen, and Eva Leslie. 2002. Environment factors associated with adults' participation in physical activity: A review. *American Journal of Preventive Medicine* 22(3):188–199.

InsideGNSS. 2008. New GPS Standard Positioning System (SPS) Performance Standard.10.

Karimanzira, D., P. Otto, and J. Wernstedt. 2006. Application of machine learning methods to route planning and navigation for disabled people. In *The 25th IASTED International Conference Lanzarote*. Canary Islands, Spain.

Karimi, Hassan A., Benjamin Zimmerman, Alper Ozcelik, and Duangdueng Roongpiboonsopit. 2009. SoNavNet: A framework for social navigation networks. In *International Workshop on Location Based Social Networks (LBSN'09)*. Seattle, Washington.

Kasemsuppakorn, Piyawan, and Hassan A. Karimi. 2009a. Pedestrian Network Data Collection through Location-Based Social Networks. In *5th International Conference on Collaborative Computing: Networking, Applications and Worksharing*. Washington, DC.

Kasemsuppakorn, Piyawan, and Hassan A. Karimi. 2009b. Personalised routing for wheelchair navigation. *Journal of Location Based Services* 3(1):24–54.

Kothuri, Ravi, Albert Godfrind, and Euro Beinat. 2007. *Pro Oracle Spatial for Oracle Database 11g*. Berkeley, CA: APress.

Lo, Chor Pang, and Albert K. W. Yeung. 2006. Digital representation and organization of geospatial data. In *Concepts and Techniques of Geographic Information Systems* (2nd ed.), edited by Chor Pang Lo and Albert K.W. Peung. Upper Saddle River, NJ: Prentice Hall.

Mayerhofer, Bernhard, Bettina Pressel, and Manfred Wieser. 2008. ODILIA: A mobility concept for the visually impaired. In *Computers Helping People with Special Needs*, edited by Klaus Miesenberger, Joachim Klaus, Wolfgang Zagler and Arthur Karshmer. Berlin: Springer.

Metropolitan Transportation Commission Planning Section. 2006. MTC studies travel patterns of transit station area residents.

OpenStreetMap (OSM). Public GPS Traces 2010. Available from http://www.openstreetmap.org/traces.

Pedestrian and Bicycle Information Center. 2009. *Sidewalks and Walkways*. Available from http://www.walkinginfo.org/engineering/roadway-sidewalks.cfm.

Scharl, Arno, and Klaus Tochtermann. 2007. The Geospatial Web: How geobrowsers, social software and the Web 2.0 are shaping the network society. London: Springer.

Sinnot, Roger W. 1984. Virtues of the haversine. *Sky and Telescope* 68(2):158.

Sobek, Adam D., and Harvey J. Miller. 2006. U-Access: A web-based system for routing pedestrians of differing abilities. *Journal of Geographical Systems* 8:269–287.

U.S. Geological Survey (USGS). 2009. The National Map: Orthoimagery. Available from http://nationalmap.gov/ortho.html.

Walter, V., M. Kada, and H. Chen. 2006. Shortest path analyses in raster maps for pedestrian navigation in location based systems. In *International Symposium on Geospatial Databases for Sustainable Development*. Goa, India: ISPRS Technical Commission IV.

Washington State Legislature. 2009. RCW 46.04.400 Pedestrian 2003. Available from http://apps.leg.wa.gov/RCW/default.aspx?cite=46.04.400.

Wiedemann, Christian. 2003. External evaluation of road networks. *ISPRS Archieves* XXXIV (Part 3/W8):93–98.

Williams, Ed. 2009. Aviation formulary v. 1.44. Available from http://williams.best.vwh.net/avform.htm.

Worboys, Michael, and Matt Duckham. 2004. *GIS: A Computing Perspective* (2nd ed.). Boca Raton, FL: CRC Press.

8 A CyberGIS Environment for Analysis of Location-Based Social Media Data

Shaowen Wang, Guofeng Cao,
Zhenhua Zhang, Yanli Zhao,
Anand Padmanabhan, and Kaichao Wu

CONTENTS

ABSTRACT

Social media (e.g., Twitter) data, often massive, dynamic, and represented in unstructured forms, contain tremendous quantities of location information for understanding spatial dynamics of geospatial processes. With the dramatic increase of volume and complexity of such data, geographic information systems (GIS) capable of data-intensive and high-performance spatial analytics have become increasingly important to take advantage of this type of data sources. Conventional GIS and spatial analysis approaches, however, are rather limited in providing such capabilities. Cyberinfrastructure, which represents integrated computing, information, and communication technologies for high performance, distributed and coordinated knowledge development, and discovery, provides a suitable means to address this challenge. Within such a context, this research describes a CyberGIS environment—a new type of GIS based on cyberinfrastructure, for effective collection, access, and analysis of location-based social media data. This CyberGIS environment seamlessly integrates a set of spatial analytical services that allow users to interactively explore spatiotemporal patterns of massive social media data. Case studies including detection of early outbreaks of flulike diseases and mapping flows of individual movement trajectories based on Twitter data are conducted to evaluate the design and implementation of the CyberGIS environment. The CyberGIS implementation encompasses a scalable architecture for managing and analyzing Twitter data, a highly interactive online user interface, and a set of integrated spatial data exploration and analysis services. These services allow for flexible data access and on-demand computing of multiple spatial analyses while enabling a number of users to simultaneously investigate spatiotemporal patterns without being concerned about managing the complexity of cyberinfrastructure access and social media data. This chapter demonstrates the CyberGIS environment in connecting location-based social media services with cyberinfrastructure for reaping the benefits of both.

8.1 INTRODUCTION AND BACKGROUND

Social media, such as social networks (e.g., Facebook), blogs and microblogs (e.g., Twitter), and photo-, audio-, and video-sharing services (e.g., YouTube and Flickr), can be understood as "a group of Internet-based applications that are built on the ideological and technological foundations of Web 2.0, and that allow the creation and exchange of user generated content" (Kaplan and Haenlein 2010). These online applications and associated data generated have been experiencing a spectacular rise in popularity; over a short period, hundreds of millions of users have been attracted to these services generating a massive amount of social media data in an unprecedented speed, scale, and extent.

Twitter, for example, has rapidly gained approximately 140 active million users worldwide, generating 340 million so-called tweets, that is, individual user posts, every day as of March 2012. Although each tweet is trivial and limited to only 140 characters, the aggregate of millions of tweets may provide a rather realistic representation of landscapes for topics of interest. Same as conventional data sources, this new modality of social media data has become increasingly important to the development of human knowledge. In fact, the Library of Congress has begun archiving Twitter feeds. Extensive studies with significant societal impacts have been conducted by capitalizing on these social networking and media data, ranging from predictions of the stock market (Bollen et al. 2011) to predictions of earthquakes (Sakaki and Okazaki 2010), from disaster responses (Goodchild and Glennon 2010) to infectious disease tracking (Signorini and Segre 2011), from measuring public opinion and political sentiment without explicit surveys (O'Connor and Balasubramanyan 2010) to winning the DARPA's (U.S. Defense Advanced Research Projects Agency) network challenge (Pickard et al. 2011).

Social media has also been recognized as proxies to understand geography. Intentionally or unintentionally, people are sharing their whereabouts when using social media services (Hecht et al. 2011). Particularly, with widespread use of location-aware mobile devices and continuing improvements of location-based services, location-based social media data are increasingly available and becoming a crucial attribute of social media. Such massive, dynamic, georeferenced data, despite quality issues such as noises and possible spams, offer an unprecedented opportunity to understand and exploit spatiotemporal microdynamics across multiple scales in our complex social systems, which were previously considered not possible. Gaining insights and desirable knowledge from these big geospatial data, however, poses a wide range of fundamental challenges pertaining to geographic information science (GIScience).

First, location-based social media data are often "big" and continuously "coming," considering the aforementioned case of daily new tweets and even extending the desirable time window to a number of months or years. The magnitude of this data volume is well beyond the capability of any mainstream geographic information systems (GIS). Especially, data analytics may not be achievable within a reasonable amount of time without resorting to high-performance computing strategies.

Second, social media data are generated dynamically and continuously. Users of social media services frequently update or change their status and locations, and for certain emergency events, volunteers can rapidly contribute their knowledge and experiences. These real-time or near-real-time crowdsourcing data, complemented with official and authoritative data sources, become particularly valuable in time-critical cases, for example, disaster response (Goodchild and Glennon 2010) and disaster relief (Gao et al. 2011). Conventional GIS, however, usually represent reality as static forms instead of dynamic processes (Goodchild 2004), although recent efforts have been made to start shifting this scene. To account for dynamic information often requires spatiotemporal analytics, which are often computationally challenging, particularly in time-critical cases where instant responses are required, and this once again highlights the need for high-performance computing.

Third, in contrast to well-structured conventional geospatial data sources, social media data are often produced in unstructured forms. Extra efforts, such as data mining techniques, are often necessary to make the data meaningful and sensible. In addition, despite ownership controversies of social media data, social media services usually do not provide direct access to all the data being produced, which causes data access to be a nontrivial problem. Researchers would have to come up with their own ad hoc mechanisms to obtain data of their interests, usually encoded in a form of live streams, via designated access interfaces provided by social media services. Issues of uncertainty and noises further compound this data access challenge, which hinders applications of these data sources to broader fields.

Cyberinfrastructure was termed to refer to the integration of high-performance computing, information, and communication technologies as well as related human expertise for coordinated knowledge development and discovery (Atkins et al. 2003). Worldwide research communities have embraced the powerful synthesis vision that cyberinfrastructure represents, and its enormous potential to accelerate innovations and enable discoveries in science and engineering. A CyberGIS vision, a new GIS modality based on cyberinfrastructure, has been recently developed to advance geospatial technologies and scientific problem solving based on the synthesis of cyberinfrastructure, GIScience, and spatial analysis (Wang 2010). Early research and development of CyberGIS have demonstrated its great potential to address fundamental challenges of cyberinfrastructure and GIScience (Wang 2010; Wright and Wang 2011; Anselin and Rey 2012), particularly computational ones tied to the popular notion of big data.

With an open architecture and concrete implementation, this chapter describes a CyberGIS environment for efficient collection, management, access, analysis, and visualization of location-based social media data, Twitter feeds in particular, with a focus placed on addressing the aforementioned challenges. By seamlessly integrating a system of collecting and managing location-based Twitter data, a suite of spatiotemporal analytical services, and a hybrid cyberinfrastructure environment with high-performance computational resources, this user-centric CyberGIS environment provides an effective means to explore spatiotemporal patterns of massive quantities of Twitter feeds.

Specifically, a Twitter feeds crawling service and a spatiotemporal database are developed to collect and manage Twitter feeds in near real-time. Based on collected location-based Twitter feeds, a set of spatiotemporal data exploratory and analysis services as well as a highly interactive online user interface are developed to take advantage of high-performance cyberinfrastructure resources through GISolve, a geospatial middleware toolkit that harnesses cyberinfrastructure for geospatial problem solving (Wang et al. 2005). Case studies, including detecting early outbreaks of flulike diseases and mapping moving trajectories of individuals, are conducted to showcase the capabilities of the CyberGIS environment while demonstrating the benefits of synergistically linking CyberGIS and location-based social media services.

In the remainder of this chapter, architecture and design are first described in Section 8.2 with an introduction to the Twitter feeds collection service. Section 8.3 discusses two representative spatiotemporal analytics of location-based social media data within a CyberGIS environment, including spatiotemporal cluster detection and flow mapping analyses. Case studies are detailed in Section 8.4 to demonstrate the utility of the CyberGIS environment. Section 8.5 summarizes the chapter and provides a brief discussion on future work.

8.2   ARCHITECTURE AND DESIGN

Although it is millions of users that are generating massive social media contents, social media services, as hosts of these data, usually limit direct or full access to these contents. In Twitter feeds, there is a commercial alternative for full access, but it may be too costly to most researchers. Therefore, one often has to build a custom application to collect desirable data by leveraging open access interfaces provided by social media services. To enable analysis of social media data, this chapter describes a generic Twitter feeds collection service using the Twitter's streaming

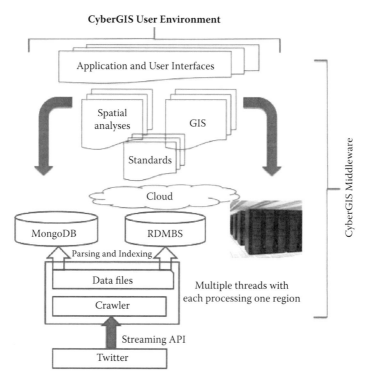

FIGURE 8.1   **(See color insert.)** CyberGIS architecture.

access application programming interface (API) for organizing location-based data as part of the CyberGIS environment (Figure 8.1).

Twitter, as one of the most popular microblogging social media services, enables users to post text-based message updates and has gained enormous popularity in recent years. Twitter provides access to part of its corpus of textual data through two main APIs: representational state transfer (REST) and streaming, which can be openly exploited. Twitter REST API allows for access to core corpus data while providing ability to run searches against an index of recent tweets. The streaming API, on the other hand, is intended for long-lived connections and, thus, provides near-real-time high-volume tweets that are sampled and can also be filtered by keywords or geographical boundary of interest. However it is worth noting that Twitter typically throttles data that match the filter criteria based on the total tweets (see https://dev.twitter.com/docs/faq#6861).

A crawler has been developed in Java based on the streaming API. The data are mainly stored in MongoDB, a scalable, high-performance, and open source NoSQL database. By continually streaming Twitter data for over the past 8 months, and extracting geospatial data of the continental United States, the crawler has been collecting approximately 1.35 million tweets per day, and amassed over 200 GB data. The tweets along with their spatial context stored in the MongoDB are easily accessible through the CyberGIS environment for large-scale geospatial analysis. The CyberGIS environment includes two key complementary components: middleware and user environment. A primary function of the middleware is to exploit spatial characteristics of GIS data and analytical operations for enhancing computational performance, and to manage the complexity of accessing cyberinfrastructure resources (e.g., high performance computers) and services (Wang 2010). The user environment is built on component-based application and user interfaces through the use of cutting-edge Web 2.0 technologies and service-oriented architecture, and provides users with friendly entry points for performing spatial analysis enabled by the middleware (Wang et al. 2009).

8.3   SPATIOTEMPORAL DATA ANALYSIS

As social media data contain rich spatial and locational contexts, various studies from different perspectives (e.g., data mining and visual analytics) have been conducted on the analysis of such data with a particular emphasis on their spatiotemporal aspects. Mei et al. (2006), for example, analyzed weblogs in a spatial and temporal context, and proposed a probabilistic approach to discovering and summarizing the spatiotemporal patterns of topics and subtopics in weblogs. Backstrom et al. (2008), based on georeferenced search engine query logs, studied the spatial variations of queries and the geographical shifting over time. More recently, Sakaki and Okazaki (2010) developed a probabilistic spatiotemporal model to estimate earthquake centers and typhoon trajectory by monitoring Twitter feeds in real time. With Senseplace2, MacEachren et al. (2011), from a perspective of visual analytics, demonstrated a geovisual analytic system where Twitter feeds are indexed by space, time, and themes to support situational awareness in crisis management.

Despite significant progress that has been made recently, computational performance challenges in spatiotemporal analysis of social media data, however, have not been adequately addressed. Such challenges often become rather prominent in empirical research, particularly when dealing with large volumes of social media data sets with high computational demanding methods (e.g., Monte Carlo simulation), which is often the case in spatiotemporal analytics of social media data. The CyberGIS environment developed in this research suggests a promising means to address these computational challenges. Within such an environment, two social media data analyses—exploratory spatial clustering detection of disease risk and flow mapping—are detailed in this section.

8.3.1   SPATIAL CLUSTER DETECTION OF DISEASE RISK

Data elements in social media, Twitter feeds in our particular case, are often represented as points (or events) indexed based on various space and time settings. Spatial cluster detection based on such data often corresponds to finding spatial regions that contain an unusually high or low concentration of events, which is of long traditional interest in spatial analysis and related fields such as geography, public health, and spatial statistics. Numerous methods have been developed in this context. Particularly in exploratory spatial data analysis (ESDA), one set of these methods was developed (see Murray, 2000, and Murray and Estivill-Castro, 1998), without prior knowledge of clustering distribution, to seek an optimal mutually exclusive partition of space by minimizing with-in cluster differences while maximizing between-cluster differences. Another broad set of statistical approaches is based on the standard Monte Carlo procedure for significance testing to identify anomalously high counts of point patterns. Typical examples of this approach include the geographical analysis machine (GAM) (Openshaw et al. 1987; Openshaw 1998) and spatial scan statistics (Kulldorff and Nagarwalla 1995; Kulldorff 1997). Among these methods, kernel density estimation (KDE) (Silverman 1986), particularly with adaptive kernel bandwidth (Brunsdon 1995) and based on relative risk functions (Bithell 1990, 1991), has attracted much attention in disease risk mapping because of the minimal underpinning assumptions in theory, and simplicity and flexibility in practice.

In this section, a recently proposed relative risk-based KDE method that can account for inhomogeneous background by an adaptive kernel band width (Shi 2010) is adapted for indicating clusters with unusually high or low disease risk. As with GAM and other spatial exhaustive sampling approaches, the method is embedded into a standard Monte Carlo simulation procedure for further significance testing. Consider N independent, identically distributed observations $\mathbf{x}_1, \ldots, \mathbf{x}_N$ in a two-dimensional geographic space. Given a location, $\mathbf{x}$, density $f(\mathbf{x})$ may then be estimated as follows:

$$\hat{f}(\mathbf{x}) = \frac{1}{N} \sum_{i=1}^{N} \frac{1}{h(\mathbf{x})^2} K\left(\frac{\mathbf{x} - \mathbf{x}_i}{h(\mathbf{x})}\right) \tag{8.1}$$

where $\hat{f}(\mathbf{x})$ is the estimated density at location $\mathbf{x}$; $K(\cdot)$ indicates a kernel function (typically a radically symmetric) governed by bandwidth $h(\mathbf{x})$. Note that $h(\mathbf{x})$ could be varied for different locations accounting for the inhomogeneous background information.

By introducing a method of kernel density ratio between cases and control, corresponding to disease cases and population at risk in our discussion, Bithell (1990) described the concept of spatial relative risk. If densities for disease cases and population at risk are estimated through independent KDE processes (Silverman 1986), the relative disease risk, $\hat{r}(\mathbf{x})$, at location $\mathbf{x}$ can be estimated by Equation (8.2). Readers are referred to Shi (2010) for detailed derivation of $\hat{r}(\mathbf{x})$:

$$\hat{r}(\mathbf{x}) = \frac{\sum_{i=1}^{N} K\left(\frac{\mathbf{x} - \mathbf{x}_i}{h(\mathbf{x})}\right)}{P_k} \tag{8.2}$$

where P_k is the population size within the current kernel bandwidth $h(\mathbf{x})$ for the target location $\mathbf{x}$. Fixed bandwidth $h(\mathbf{x})$ independent of the population distribution could lead to varying population sizes for different locations $\mathbf{x}$, and therefore yields inconsistency regarding estimation errors of disease risk (Shi 2010; Cai et al. 2012). To address this problem, an adaptive bandwidth strategy through which $h(\mathbf{x})$ could vary according to the inhomegeneous background popuation distribution (Shi 2010) was adopted in this analysis to maintain the stability of populuation within each $h(\mathbf{x})$, and thus the estimation errors of disease risk. Based on such an adaptive KDE-based method, the risk of vulnerability to a certain disease could be mapped in a study area given the distribution of population and occurrences of this disease, and clusters could be identified as regions with unusually high or low risk values. As suggested in common statistical ESDA-based cluster detection methods (e.g., Openshaw et al. 1987; Anselin 1995), this disease risk estimation method is further integrated into a random labeling Monte Carlo simulation procedure to test the significance of the detected clusters. Given a study area, this procedure can be summarized as follows:

Step 1—Define a lattice of grid points that covers the study area.

Step 2—Compute values of bandwidth for each grid point, so that within each bandwidth, there is a constant size of population size.

Step 3—Estimate disease risk at each grid point by applying the risk estimator (Equation 8.2) to the occurrences of disease.

Step 4—Generate simulated disease occurrences by randomly selecting a subset of population, under the assumption that each individual within the population at risk has an equal chance of being an occurrence of the disease. Repeat step 3 by replacing the disease occurrences with simulated ones.

The entire procedure is repeated to generate a reference distribution of disease risks for each grid, based on which the p-value for the observed disease risk can be estimated for each grid point. For example, if the observed risk at a grid point computed from step 3 is larger than 995 of the 1000 simulated risks, then the p-value of the observed risk, or the probability that the observed risk, is smaller than the simulated risks at this grid point is 0.005.

To achieve the computational feasibility of the proposed procedure (adaptive KDE method and Monte Carlo simulation) on the significant size of location-based social media data set that we have been collecting, the CyberGIS middleware is employed to efficiently harness enormous computational resources provided by the National Science Foundation Extreme Science and Engineering Discovery Environment (XSEDE).

8.3.2 FLOW MAPPING

Geographical movement (such as movement of objects, materials, and information; or migration of people, animals, and birds) has been extensively studied in many research fields, such as geography, sociology, and public health. One of the cartographic methods most often resorted to depict and represent geographical movement is arguably flow mapping, in which each edge represents a movement (flow) between pairwise interacting geographical regions and thickness of each edge is corresponding to the flow magnitude on this edge. Location-based social media data provide real-time or near-real-time individual-level moving trajectories, and thus an appealing opportunity to investigate geographical movements, people migration in particular, across multiple spatiotemporal scales, from macro migration trends across the globe to characteristics of individual daily mobility behaviors. Within the CyberGIS environment, an interactive flow mapping service for location-based

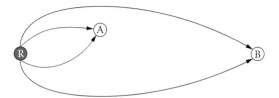

FIGURE 8.2   An illustration of spiral tree with root R and two nodes A and B, and corresponding spiral regions.

Twitter data is described in this section to support exploratory cluster detection of people migration across multiple spatiotemporal scales.

The idea of flow mapping could be dated back to the mid-1800s and the French engineer Charles Minard was one of the first to use hand-drawn flow maps to depict social movement phenomena. In the modern era, Tobler (1987) introduced a generic method to produce flow maps with the assistance of computers. Without taking advantage of the edge bundling and edge routing techniques implied in the hand-drawn methods, Tobler (1987) could possibly yield results with visual clutter, such as overlaps and crossings between edges. Based on a hierarchical clustering strategy, Phan et al. (2005) developed an algorithm to create single-source flow maps with bundled edges. Crossings of edges, however, could not be completely avoided by the adopted iterative ad hoc method for edge routing. Most recently, Verbeek et al. (2011) presented a 2-approximation optimal spiral tree algorithm for single-source flow mapping by taking advantage of spiral tree features: (1) if node A is within the spiral region of node B (defined by the boundary in red lines), the entire spiral region of node A (defined by the boundary in blue lines) should be contained in the spiral region of B, as illustrated in Figure 8.2; and (2) a right spiral boundary never crosses another right spiral boundary. The same features are valid for a left spiral boundary. The detailed description of the algorithm can be found in Verbeek et al. (2011).

To study the spatiotemporal patterns of human movement based on Twitter data, the flow mapping analysis decomposes the geographical space of interest into regions represented by cells with a specified spatial resolution. As an illustrative example, the left map panel of Figure 8.3 represents the spatial distribution of Twitter feeds over a span of 6 months that have been collected by the crawler introduced in the previous section; each number in each cell represents the number of Twitter feeds posted within the cell boundary and a *yellow* dot represents the center of each cell, derived by averaging geographic coordinates of every tweet location. Given a certain time window (temporal resolution), Twitter users in a region (spatial cell, indexed by the center of tweets locations) could frequently move to others. By monitoring such movements of Twitter users in a specified spatiotemporal resolution, the spiral-tree-based flow-mapping algorithm (Verbeek et al. 2011) was applied in the flow mapping service for exploratory cluster detection of people movements. To facilitate the access to this flow mapping service, a Web-based application is developed within the CyberGIS user environment, through which users can interactively specify the configuration settings, such as study areas, and spatial and temporal scopes of interest, as illustrated in Figure 8.3.

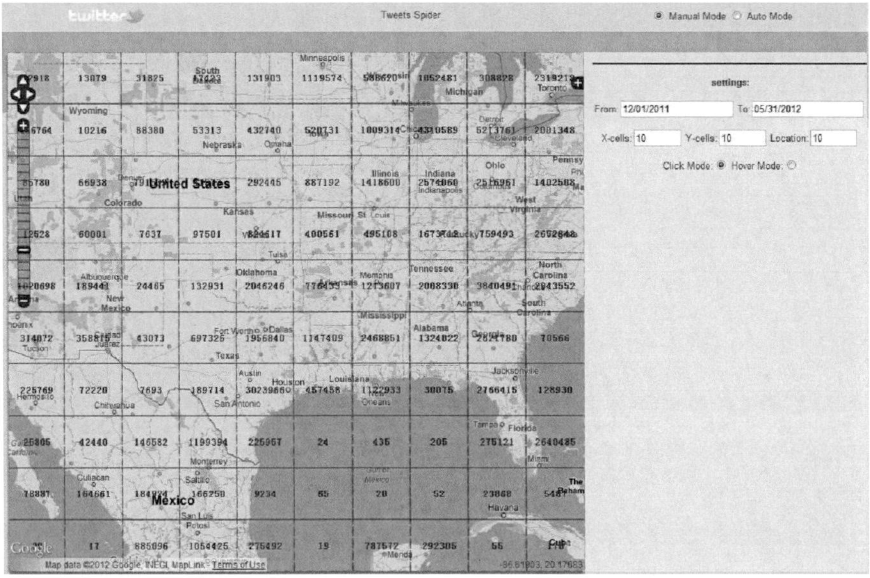

FIGURE 8.3   (See color insert.) User interface of the flow mapping service.

8.4   CASE STUDIES

8.4.1   Detection of Flulike Diseases

Recent studies have shown strong correlation between flu instance reports on social media and national influenza rates (Achrekar et al. 2011). In this case study, we hypothesize that social media data could be exploited to provide an early indication of spatiotemporal patterns of flu risk. To evaluate the validity of this hypothesis, we extract geolocated flu-related messages from the Twitter feeds that have been collected through the crawler, and then analyze these messages to detect daily spatial patterns of flu risk in the United States, through the use of the spatial cluster detection method described in Section 8.3.1.

8.4.1.1   Data

We aggregated two types of data sets on a daily basis using the Twitter location data service: raw tweets and flu-related tweets. Raw tweets represent all the Twitter feeds posted in the conterminous United States that we have been collecting, and flu-related tweets are those filtered by a list of predefined flu-related keywords while collecting Twitter streams (Ginsberg et al. 2009; Lampos et al. 2010; Signorini and Segre 2011; Smrz and Otrusina 2011). In addition, we only use tweets with location information, and each tweet is associated with a geographic coordinate. In our analysis, we assume that one flu-related tweet represents a flu occurrence, and one raw tweet represents an individual within population at risk. Figure 8.4 shows distribution of raw tweets and flu-related tweets aggregated on four different days (January 24, January 31, February 7, and February 14 of year 2012) in the conterminous United States.

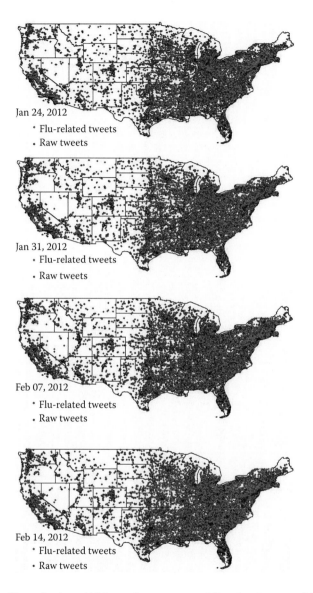

FIGURE 8.4   (See color insert.) Maps of raw tweets and flu-related tweets of four different sample days in the conterminous United States.

8.4.1.2   Results

The data sets of the four days (Figure 8.4) are used for analyzing daily spatial cluster change of flu risks in the conterminous United States. The parameters for the spatial cluster detection service are configured as follows: the resolution of output lattice is 5 kilometers, the constant population size P_k in Equation (8.1) is 1000, 999 iterations in Monte Carlo simulation, and the kernel function is specified as follows with $d = \dfrac{\mathbf{x} - \mathbf{x}_i}{h(\mathbf{x})}$:

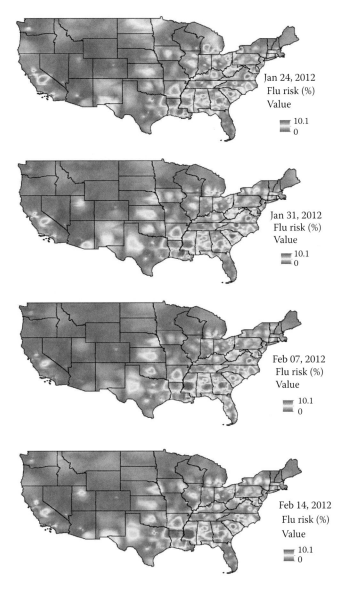

FIGURE 8.5 **(See color insert.)** Maps of estimated flu-risk in the conterminous United States of four different sample days.

$$\begin{cases} K(d) = \left(1 - d^2\right) & \text{if } d < 1 \\ K(d) = 0 & \text{otherwise} \end{cases}$$

Results of running the service are demonstrated in Figure 8.5 and Figure 8.6. Figure 8.5 represents maps of estimated observed flu risks on the four different days, and Figure 8.6 illustrates significantly high flu-risk areas where observed flu risk is

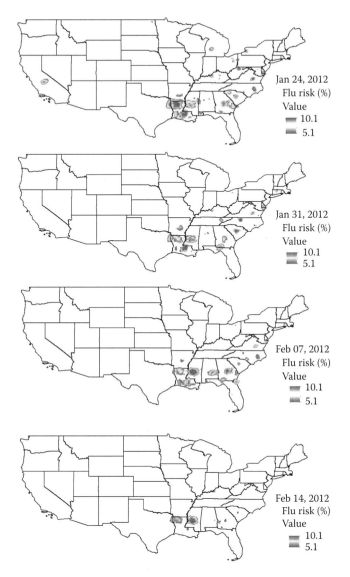

FIGURE 8.6   **(See color insert.)** Maps of significantly high flu-risk areas in the contermi-nous United States of four different sample days: estimated flu-risk >5.1% and *p*-value <0.005.

larger than 5.1% and *p*-value is smaller than 0.005. As the results clearly show, high flu-risk areas were mainly distributed in the mid-south of the United States, though minor spatial pattern changes over time have been observed. Further domain knowl-edge and investigation are necessary to provide explanation of such spatiotemporal patterns, and this is beyond the scope of this study.

In summary, this case study demonstrates the great complement between the Twitter data service and the CyberGIS-based spatial cluster detection service in detecting daily spatial pattern changes of flu risk across the United States. The output of the proposed service is not only timelier than conventional flu reports, which usually take at least multiple days, it also could generate detailed spatiotemporal pattern information that official flu reports have difficulty providing. These capabilities demonstrate the potential of this service as an early warning system of flu risk and a promising complement to traditional disease surveillance.

8.4.2   EXPLORATORY CLUSTER DETECTION OF MOVING TRAJECTORIES

Individual-level moving trajectories are important data sources for investigation of human mobility characteristics and people migration. In public health, for example, such collection of trajectories may help reach beyond the cluster detection of flulike diseases and provide further valuable information for transmission traces of flulike diseases if movement patterns of infected individuals could be identified as we did in the previous case study. To demonstrate the advantages of the flow mapping service described in Section 8.3.2, we focus on the Urbana–Champaign–Bloomington area of Illinois in the United States, and investigate, across multiple spatiotemporal scales, the movement patterns based on individual-level moving trajectories implied in Twitter feeds collection over a span of 6 months (December 1, 2011 to May 31, 2012).

8.4.2.1   Data

For the conterminous United States, a collection of 178,241,135 location-based tweets have been collected from December 1, 2011 to May 31, 2012, and approximately 60% of them have exact geographic coordinates (longitude–latitude) while the rest 40% of the collection have approximate platial information (e.g., city name). Based on such a data set, one could specify the study area and spatiotemporal scales of interest to investigate people movement patterns through the use of the flow mapping service described in Section 8.3.2.

8.4.2.2   Results

In the first analysis, we focus on the area of the University of Illinois as displayed on the map in Figure 8.7, and represent this area with a grid of cells with the cell size of approximately 500 meters. As one of the major residential communities close to the university, Orchard Downs (OD) family and graduate housing is located in the southwest campus area and the residential area is roughly covered by the cell highlighted with red color (Figure 8.7). From December 1, 2011 to January 1, 2012, moving trajectories of OD residents are analyzed with the proposed CyberGIS-based flow mapping service. From the output flow map (Figure 8.7), one can clearly see that the OD residents tend to travel to the north side (up) of the area, which is not surprising considering that the south sides of this community are mainly for agricultural uses. The thick blue edges of Figure 8.7 represent prominent movement clusters from OD to the university campus area, which is reasonable considering most of the OD residents usually work or take classes on the university campus.

   In the second analysis, we extend the spatial extent of the study area to Urbana–Champaign–Bloomington, as illustrated in Figure 8.8, and the temporal extent of the period from January 1, 2011 to May 31, 2012. Similarly, this study area is represented with a grid of cells with the cell size of approximately 6000 meters, and

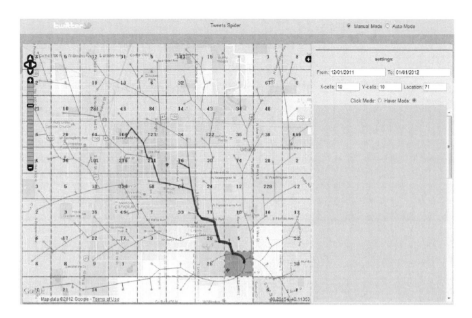

FIGURE 8.7   (See color insert.) Sample flow mapping result I: Movement patterns of Orchard Downs residents in the area of the University of Illinois from Dec. 1st 2011 to Jan. 1st 2012.

residential areas of Champaign and Urbana are roughly covered in two adjacent cells. Based on such settings and following the similar fashion in the first analysis, we derive the movement clusters of people from Urbana based on the individual moving trajectories implied in the Twitter feeds collections. From Figure 8.8, one can notice two prominent clusters represented by thick edges with pink color. The thickest edge is from Urbana to Champaign and represents that people from Urbana travel most frequently to Champaign. The second thickest edge is from Urbana to Bloomington and represents that, among the places on this map, people from Urbana travel second most frequently to Bloomington. Both of these cases can be explained considering that Champaign is the closest town to Urbana, and Bloomington is the largest city in the study area.

As mentioned, based on the flow mapping service, one could investigate the movement patterns of people in the area of interest with specified spatial and temporal scales. The promising preliminary results, however, demonstrate the potential of the CyberGIS-based flow mapping service for location-based social media data in movement pattern detection, and more importantly, showcase the natural complement between social media services and the CyberGIS environment.

8.5   CONCLUDING DISCUSSION

Location-based social media data provide a tremendous opportunity to gain better understanding of complex spatiotemporal dynamics of social and environmental systems by enabling unprecedented knowledge discovery from individual-level dynamic observations. Conventional GIS, however, are not in a position to deal with

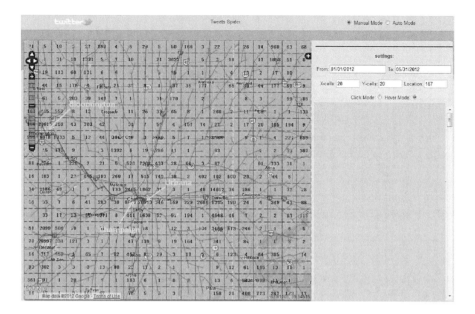

FIGURE 8.8   (See color insert.) Sample flow mapping result II: Movement patterns of Urbana residents in the area of Urbana–Champaign–Bloomington, Illinois from Jan. 1st 2012 to May 31st 2012.

these massive, dynamic, and nonstructured data sources. CyberGIS, as a new GIS modality based on synthesizing cyberinfrastructure, GIS, and spatial analysis, provides a promising means to advance and transform data-intensive geospatial sciences and technologies. Within this context, this chapter described a CyberGIS environment for analyzing location-based social media data. Through a seamless integration of various cyberinfrastructure resources and services based on CyberGIS middleware, a social media data collection service based on Twitter, and two spatio-temporal exploratory analyses, this environment provides a user-friendly and highly interactive approach to harnessing (e.g., querying and analyzing) location-based social media data.

Two well-known spatial analysis methods were adopted for demonstrating the capability and utility of the CyberGIS environment. Both methods require CyberGIS support to be applied to massive social media data. The kernel density estimation method focuses on gaining overall trends from individual-level observations while the flow mapping method represents a classic multiscale analytical cartographic technique. For treating massive spatial data, both methods posed computational challenges that were eventually resolved using cyberinfrastructure enabled by the CyberGIS environment.

Corresponding to the two analytical methods, two case studies were designed to evaluate the CyberGIS environment. The first case study aims to show the insights that can be gained from analyzing location-based social media data for revealing spatiotemporal patterns of disease risks. These insights otherwise difficult, or even impossible to acquire, especially for the purpose of capturing early indications of infectious disease spread. The second study is designed for the same purpose but focuses on a multiscale cartographic approach for depicting movement patterns extrapolated from individual-level movement trajectories. Results from both case studies are preliminary, and validation and further experiments will be conducted as part of our future work. These two complementary studies have demonstrated the significance of the CyberGIS environment for scalable knowledge discovery based on massive social media data.

The CyberGIS user environment allows multiple users to simultaneously perform the spatial analyses of the location-based social media data. Users do not need to know any specific details about how to access cyberinfrastructure or retrieve social media data, while being able to focus on understanding the configurations of the methods and related results. As the focus of this research is placed on the methods and integration of the social media data analytics within the CyberGIS environment, thorough evaluations of computational workflow and performance will be further investigated in the future work. Our CyberGIS approach demonstrates tremendous potential for a large number of users to gain in-depth knowledge from massive location-based social media data enabled by cyberinfrastructure.

ACKNOWLEDGMENTS

This material is based in part upon work supported by the U.S. National Science Foundation under grant numbers OCI-1047916 and BCS-0846655. Any opinions, findings, and conclusions or recommendations expressed in this material are those of the authors and do not necessarily reflect the views of the National Science Foundation. We are grateful for the support of a visiting scholarship by the K. C. Wong Education Foundation, which has benefited this research. This work used the Extreme Science and Engineering Discovery Environment (XSEDE), which is supported by the National Science Foundation grant number OCI-1053575.

REFERENCES

Achrekar, H., Fang, Z., Li, Y., Chen, C. X., Liu, B., and Wang, J., 2011. A spatio-temporal approach to the discovery of online social trends. *Combinatorial Optimization and Applications, Lecture Notes in Computer Science*, 6831, pp. 510–524.

Anselin, L., 1995. Local indicators of spatial association—LISA. *Geographical Analysis*, 27, pp. 93–115.

Anselin, L., and Rey, S., 2012. Spatial econometrics in an age of CyberGIScience, *International Journal of Geographical Information Science*, 26(12), pp. 2211–2226.

Atkins, D. E., Droegemeier, K. K., Feldman, S I., Garcia-Molina, H., Klein, M L., Messerschmitt, D. G., Messina, P., Ostriker, J. P., and Wright, M. H., 2003. Revolutionizing Science and Engineering through Cyberinfrastructure: Report of the National Science Foundation Blue-Ribbon Advisory Panel on Cyberinfrastructure, National Science Foundation.

Backstrom, L., Kleinberg, J., and Kumar, R., 2008. Spatial variation in search engine queries. *Proceedings of the 17th International Conference on World Wide Web*.

Bithell, J., 1990. An application of density estimation to geographical epidemiology. *Statistics in Medicine*, pp. 691–701.

Bithell, J., 1991. Estimation of relative risk functions. *Statistics in Medicine*, 10, pp. 1945–1951.

Bollen, J., Mao, H., and Zhang, X., 2011. Twitter mood predicts the stock market. *Journal of Computational Science*, 2(1), pp. 1–8.

Brunsdon, C., 1995. Estimating probability surfaces for geographical point data: An adaptive kernel algorithm. *Computers & Geosciences*, 21, pp. 877–894.

Cai, Q., Rushton, G., and Bhaduri, B., 2012. Validation tests of an improved kernel density estimation method for identifying disease clusters. *Journal of Geographical Systems*, 14(3), pp. 243–264.

Gao, H., Barbier, G. & Goolsby, R., 2011. Harnessing the crowdsourcing power of social media for disaster relief. Intelligent Systems, IEEE.

Ginsberg, J., Mohebbi, M. H., Patel, R. S., Brammer, L., Smolinski, M. S., and Brilliant, L., 2009. Detecting influenza epidemics using search engine query data. *Nature*, 457, pp. 1012–1014.

Goodchild, M., 2004. GIScience, geography, form, and process. *Annals of the Association of American*, 94(4), pp. 709–714.

Goodchild, M., and Glennon, A., 2010. Crowdsourcing geographic information for disaster response: A research frontier. *International Journal of Digital Earth*, 3(3), pp. 231–241.

Hecht, B., Hong, L., and Suh, B., 2011. Tweets from Justin Bieber's heart: The dynamics of the location field in user profiles. In Proceedings of the 2011 Annual Conference on Human Factors in Computing Systems, New York, New York.

Kaplan, A., and Haenlein, M., 2010. Users of the world, unite! The challenges and opportunities of Social Media. Business Horizons, 53, pp. 59–68.

Kulldorff, M., 1997. A spatial scan statistic. *Communications in Statistics—Theory and Methods*, 26(6), pp. 1481–1496.

Kulldorff, M., and Nagarwalla, N., 1995. Spatial disease clusters: Detection and inference. *Statistics in Medicine*, 14, pp. 799–810.

Lampos, V., Bie, T. D., and Cristianini, N., 2010. Flu detector-tracking epidemics on twitter. *Machine Learning and Knowledge Discovery in Databases, Lecture Notes in Computer Science*, 6323, pp. 599–602.

MacEachren, A. M., Jaiswal, A., Robinson, A. C., Pezanowski, S., Savelyev, A., Mitra, P., Zhang, X., and Blanford, J., 2011. SensePlace2: GeoTwitter analytics support for situational awareness. In *IEEE VAST*, pp. 181–190.

Mei, Q., Liu, C., and Su, H., 2006. A probabilistic approach to spatiotemporal theme pattern mining on weblogs. In *Proceedings of the 15th International Conference on World Wide Web*, pp. 533–542.

Murray, A., 2000. Spatial characteristics and comparisons of interaction and median clustering models. *Geographical Analysis*, 32, pp. 1–18.

Murray, A., and Estivill-Castro, V., 1998. Cluster discovery techniques for exploratory spatial data analysis. *International Journal of of Geographical Information Science,* 12, pp. 431–443.

O'Connor, B., and Balasubramanyan, R., 2010. From tweets to polls: Linking text sentiment to public opinion time series. In *Proceedings of the Fourth International AAAI Conference on Weblogs and Social Media*, pp. 122–129.

Openshaw, S., 1998. Building automated geographical analysis and explanation machines. In *Geocomputation: A primer*, P. A. Longley, S. M. Brooks, R. McDonnell, and B. MacMillan, eds. Chichester, UK: Wiley.

Openshaw, S., Charlton, M., Wymer, C., and Craft, A., 1987. A Mark I geographical analysis machine for the automated analysis of point data sets. *International Journal of Geographical Information Systems*, 1, pp. 335–358.

Phan, D., Xiao, L., Yeh, R., Hanrahan, P., and Winograd, T., 2005. Flow map layout. In *Proceedings of the 2005 IEEE Symposium on Information Visualization*, pp. 219–224.

Pickard, G., Pan, W., Rahwan, I., Cebrian, E., Crane, R., Madan, A., and Pentland, A., 2011. Time critical social mobilization. *Science*, 334(6055), pp. 509–512.

Sakaki, T., and Okazaki, M., 2010. Earthquake shakes Twitter users: Real-time event detection by social sensors. In *Proceedings of the 19th International Conference on World Wide Web*, pp. 851–860.

Shi, X., 2010. Selection of bandwidth type and adjustment side in kernel density estimation over inhomogeneous backgrounds. *International Journal of Geographical Information Science*, 24, pp. 643–660.

Signorini, A., and Segre, A., 2011. The use of Twitter to track levels of disease activity and public concern in the US during the influenza A H1N1 pandemic. PLoS One, 6(5): e19467.

Silverman, B. W., 1986. *Density Estimation for Statistics and Data Analysis*. London: Chapman & Hall.

Smrz, P., and Otrusina, L., 2011. Finding indicators of epidemiological events by analysing messages from twitter and other social networks. In *Proceedings of the Second International Workshop on Web Science and Information Exchange in the Medical Web*, pp. 7–10.

Tobler, W., 1987. Experiments in migration mapping by computer. *Cartography and Geographic Information Science*, 14(2), pp. 155–163.

Verbeek, K., Buchin, K., and Speckmann, B., 2011. Flow map layout via spiral trees. *IEEE Transactions on Visualization and Computer Graphics*, 17, pp. 2536–2544.

Wang, S., 2010. A cyberGIS framework for the synthesis of cyberinfrastructure, GIS, and spatial analysis. *Annals of the Association of American Geographers*, 100(3), pp. 535–557.

Wang, S., Armstrong, M., and Ni, J., 2005. GISolve: A grid-based problem solving environment for computationally intensive geographic information analysis. In *Proceedings of the 14th International Symposium on High Performance Distributed Computing (HPDC-14)–Challenges of Large Applications in Distributed Environments (CLADE) Workshop*, pp. 3–12.

Wang, S., Liu, Y., Wilkins-Diehr, N., and Martin, S., 2009. SimpleGrid toolkit: Enabling geosciences gateways to cyberinfrastructure. *Computers & Geosciences*, 35(12), pp. 2283–2294.

Wright, D., and Wang, S., 2011. The emergence of spatial cyberinfrastructure. *Proceedings of the National Academy of Sciences*, 108(14), pp. 5488–5491.

Section 3

Services

9 Multisensor Map Matching for Pedestrian and Wheelchair Navigation

Ming Ren and Hassan A. Karimi

CONTENTS

ABSTRACT

For pedestrian or wheelchair navigation, availability and accuracy of GPS signals may not be good enough for localization of pedestrians and wheelchair users on sidewalks due to the presence of high-rise buildings in urban areas. To overcome these shortcomings of GPS, in this chapter multisensor map matching is discussed. To that end, an algorithm that integrates accelerometer, compass, and vision data was developed to fill in signal gaps in GPS positions. Experiments were conducted to evaluate the developed algorithm using real field test data (GPS coordinates and other sensors data). The experimental results show that the developed algorithms can provide high-quality and uninterrupted localization estimation in pedestrian and wheelchair navigation services. The algorithm is applicable to pedestrian and wheelchair navigation applications requiring seamless localizations.

9.1 INTRODUCTION

The positioning technologies used in navigation systems have undergone a major evolution over the last few years. Several positioning technologies and techniques are currently being used or under research worldwide. These include Global Navigation Satellite System (GNSS), WiFi positioning system (WPS), cellular positioning system (CPS), and dead reckoning (DR) (Rizos et al., 2005; LaMarca et al., 2008). Among these positioning technologies, GNSS, such as GPS, are widely used in outdoor navigation systems. However, one shortcoming of these satellite-based positioning technologies is degradation of accuracy due to the environments and signal blockage caused by structures such as high-rise buildings.

Compared to vehicles, positioning pedestrians or wheelchairs using GNSS receivers is more challenging in that pedestrians move in low speeds and often close to buildings, where signals are easily disrupted and multipath reflections decrease accuracy. As an alternative, to enhance positioning quality, integrating GNSS data with other types of positioning data has become a trend (Hasan et al., 2009).

Sensor-integrated positioning estimates positions through sensory fusion of GPS and additional sensors like motion sensors or vision sensors. These additional sensors are used in the measurement of relative movement distance. Sensor data fusion is based on the combination of different sensor signals to calculate more accurate locations in navigation systems. For instance, in an inertial navigation system (INS), data from inertial positioning sensors like an accelerometer or a gyroscope are used to estimate changes in position over time. The measurement unit is called odometry, which is used to interpret data received from the movement of actuators to determine position replacement over time. Likewise, often equipped on wheelchairs, devices like rotary encoders help measure wheel rotations (Ohno et al., 2004). Similarly, visual odometry estimates traveled distance based on sequential camera images by using computer vision techniques (Hagnelius, 2005). Furthermore, with advances in mobile computing GPS, WiFi, 3G or 4G connection, cameras, accelerometers, compasses, and gyroscopes have become standard technologies on smartphones.

Availability of such technologies on smartphones makes it possible to provide enhanced localization through positioning data fusion.

To take advantage of such advanced technologies on smartphones, a novel map matching algorithm that integrates computer vision and an accelerometer with GPS will be presented in this chapter. The algorithm will provide users with uninterrupted location estimation in pedestrian and wheelchair navigation services. The algorithm was tested on a smartphone/server architecture.

9.2   BACKGROUND

As the most popular GNSS, GPS positional accuracy ranges from a few meters to 10 m in stand-alone mode and a submeter to a few meters in differential mode (i.e., differential GPS or DGPS). However, GPS often suffers from availability and accuracy issues. Due to obstructions, a sufficient number of satellites may not be available for a short period of time. In urban areas, especially downtown areas, GPS signals could be very weak compared with those in rural areas. Accuracy of GPS data also can be influenced by weather and fluctuate in the same location over time. As an alternative, to overcome these shortcomings, GPS usually is integrated with other positioning technologies to bridge the absence of satellite signals.

In pedestrian and wheelchair navigation services, the positioning technologies that can be integrated with GPS include WPS, CPS, VPS, DR, and INS (see Rizos et al., 2005; Retscher et al., 2006). They can be categorized into absolute and relative positioning as shown in Table 9.1.

WPS has grown rapidly in recent years and provides reasonable positional accuracy, but it suffers from limited signal coverage. WPS cannot locate targets when they are out of range of WiFi signals. WPS accuracy also depends on WiFi hotspot databases, which are built by fingerprinting wireless access points and must be constantly updated to keep up with WiFi hotspot changes (LaMarca et al., 2008). Moreover, only a few commercial companies build databases of sufficient size to be used for WiFi positioning. Building and maintaining such a database requires accurate and reliable fingerprinting techniques.

In contrast to WPS, CPS has good signal coverage in urban areas, but it is less accurate. Kitching (2000) proposed integrating GPS and CPS at two levels: (1) at the measurement data level and (2) at the infrastructure level. Although the addition of cellular network base stations (BSs) can improve the horizontal accuracy of GPS positions, a number of infrastructure modifications are required to enable the

TABLE 9.1
Positioning Technologies for Integration with GPS

Outdoor Positioning	Type	Accuracy	Coverage
WPS (WiFi)	Absolute	High; ±1 to 3 m	Limited
CPS (Cellular)	Absolute	Low; ±50 to 100 m	Good
VPS (Vision based)	Absolute/Relative	High	Middle distance
DR (Odometry + INS)	Relative	Low; ± 20 to 50 m per 1 km	—

cellular ranging measurements suitable in a positioning solution. Some commercial companies, like Qualcomm, are currently working on such integrations.

Integrating DR sensors, which are based on relative positioning techniques, is another alternative to overcome GPS errors in navigation systems and services. DR sensors, for example, gyroscope and accelerometer, obtain the traveled distance from velocity and acceleration measurements, and estimate direction of motion or heading and height difference.

A wheelchair's position can be estimated based on distances measured with odometer devices mounted on both wheels of the wheelchair. Accelerometers can provide relatively high positional accuracy in a relatively short time and, due to their bias drift, the position error will grow over time. For pedestrians, positioning data can come from accelerometer measurements based on an INS or from a step-counter and step-length estimator from a typical pedometer. Accelerometer measurements used for pedestrians have the same problem as in wheelchairs, that is, position errors would be accumulated over time. Regarding pedestrian dead reckoning (PDR), the positional accuracy in the pedometer–GPS integration relies mainly on estimations of the number of steps (counted by the accelerometer) and the length of the steps (calculated by the pedometer). For a pedometer to measure distance, the average step length of a user must be measured, assuming users walk at a consistent pace.

Computer vision also can be used to compute distance and estimate indoor and outdoor locations, which can be integrated with GPS to provide continuous positions. Vision-based positioning is an active research topic (Chen and Shibasaki, 1999; Henlich, 1997; Koller et al., 1997; Malis et al., 2002; Tardif et al., 2008) and compared to the aforementioned positioning technologies, it has several advantages as it is less influenced by the environment. VPS can help positioning in areas with no good GPS signals. In addition, using a camera as a sensor for vision-based positioning is relatively inexpensive and practical.

As all types of odometry may suffer from accuracy problems, visual odometry must also deal with errors that can occur through the accumulation of data on the continuous motion of objects. In spite of this, as compared to the type of odometry that uses motion sensors, visual odometry can be more accurate (Hagnelius, 2005). For this reason, visual odometry is chosen to implement relative positioning for pedestrian and wheelchair navigation described in this chapter. GPS positions then can be integrated with vision-based positioning results for map matching user's locations on the sidewalk.

Previous work on VPS mainly consists of research on automatic driving systems (Castro et al., 2001) and pedestrian navigation systems (Fritz et al., 2006; Steinhoff et al., 2007). Both systems use vision-based positioning techniques to provide either absolute or relative positioning. Absolute positioning computes the absolute position of an object by measuring distances from it to other known objects, like buildings, on the basis of recognizing known objects from images. Relative positioning calculates incremental positions by measuring movement or rotation/orientation step by step.

In pedestrian and wheelchair navigation systems, landmarks are used as references for absolute positioning. Given known locations of landmarks and estimated distances from users to those landmarks, pedestrian navigation systems can provide users with their absolute positions. Locations of known landmarks are recorded in

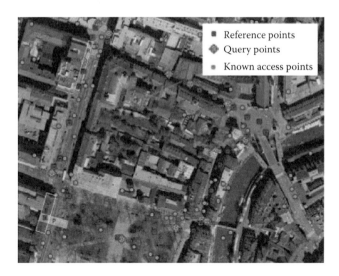

FIGURE 9.1   Example image with landmarks. (After Steinhoff, U., D. Omercevic, R. Perko, B. Schiele, and A. Leonardis, 2007, How computer vision can help in outdoor positioning, *LNCS* 4794, pp. 124–141.)

the database server in advance. An example scenario is as follows. A user on a tour captures pictures and sends them to the server. Once landmarks shown in the pictures are matched with the known landmarks in the server, the tour system can estimate the user's location by retrieving the known landmark's location in the database server and computing the distance from the user to the known landmarks. Figure 9.1 shows an example image with landmarks. In this figure, the red points on the image are landmarks, which are used as reference points. When a user travels on a tour, their locations are computed and marked on the image, which are shown as query points. Steinhoff et al. (2007) concluded that the absolute positions obtained for the way-points in the tour system were more accurate than GPS.

By contrast, automatic driving systems use a relative positioning approach in their vision-based positioning. By tracking visual features observed from a moving camera, relative positioning uses the transformation relationship between an image coordinate system and a world coordinate system to estimate a vehicle's location. To accomplish this, correspondence between image pairs or sequences must be obtained. Once an image set has been matched, bundle adjustment techniques can be used to compute the camera position, which is a surrogate for the vehicle's position. With the location of a starting point, the automatic driving system can also mark a vehicle by its absolute positions on a map by adding relative movements, which is called visual odometry in robotics (Olson et al., 2001; Levin et al., 2004). Tardif et al. (2008) presented a system for motion estimation of a vehicle using an omni-directional camera that successfully performed high precision camera trajectory estimation in urban scenes with a large amount of clutter. Other systems prefer to use two cameras as a pair to improve accuracy (Olson, 2001; Hartley and Zisserman, 2004; Kitt et al., 2010).

9.3 MULTISENSOR POSITIONING USING MONOCULAR VISUAL ODOMETRY

This section presents a positioning algorithm by integrating vision, an accelerometer, and GPS on smartphones. Different from previous works on visual odometry (e.g., Davide, 2008; Kitt et al., 2010), in this algorithm only one camera is available for use on smartphones. Unlike using a stereovision, monocular visual odometry has to deal with the ambiguity problem of scale factors from a single image in order to reconstruct the 3D structure of the real world (Hakeem et al., 2006; Esteban et al., 2010). To address the problem of scale factors in monocular visual odometry, an accelerometer is utilized to measure the distance between a pair of consecutive frames to calculate the scale factor between them. More important, consecutive frames have to be extracted from real-time video capture where accelerometer is used to assist in frame extraction.

In brief, an accelerometer is used in this algorithm with two roles. One is to identify user's movement pattern for video frame extraction. The other is to use the acceleration data collected between the first extracted frame and the second extracted frame to calculate the distance of movement for scale factor estimation.

9.3.1 ACCELEROMETER-ASSISTED MONOCULAR VISUAL ODOMETRY FOR MOTION ESTIMATION

Figure 9.2 shows how monocular visual odometry works for positioning in pedestrian and wheelchair navigation systems.

In this figure, a videostream is obtained from a camera, and a sequence of images is taken as the user moves. Images are shown on image planes from C_0 to C_4. A different set of image features corresponding to 3D objects is used to compute the motion between consecutive frames. For instance, images of a street scene in the figure are taken in consecutive frames C_0 to C_2, which have overlapping objects, like buildings. Some features of objects are marked as circles on the image planes, and the common features will be used for camera pose estimation in the motion.

The first step of frame-to-frame motion estimation is to extract a set of salient features that are present in each frame. Scale-invariant feature transform (SIFT) is employed to extract local features and build descriptors as feature vectors. In the second step, the features across consecutive frames are matched using a nearest neighbors search and a minimum distance threshold in the SIFT descriptor space, obtaining matches between frames Fr_i and Fr_{i+1}. In the third step, the normalized 8-point algorithm is used to compute the frame-to-frame motion, as described by Hartley and Zisserman (2004), due to its computational simplicity. Outliers are then removed between Fr_i and Fr_{i+1} frames using RANdom SAmple Consensus (RANSAC), and the final motion is recomputed using only the set of inliers. This yields a fundamental matrix F that describes the camera motion.

In the next step, given that the camera was calibrated beforehand, we already know the calibration matrix K. Therefore, we can obtain the essential matrix by applying E = K'*F*K. In the last step, the frame-to-frame rotation matrix $R_{i-1,i}$

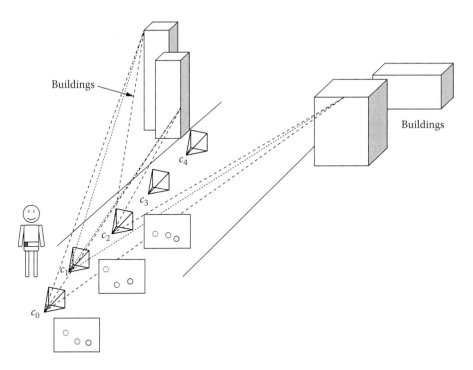

FIGURE 9.2   Frame-to-frame motion estimation.

and the translational vector $T_{i-1,i}$ are eventually obtained using the method by Horn (2004). This yields four possible solutions, from which we select the one with more inliers in front of both cameras. Furthermore, a scale factor for each translational vector must be calculated to recover the overall camera pose. Figure 9.3 illustrates the entire procedure.

9.3.1.1   Fundamentals of Visual Odometry for Motion Estimation

To better understand the relationship between the image plane and the 3D modeling of the real world, vision geometry related notations are first defined:

(X, Y, Z) are the coordinates of a 3D point in the world coordinate space.
(u, v) are the coordinates of the projection point in pixels.
(x, y, z) are the coordinates of a 3D point in the image coordinate system.
K represents a camera projection matrix, which is a matrix of intrinsic parameters that do not depend on the scene viewed. The matrix represents the quality of each camera, so once K is estimated, it can be reused as long as the same camera is used.
(cx, cy) is a principal point, which is usually at the image center.
(fx, fy) are the focal lengths expressed in pixels.
[R|T] represents a matrix of extrinsic parameters. This is a joint rotation-translation matrix, where R is the rotation matrix and T is the translation matrix.

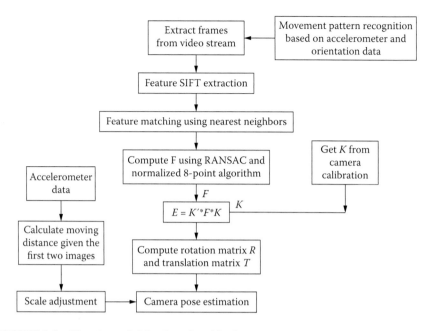

FIGURE 9.3   Flowchart of vision-based positioning.

Based on these definitions, Equation (9.1) shows the relationship between the image and 3D scene in the image coordinate system. Equation (9.2) shows the relationship between the image coordinate system and the world coordinate system. Given Equations (9.1) and (9.2), Equation (9.3), which shows the relationship between the image pixels on the image plane and the corresponding 3D points in the world coordinate system, can be calculated.

$$
\begin{bmatrix} u \\ v \\ w \end{bmatrix} = \begin{bmatrix} fx & s & cx & 0 \\ 0 & fy & cy & 0 \\ 0 & 0 & 1 & 0 \end{bmatrix} \begin{bmatrix} x \\ y \\ z \\ 1 \end{bmatrix} = K \begin{bmatrix} x \\ y \\ z \\ 1 \end{bmatrix} \tag{9.1}
$$

$$
\begin{bmatrix} x \\ y \\ z \\ 1 \end{bmatrix} = \begin{bmatrix} R_{11} & R_{12} & R_{13} & 0 \\ R_{21} & R_{22} & R_{23} & 0 \\ R_{31} & R_{33} & R_{33} & 0 \end{bmatrix} \begin{bmatrix} X \\ Y \\ Z \\ 1 \end{bmatrix} + \begin{bmatrix} T_1 \\ T_2 \\ T_3 \\ 1 \end{bmatrix} = \begin{bmatrix} R & T \\ 0_3^T & 1 \end{bmatrix} \begin{bmatrix} X \\ Y \\ Z \\ 1 \end{bmatrix} \tag{9.2}
$$

$$
\begin{bmatrix} u \\ v \\ w \end{bmatrix} = K \begin{bmatrix} I_3 | 0_3 \end{bmatrix} \begin{bmatrix} R & T \\ 0_3^T & 1 \end{bmatrix} \begin{bmatrix} X \\ Y \\ Z \\ 1 \end{bmatrix} \tag{9.3}
$$

Set

$$P = K\begin{bmatrix} I_3 | 0_3 \end{bmatrix} \begin{bmatrix} R & T \\ 0_3^T & 1 \end{bmatrix},$$

we have

$$\begin{bmatrix} u \\ v \\ w \end{bmatrix} = P \begin{bmatrix} X \\ Y \\ Z \\ 1 \end{bmatrix}.$$

Given an image represented by $\begin{bmatrix} u \\ v \\ w \end{bmatrix}$, if we have $\begin{bmatrix} X \\ Y \\ Z \\ 1 \end{bmatrix}$, which are the coordinates of object points in the real world, it is then possible to calculate matrix P. Matrix P is mainly made up of two matrices, K and $\begin{bmatrix} R & T \\ 0_3^T & 1 \end{bmatrix}$. Where K is the intrinsic matrix and $\begin{bmatrix} R & T \\ 0_3^T & 1 \end{bmatrix}$ is the extrinsic matrix. Since all parameters in K are fixed when the same camera is used, if K is known, then the rotation matrix R and translation matrix T can be calculated after P is estimated.

The problem of estimating the trajectory of a user's movement can be defined as the trajectory of both the rotation matrix $R_{i-1,i}$ and the translational vector $T_{i-1,i}$ in a given frame, as well as the characterization of the relative movement between two consecutive frames (see Figure 9.4).

Matrix K must be obtained before the camera pose recovery can proceed. To obtain the intrinsic matrix K, camera calibration must be performed. Since offline camera calibration is more accurate than online camera calibration, we estimate K by using offline camera calibration.

9.3.1.2 Camera Calibration

Camera calibration finds the essential parameters of the camera that affect the imaging process. Specifically, with the definition of matrix K in Equation (9.1),

that is $K = \begin{bmatrix} fx & s & cx & 0 \\ 0 & fy & cy & 0 \\ 0 & 0 & 1 & 0 \end{bmatrix}$, camera calibration is to estimate all parameters in

matrix K, which involves calibrating the position of image center, which may not be at the image's true center, estimating the focal length, using different scaling factors for row pixels and column pixels, and accounting for any skew factor and lens distortion (pin-cushion effect). In camera calibration, by taking pictures of a known object and by knowing the coordinates of given object points in the real world, it is possible to obtain internal camera parameters through optimization algorithms.

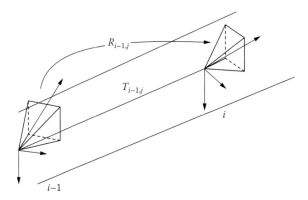

FIGURE 9.4 Estimation of rotation matrix $R_{i-1,i}$ and translational vector $T_{i-1,i}$ in the motion between video frame Fr_{i-1} and Fr_i.

To implement camera calibration, the camera calibration toolbox (http://www.vision.caltech.edu/bouguetj/calib_doc/) was used.

9.3.1.3 Video Frames Extraction Based on User's Movement Pattern Recognition

Before image analysis and position estimation are performed, frames must be extracted from the video. A major criterion for frame extraction is to ensure that there are overlaps between consecutive images in order to perform feature matching. Furthermore, to obtain precise feature matching, dispersedly distributed overlapped features in the images are highly preferred. The objective is to select those frames that will appropriately meet the requirement of image feature extraction and matching. Moreover, the overlapped features extraction and matching in consecutive frames will further impact the camera pose estimation.

The key to frame extraction is to decide the points in time at which the extraction should be performed. Recognition of user's movement patterns can aid in this determination. For example, there is no need to extract frames if a user is stopped at a traffic light. Conversely, if a user is making a turn into the next segment of the sidewalk, more frequent frame extraction is required than would be required in the time period when a user is moving straight. This is because turning makes adjacent frames more likely to lose overlapped features. This could cause problems in matching features and may eventually decrease the accuracy of geometric calculation for the camera pose estimation.

For pedestrians, outdoor activities in navigation services can be classified into four modes, which are no movement, walking, running, and turning. In the computer vision context, the four different movement modes, which are no movement (operating at zero speed), walking (operating at a low speed), running (operating at a relatively high speed), and turning (operating with change in viewpoint of images), require varying frame extraction intervals. To extract frames appropriately, with changes of speed and changes of viewpoint in the movements, frame extraction intervals are set up based on different modes of movements.

A study by Richard et al. (1999) indicates that the mean walking speeds are 1.51 m/sec for younger pedestrians and 1.25 m/sec for older pedestrians. Since a speed of 1.51 m/sec is considered an average speed for walking, 1.51 m/sec is used as a baseline in order to set up frame extraction rates that correspond to different movement modes. The number of standard video frames in 1 second is 30. Taking walking mode as an example, if the standard pedestrian's walking distance in 1 second is 1.51 m, the frame extraction rate for a 2 m distance interval should be $(2/1.51) \times 30$, which is about 1 frame per 40 frames. Since frame extraction rate is proportional to movement speed, using 1 frame per 40 frames in frame extraction as a baseline, the frame-extraction rate in running mode is set as 1 frame per 30 frames, which corresponds to the relatively higher running speed considered in our experiments. In turning mode, to keep overlapped features in adjacent frames as much as possible, the frame extraction rate is set as a half of the frame extraction rate in walking mode. In summary, the following rules for frame extraction are made corresponding to the four modes of movements:

1. When a user is walking in a straight path, frame extraction rate is 1 frame per 40 frames.
2. When a user is running, frame extraction rate is 1 frame per 30 frames.
3. When a user is making a turn, frame extraction rate is 1 frame per 20 frames.
4. When a user is not moving, there is no need to extract frames.

We use accelerometer data for user's movement pattern recognition. Four features were extracted from each of the three axes in the accelerometer, giving a total of twelve attributes. Table 9.2 shows the attributes obtained by feature extraction from three-axis accelerometer data.

To implement user's movement pattern recognition, a decision tree classification is employed for feature selection and further utilized for recognition. Figure 9.5 shows five attributes out of twelve selected based on training data that we collected and labeled. The selected features are My, Sy, Sz, Ez and Corrxy corresponding to those in Table 9.1. The leaf nodes in the decision tree represent the four modes, no movement, walking, running and turning.

A smartphone carried on by a user who is moving on a sidewalk can collect motion data (through accelerometer) and take live video streams (through camera). After matching the user's movement mode with one of the four movement modes, appropriate image frames can be extracted by following the rules described earlier. The next section describes how we employ motion estimation from frame to frame.

TABLE 9.2
Twelve Attributes by Feature Extraction from Three-Axis Accelerometer Data

Feature	Mean			Standard Deviation			Energy			Correlation		
Attribute	x	y	z	z	y	z	x	y	z	Corrxy	Corrxz	Corryz

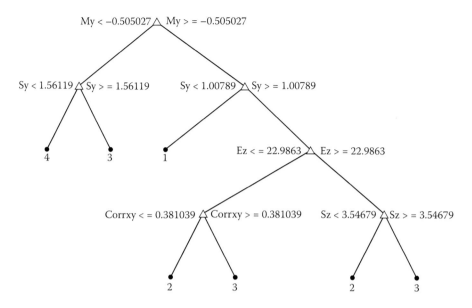

FIGURE 9.5 Movement recognition decision tree.

9.3.1.4 Feature Extraction for Map Matching in Pedestrian Navigation

After the images are obtained, feature extraction is the next step in estimating motion. With a prior knowledge about man-made environments on streets, such as rectangular objects with dominant planes (Ohnishia and Imiya, 2006) like buildings, objects matching in the vision-based map matching can make use of some of the special characteristics of street-view images. In a sequence of street-view images, sky and ground are both viewed as backgrounds due to their stable and static characteristics, whereas other objects like buildings and cars are unique or diverse, so they are more helpful to use in location identification. Unique objects in urban environments include:

- Buildings
- Vehicles (e.g., cars, bikes, strollers)
- Pedestrians
- Vegetation (e.g., trees, flowers, bushes)
- Urban furniture (e.g., city lights, telephone poles, parking meters, benches)
- Signs and banners

Note that signs and banners can be used to recognize specific locations only when optical character recognition (OCR) technology is applied to recognize characters in images, while some types of vegetation and urban furniture may appear in multiple locations and images. Vehicles and pedestrians are moving objects that are not stable in locations, and as a result, they are not considered to be reliable features to use for feature matching. Buildings are the most stable and distinctive objects for location estimation in pedestrian and wheelchair navigation. Inspired by human cognitive mechanisms that daily navigation strongly relies on landmark information, overlapped landmarks in the image sequence are features extracted as interest points. When a user is moving, viewpoints and objects in motion sequences must change with the movements of the user. Finally, due to its sensitivities to changes in viewpoints, scales, lighting, and environment, global features (such as color histogram, texture, and edge) are not suitable for location estimation. With change of distance and viewpoint during movements, features with rotation-invariance and scale-invariance are needed. After analyzing various features discussed by researchers (e.g., Wang et al., 2004; MOBVIS, 2006), local features are chosen both for object recognition and for subsequent location estimation. Of all the current local feature extraction algorithms, SIFT is the most effective algorithm for street-view images (Deselaers et al., 2007).

The SIFT descriptor transforms image data into scale-invariant coordinates that are relative to local features. The SIFT descriptor is a well-known method in computer vision for its capabilities in robust matching to the database records, despite viewpoint, illumination, and scale changes in images. SIFT is suitable for object recognition in urban environments where illumination and scale changes usually degrade performance (Lowe, 1999, 2004).

The major computations to generate SIFT features consist of the following:

1. Scale-space extrema detection. This is the initial preparation. Of all scale levels and their corresponding image locations, a difference-of-Gaussian function is used to identify potential interest points, which are invariant to scale and orientation.
2. Keypoint localization. At each candidate location, a detailed model is used to determine location and scale. A technique similar to the Harris Corner Detector (Derpanis, 2004) is used in SIFT. Keypoints are selected by eliminating some instable candidates, like edges and low contrast regions in terms of their stability.
3. Orientation assignment. One or more orientations are assigned to each keypoint location, based on local image gradient directions. Image data are processed and transformed relative to the assigned orientation, scale, and location for each feature. This effectively cancels out the effect of transformation.
4. Keypoint descriptor. With scale and rotation invariance in place, local image gradients are measured at the selected scale in the region around each keypoint. This helps identify unique features, allowing for significant levels of local shape distortion and changes in illumination.

9.3.1.5   Scale Adjustment

With only one camera, the baseline between two instants is unknown and the scale factor of reconstruction is ambiguous. To address this ambiguity problem, an accelerometer is used to assist with estimating the scale factor. Figure 9.6 shows the motion estimation for consecutive frames.

When the distance between cameras C_0 and C_1 is normalized to 1, the location of the third camera, C_2, can be estimated across the translation direction. All scales

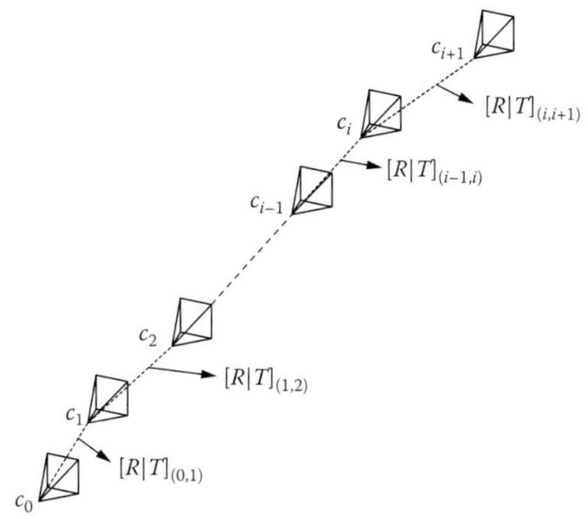

FIGURE 9.6 Scale adjustment.

between consecutive images are adjusted based on the normalized distance $d(C_0, C_1)$. To solve the ambiguity of the scale factor in translational vector, an accelerometer is used to measure the distance between the first frame and the second frame, that is, $d(C_0, C_1)$. The integral of acceleration over time from t_0 to t_1 will yield velocity, and the integral of velocity over time results in distance. Since accelerometers return data in units of the gravitational constant, that is, g, acceleration values need to be multiplied by 9.81 to convert to meter per second squared (m/s^2). During this process, errors may accumulate in the integral calculation. As a result, we will only use the accelerometer to calculate first-step distance, $d(C_0, C_1)$, which measures the movement from the first frame to the second one.

Once the distance between C_0 and C_1 is obtained, the translation is only determined up to the scale ratio between each pair of consecutive frames. The ratio between these distances must be calculated before the camera's pose can be reconstructed.

To calculate this distance ratio, we first calculate the motion between three consecutive frames using frame-to-frame feature matches. This produces two different motion estimations: [R|T](i,i+1) and [R|T](i+1,i+2). The quality of this motion estimation is greater than the motion estimation of [R|T](i,i+2), due in part to the larger number of matches. These two motions are translated into two different scale factors: s(i,i+1) and s(i+1,i+2).

Given the motion estimation of the camera and the reconstructed 3D points of three-frame matches, we can establish the following relation:

$$ K^{-1} \begin{bmatrix} u \\ v \\ w \end{bmatrix} = \begin{bmatrix} R \mid s_i T \end{bmatrix} \begin{bmatrix} X \\ Y \\ Z \\ 1 \end{bmatrix} \tag{9.4} $$

where s_i is the scale ratio that relates the translation between cameras i and i+1 and cameras i+1 and i+2. The ratio $s_i = s(i,i+1)/s(i+1,i+2)$ is calculated using matches across all three frames and a linear system of equations as in the P6P DLT algorithm.

9.3.2 INTEGRATED POSITIONING AND MAP MATCHING

For pedestrian and wheelchair outdoor navigation, when high-quality GPS data are available, users' location can be solely measured based on GPS data. When pedestrians move into areas with poor or no GPS signals, the monocular visual odometry is needed to continue tracking user's location movement. Therefore, whether visual odometry is required or only GPS data are needed is determined by the quality of GPS data. The quality of GPS data can be detected in real time as users move.

If GPS horizontal accuracy measured in meters is below a threshold, GPS signals are considered as good quality, where absolute positions can be obtained. Conversely, if GPS accuracy is above a threshold, GPS signals are considered as poor quality, where a vision-based positioning algorithm is applied to calculate the relative distance in user's movement. Figure 9.7 shows the flowchart of the vision-based positioning results combined with GPS-based map matching results to obtain user's absolute locations. To enhance the accuracy of map matching, orientation data from compasses are used to aid map matching while users are making turns.

9.3.2.1 Integrated Map Matching

Given the information from the camera pose recovery, the relative displacement combined with GPS historical positions and orientations in the movement are used to perform map matching on a sidewalk network. Figure 9.8 shows the overview of the map matching algorithm, which integrates the camera pose recovery results with the GPS historical trajectory and orientation data on the sidewalk network.

In areas with poor GPS signals, the camera is used to capture images for measuring continuous user's movement distance. GPS historical data provides the starting

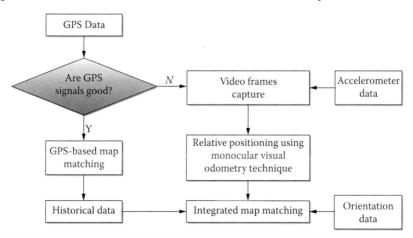

FIGURE 9.7 Flowchart of multisensor map matching algorithm using monocular visual odometry.

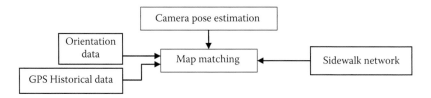

FIGURE 9.8 Flowchart of map matching algorithm.

positions at the time when the camera is triggered to be active. Camera pose estimation is performed to obtain the relative displacements between consecutive image planes. Orientation data, as indicated by the compass, are integrated with estimated positions to help map matching as users move about the environment.

9.3.2.2 Coordinate System Conversion for Tracking

To integrate data from different sensors in multisensor map matching, four coordinate systems are involved. These are a 2D image coordinate system, a 3D camera coordinate system, a world coordinate system, and a map coordinate system. In the integration process, GPS positions and sidewalk map data are in the world coordinate system, presented by longitudes and latitudes in the WGS-84 projection system. In the camera's pose estimation, image sequences are extracted from real-time video streams, and image feature extraction and computation are conducted in the 2D plane coordinate system. Image features are further reconstructed in the 3D camera coordinate system. Therefore, 2D image plane coordinates are transformed and presented in the 3D camera coordinate system, as shown in Equation (9.1), and are further translated to a 3D world coordinate system, as shown in Equation (9.3). Eventually, all the positioning data and map matching results are shown in the map coordinate system. This requires a conversion between the 3D world coordinate system and the 2D map coordinate system, from WGS-84 to Universal Transverse Mercator (UTM) (Grewal et al., 2002), in order to track user's locations on the 2D map.

9.4 EXPERIMENTATIONS AND ANALYSIS

To validate the multisensor integrated map matching algorithm, experiments were conducted on a smartphone/server architecture. Multisensor data including camera, accelerometer, compass, and GPS data were collected from Samsung GT I9000 Galaxy S smartphone by walking on the sidewalk of the main campus of the University of Pittsburgh.

9.4.1 CAMERA CALIBRATION

The camera was calibrated beforehand by applying "Camera Calibration Toolbox for Matlab." Twenty 720 × 480 photos of a black-and-white checkerboard were taken from different angles by the Samsung phone. Figure 9.9 shows one of these photos,

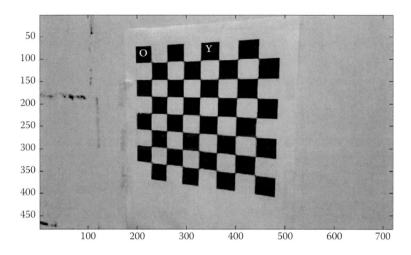

FIGURE 9.9   **(See color insert.)** A checkerboard to calibrate camera.

on which each corner of the grid on the checkerboard is selected as the featured point and is marked with a red cross.

Camera internal parameters were estimated in the camera calibration. K is the intrinsic matrix, as shown in Equation (9.5).

$$K = \begin{pmatrix} 686.646920000000 & 0 & 359.500000000000 \\ 0 & 687.467270000000 & 239.500000000000 \\ 0 & 0 & 1 \end{pmatrix} \quad (9.5)$$

9.4.2 EXPERIMENTAL RESULTS

Several video clips were taken on the main campus of the University of Pittsburgh. Buildings, trees, pedestrians, cars, urban furniture, and signs were the most common objects captured from the street view.

Using video as an example, the videostream was recorded as a user was walking on the sidewalk in front of the School of Information Sciences at the University of Pittsburgh. A sequence of frames was captured in the videostream. No particular attention was given to the distance between frames, since frames are automatically extracted based on the user's movement pattern. These captured frames are saved as images for further image processing and feature extraction. Figure 9.10 shows a sequence of images extracted from a video that was collected in motion.

FIGURE 9.10   **(See color insert.)** A sequence of images extracted from a video.

FIGURE 9.11   (See color insert.) SIFT features of a streetview image.

9.4.2.1   SIFT Feature Extraction and Feature Matching

The SIFT algorithm is first applied to extract features from images. Figure 9.11 shows an example of SIFT feature extraction from one image taken on the campus. In the figure, the length of the arrow represents the scale of the extracted SIFT features and the arrow's direction represents the extracted features' dominant direction. Figure 9.12 shows two image frames that were extracted from the video and with their SIFT features extracted. The lines that link two features in the two images show the correspondences that occur after feature matching is performed. Table 9.3 presents the number of feature points extracted from each image and the number of matched points in both images.

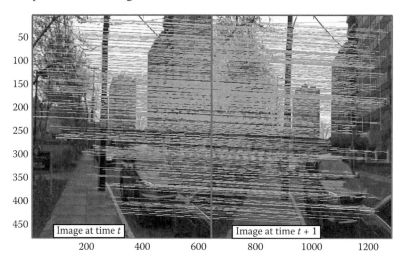

FIGURE 9.12   (See color insert.) Matched SIFT feature points.

TABLE 9.3
Keypoints and Matched Points

Image Sequence	Feature Points	Matches
Image 1	2575	737
Image 2	2500	737

FIGURE 9.13 **(See color insert.)** Feature points in Image 1 versus Epipolar lines in Image 2.

In Table 9.3, an image pair in one image sequence is taken as an example, 2575 SIFT feature points are extracted from Image 1 and 2500 SIFT feature points are extracted from Image 2. After feature matching, both images have 737 feature points in common corresponding to same feature points on objects in the real world. Similarly, feature extraction and feature matching are implemented between all the continuous image pairs taken in the experiment. Taking 100 images as samples in the experiment, 87.5% features in the images are matched correctly. Since images are taken with changing viewpoints, the high quality of feature extraction and accuracy in feature matching indicate that SIFT is insensitive to changes of viewpoints, which is appropriate in our vision-based positioning.

After SIFT features are extracted, the fundamental matrix F is calculated, given correspondences in an image sequence. In Figure 9.13, given matrix F, the marked corresponding feature points are shown in Image 1, while epipolar lines going through the matches are shown in Image 2.

9.4.2.2 Monocular Visual Odometry and Vision-Based Map Matching

After the fundamental matrix F is obtained, the essential matrix E can be calculated as discussed earlier. Therefore, camera positions and poses can be estimated, which provide user's locations. The video was taken starting in the front of the School of Information Sciences building, and Figure 9.14 shows the positioning results, as overlaid on Google Maps. The results show that locations are estimated quite precisely compared to the actual trajectories that are recorded by the collector.

FIGURE 9.14 **(See color insert.)** Positioning results by using visual odometry, top view (left) and street view (right).

However, like other odometry techniques, the visual odometry must overcome the same problem in accumulation of position error (Davide, 2008; Kitt et al., 2010). This problem also occurred in our experiment as error accumulates in a relatively long distance by the monocular visual odometry. Starting from the same origin and continuing the same route, as shown in Figure 9.14, a videostream of a user walking 332 m in about 4 minutes is recorded. The experiment is shown in Figure 9.14, where the user started walking from a point marked as S, and ended up the walk at another point marked as E. All the features, such as buildings, trees, cars, pedestrians, and signs emerge on the captured images along roads. Figure 9.15a shows that the estimated locations drifted over time when only the monocular visual odometry technique was applied. To mitigate this problem, we use geometrical and topological information in the sidewalk map to constrain the user's location in every map matching step on the sidewalk in order to reduce position error accumulation. Figure 9.15b shows the comparison of location estimation, both before and after map matching. Finally, Figure 9.15c shows the map matching results overlaid on Google Maps.

Since vision-based map matching is only needed in places where GPS signals either are not available or have poor quality, the accuracy of GPS signals is used as the criterion to determine when to start vision-based map matching. Figure 9.16 shows a sample log file from data collection. The log file includes GPS, accelerometer, and orientation data, and each GPS data point has recorded longitude, latitude, accuracy, bearing, altitude, and speed.

As Figure 9.16 shows, the collected GPS data have different accuracies in different locations at different times. The accuracies of the three GPS positions are 8.94 m, 17.89 m, and 10.0 m. Our tests show that in the multisensor map matching algorithm, if GPS accuracy is equal to or better than 10 m, then the quality of GPS data is considered to be acceptable and map matching can be performed by using only GPS data. But if GPS accuracy is worse than 10 m, vision-based map matching is needed to fill in the localization gap. In this case, the accelerometer and orientation data are used to help video frame extraction, assist map matching in movement pattern recognition, and improve the overall efficiency of map matching.

Figure 9.17 shows the experimental results that compare the GPS-based map matching results with multisensor map matching results. In Figures 9.17a and 9.17b, black points represent a GPS trajectory and red points show the GPS-based map

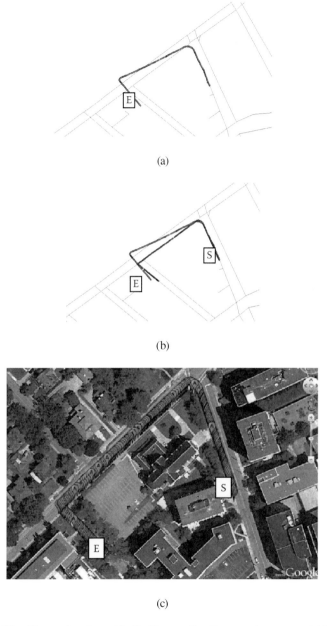

(a)

(b)

(c)

FIGURE 9.15 (See color insert.) Position estimations and map matching results. (a) Monocular visual odometry results in one route before map matching. (b) Estimated locations on one route before map matching and after map matching. (c) Map matching results overlaid on Google Maps.

```
15:20:45:463 ACCELEROMETER 0.47884035,1.9919758,9.059659
15:20:45:466 ACCELEROMETER 0.7278373,1.8579005,9.212888
15:20:45:469 ACCELEROMETER 0.9768343,1.685518,9.595961
15:20:45:474 ACCELEROMETER 1.1109096,1.4939818,10.093954
15:20:45:477 ORIENTATION 54.0,-8.0,6.0
15:20:45:513 ACCELEROMETER 1.3790601,1.3215994,10.572795
15:20:45:516 ACCELEROMETER 1.340753,1.4173675,10.649409
15:20:45:520 ACCELEROMETER 1.4939818,1.6472107,10.70687
15:20:45:525 ACCELEROMETER 1.283292,1.8770541,10.821792
15:20:45:533 ACCELEROMETER 1.091756,1.8387469,10.783484
15:20:45:536 ACCELEROMETER 1.3790601,1.4939818,10.515334
15:20:45:563 ACCELEROMETER 1.8770541,1.2449849,10.802638
15:20:45:577 GPS -
79.95218644,40.44750061,8.9442720413208,0.0,286.70001220703125
,0.0
15:20:45:582 ACCELEROMETER 2.2984335,0.8810662,11.473015
15:20:45:585 ORIENTATION 54.0,-2.0,10.0
15:20:45:610 ACCELEROMETER 0.842759,0.019153614,12.852075
15:20:45:617 ACCELEROMETER 0.2873042,0.019153614,12.852075
15:20:45:621 ACCELEROMETER -0.038307227,-0.019153614,12.469003
```

(a)

```
15:20:38:636 GPS -
79.95222581,40.44747017,17.8885440826416,0.0,286.3999938964844
,0.0
```

(b)

```
15:22:47:553 GPS -
79.95297402,40.44828781,10.0,208.6999969482422,263.70001220703
125,1.0
```

(c)

FIGURE 9.16   A sample log file to compare accuracy of GPS positions. (a) A sample log file recording GPS, accelerometer and orientation data and a highlighted GPS position with accuracy of 8.94 m. (b) A highlighted GPS position with accuracy of 17.89 m. (c) A highlighted GPS position with accuracy of 10.0 m.

matching results. Figure 9.17a shows the map matching results of all the GPS raw data, and Figure 9.17b shows the map matching results based on those GPS data with accuracy ≤10 m. In Figure 9.17b, because a section of sidewalk (along O'Hara St) has GPS accuracy worse than 10 m due to poor GPS signals, GPS-based map matching is not appropriate; this is where the vision-based map matching is performed to fill in the signal gap from GPS. The final map matching results, which are obtained by integrating GPS and vision data, are shown in Figure 9.17c. These final results prove that by using monocular visual odometry, the multisensor map matching algorithm can provide users with continuous location estimation, regardless of changes in quality of GPS data.

   In terms of time performance, vision-based map matching is a computationally intensive process, which requires high CPU and memory usage. For this, the experiments were conducted on the smartphone/server architecture. In this architecture, the server is responsible for major computations in vision-based map matching, including

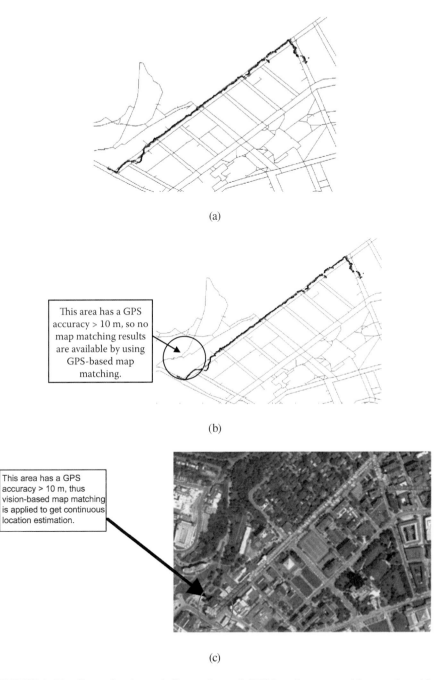

FIGURE 9.17 (See color insert.) Comparison of GPS-based map matching results with multisensor map matching results. (a) GPS-based map matching results, as compared with raw GPS data overlaid on the sidewalk map. (b) GPS-based map matching results in GPS accuracy ≤10m, compared with raw GPS data overlaid on the sidewalk map. (c) Multisensor map matching results overlaid on Google Maps.

SIFT feature extraction and feature matching and camera pose estimation. Clients (smartphones) are responsible for capturing videostreams and extracting frames from captured videostreams. Frame extraction from videostreams takes about 0.1 seconds on average. Since each image is 720 × 480 pixels and each pixel requires 8 bits of storage, data size of each image is 720 × 480 × 8 bits = 2,764,800 bits. To perform feature extraction and feature matching on the server, each image needs to be uploaded to the server, which takes about 1.53 seconds with an average of 1.8 Mbps data upload speed on the 3G networking. With this, the total time (computation time and communication time) is less than 1.7 seconds on the client side. On the server side, vision-based map matching, which involves SIFT feature extraction and feature matching, is the major computation. In the experiments, all the images extracted from videostreams have 720 × 480 pixels. In a Matlab running environment, our experiments showed that the average time of SIFT feature extraction and feature matching between two images is about 1 second. Besides this, the camera pose estimation process takes about 0.5 seconds. Therefore, the computation on the server side takes about 1.7 seconds. After adding the response time from the server to the client, the total time is 3.2 seconds. For this, we set up time intervals of 3.2 seconds to update the map matching results. The average speed of pedestrians is 1.51 m/s, so the distance moved in 3.2 seconds is below 5 m. For pedestrian and wheelchair navigation, 5 m location updates are reasonable. These time performances, on clients, over networks, and on servers, indicate that the proposed multisensor integrated map matching algorithm is suitable for pedestrian navigation applications.

9.5   SUMMARY AND CONCLUSION

To solve the issue of tracking pedestrians or wheelchair users in places that have poor or no GPS signals, this chapter presented a multisensor integrated map matching algorithm using monocular visual odometry. The algorithm provides uninterrupted localization where a monocular visual odometry technique is applied to recover the trajectory of a moving camera. Captured images rely on an accelerometer to recognize movement pattern of users to control frame extraction from videostream. Additionally, the accelerometer data help calculate the distance of the first step in the motion, which is used to estimate scale factor for camera motion estimation in the monocular visual odometry. The developed algorithm was experimented on an Android mobile phone and the results showed that it provides continuous and accurate solutions.

Conclusions can be drawn based on the results of the experiments. The visual odometry multisensor integrated map matching algorithm presented in this chapter can augment some of the drawbacks of GPS-based positioning, especially providing location estimations in places without GPS signals. The experimental results showed that the multisensor integrated map matching algorithm is both feasible and practical in providing uninterrupted location estimations in the outdoor environment. However, the positional accuracy to identify the correct sidewalk from parallel sidewalks on narrow streets still remains a challenge because the GPS accuracy is often not high enough during pedestrian and wheelchair navigation. Additionally, the user movement pattern recognition algorithm, which integrates accelerometer, compass, and GPS data, can greatly improve the efficiency of map matching on the smartphone/server architecture.

The study in this chapter can benefit other research areas such as automated wheelchair navigation and walking robots, where ensuring uninterrupted localization by map matching is one of the critical factors necessary to plan routes and achieve automatic location guidance.

REFERENCES

Castro, A. P. A., J. Demisi, S. D. Silva, and P. O. Sim. 2001. Image based autonomous navigation with fuzzy logic control, Proceedings of International Joint Conference on Neural Networks, vol. 3, pp. 2200–2205.

Chen, T., and R. Shibasaki. 1999. Development of a vision-based positioning system for high density urban areas, GISDEVELOPER.

Davide, S. 2008. Appearance-guided monocular omnidirectional visual odometry for outdoor ground vehicles, *IEEE Transactions on Robotics*, vol. 24, no. 5.

Derpanis, K. G. 2004. The Harris Corner Detector. Available:

Deselaers, T., D. Keysers, and N. Hermann. 2004. Features for image retrieval: An experimental comparison, DAGM-Symposium 2004, 228–236.

Esteban, I., Dijk, J., and F. Groen. 2010. Automatic 3D modeling of the urban landscape, International Congress on Ultra Telecommunications and Control Systems and Workshops.

Fritz, G., Seifert, C., and L. Paletta. 2006. A mobile vision system for urban detection with informative local descriptors, IEE International Conference on Computer Vision Systems.

Hagnelius, A. 2005. Visual odometry, master's thesis Umea University.

Hakeem, A., R. Vezzani, M. Shah, and M. R. Cucchiara. 2006. Estimating geospatial trajectory of a moving camera, International Conference on Pattern Recognition, vol. 2, pp. 82–87.

Hartley, R., and A. Zisserman. 2004. Multiple view geometry in computer vision. New York: Cambridge University Press.

Henlich, O. 1997. Vision-Based Positioning. Information Systems Engineering. Imperial College, Vol. 2. pp. 1–10.

Horn, B. 1990. Relative orientation, *International Journal of Computer Vision*, vol. 4, no. 1, pp. 59–78.

Kitching, D. Ian. 2000. GPS and cellular radio measurement integration, *Journal of Navigation*, vol. 53, no. 3, pp. 451–463.

Kitt, B., G. Andreas, and H. Lategahn. 2010. Visual odometry based on stereo image sequences with RANSAC-based outlier rejection scheme, IEEE Intelligent Vehicles Symposium.

Koller, D., G. Klinker, E. Rose, D. Breen, R. Whitaker, and M. Tuceryan. 1997. Real-time vision-based camera tracking for augmented reality applications, Proceedings of the Symposium on Virtual Reality Software and Technology.

LaMarca, A., and E. E. Lara. 2008. *Location systems: An introduction to the technology behind location awareness.* Morgan & Claypool.

Levin, A., and R. Szeliski. 2004. Visual odometry and map correlation. *CVPR 2004*, vol. 1, pp. 611–618.

Lowe, D. G. 1999. Object recognition from local scale-invariant features, International Conference on Computer Vision, Corfu, Greece, September.

Lowe, D. G. 2004. Distinctive image features from scale invariant features, *International Journal of Computer Vision*, vol. 60, o. 2, pp. 91–110.

Malis, E. 2002. Survey of vision-based robot control, European Naval Ship Design, Captain Computer IV Forum, ENSIETA, Brest, France, April.

MOBVIS. 2006. Software prototype and report on global and local informative features, information society technologies.

Ohnishia, N., and A. Imiya. 2006. Dominant plane detection from optical flow for robot navigation, *Pattern Recognition Letters*, vol. 27, no. 9, pp. 1009–1021.

Ohno, K., T. Tsubouchi, B. Shigematsu, and S. Yuta. 2004. Differential GPS and odometry-based outdoor navigation of a mobile robot, *Advanced Robotics*, vol. 18, no. 6, pp. 611–635.

Olson, C. F. 2001. Stereo ego-motion improvements for robust rover navigation. *ICRA'01*, vol. 2, pp. 1099–1104.

Retscher, G., and M. Thienelt. 2006. NAVIO: A navigation service for pedestrians, *Journal of Space Communication*, no. 9.

Richard, K. L., M. T. Pietrucha, and M. Nitzbur. 1996. Field studies of pedestrian walking speed and start-up time, *Journal of the Transportation Research Board*, vol. 1538, p. 27–38.

Rizos, C. 2005. Trends in geopositioning for LBS, navigation and mapping. Int. Symp. and Exhibition on Geoinformation, Penang, Malaysia, pp. 27–29.

Steinhoff, U., D. Omercevic, R. Perko, B. Schiele, and A. Leonardis. 2007. How computer vision can help in outdoor positioning, *LNCS* 4794, pp. 124–141.

Tardif, J., Y. Pavlidis, and K. Daniilidis. 2008. Monocular visual odometry in urban environments using an omnidirectional camera. IEEE IROS'08.

Wang, J., R. Cipolla, and H. Zha. 2004. Image-based localization and pose recovery using scale invariant features, *IEEE International Conference on Robotics and Biomimetics*, pp. 711–715.

10 Security and Privacy in Location-Based Services

Mohd Anwar, Amirreza Masoumzadeh, and James Joshi

CONTENTS

ABSTRACT

With the increased mobility in our modern life, location-based services (LBS) such as mobile commerce, mobile health, and location-based social networking have become essential for business, healthcare, entertainment, and even day-to-day living. At the same time, the security and privacy of LBS have become major concerns and roadblocks toward achieving the full potential of location-based applications. The location information carries the footprints of our activities, heralding our presence and absence in spatial and temporal contexts. As a result, location information needs to be protected from unauthorized parties while guaranteeing access at the appropriate granularity level by bona fide location-based service providers in order for users to receive desired location-based services. On the otherhand, resources and services of LBS providers need to be provisioned based on the location of the LBS user. In this chapter, we present the ecosystem of location information within an LBS including major entities involved, granularities, and types of location information. We then present security and privacy issues and threats that can emerge in LBSs and discuss various approaches that have been proposed to address these security and privacy concerns, including anonymization, location obfuscation, and access control and location authentication techniques. We conclude this chapter with a discussion about the research directions.

10.1   INTRODUCTION

Recently, the rapid proliferation of location-enabled wireless devices has made it possible to provide a plethora of location-based services (LBS). Supported by significant advances in mobile computing technologies, these location-based applications and their use are growing at an accelerated pace. By offering their location information, users can reap the benefits of LBS such as finding a nearest ATM machine or a restaurant, meeting friends present in a nearby location, or simply playing a location-based game. Many of us are familiar with location-based traffic report by car navigation systems such as TomTom's estimation on the delay in the current route. A 2012 report by the Pew Research Center reveals that 74% of U.S. smartphone owners use LBS, and 18% "check in" to locations.* At the same time, location information about users reveals or implies a great detail about *who they are*, *what they do*, or even with *whom they mingle*. Over time, a user's location information can help build a complete profile about him (Jin et al. 2012). As a result, LBS users can end up unknowingly or inadvertently sharing too much personal information by just sharing their location information.

Some of the location-enabled devices have become an integral part of our life. For instance, we carry our smartphones almost everywhere we go. As a result, our movement and activity can be monitored anywhere anytime (Bertino and Kirkpatrick 2009; Liu 2007). It can reveal privacy sensitive places we visit such as hospitals,

---

* http://www.factbrowser.com/facts/7727/

rallies, and courts. It can jeopardize privacy of other people such as friends and family colocated with us. Some socially sensitive relationships can even be inferred based on the trail of location information of two colocated users. For example, if Alice and Bob are co-located at the same residential address from 8 p.m. to 6 a.m. everyday then Bob is probably the spouse of Alice. Disclosure of location information not only has all the aforementioned privacy implications, it also creates security and safety concerns for the individuals. An increasing number of today's young adults carry smartphones and other location-enabled devices, and their location information may make them an easy target to child offenders. Location information not only reveals one's presence in a particular place, but also his/her absence. By collecting all the checking in tweets about members of *Foursquare*, which is a location-based social networking (LBSN) service, for instance, criminals can know that their target is not at home. This can lead to crimes such as home burglary when no one is at home. It is, thus, increasingly becoming very crucial that security and privacy issues in LBS be properly understood and appropriate solutions developed to ensure safe and secure use of LBS.

In this chapter, we present the pressing security and privacy concerns related to LBS and discuss existing approaches. We overview the general concepts related to information security and privacy, and the ecosystem of location information within an LBS including potential parties involved, granularities, and types of location information. We then present security and privacy issues and threats within an LBS environment. Then, we discuss various protection approaches that have been developed to address various security and privacy concerns, including anonymization, obfuscation, and access control. We conclude this chapter with a discussion on the research directions.

10.2   OVERVIEW: SECURITY AND PRIVACY, AND LOCATION-BASED SERVICES

In this section, we briefly overview information security and privacy concepts followed by the role of users' location information in the context of LBS.

10.2.1   INFORMATION SECURITY AND PRIVACY

10.2.1.1   Information Security

The key information security goals include *confidentiality* or *secrecy*, *integrity*, *availability*, and *accountability* (Joshi et al. 2001). The basic approaches used to achieve these goals include authentication, access control, and audit. *Authentication* refers to establishing the true identity of an entity before other activities are to be allowed within an information system; authentication is a prerequisite for governing access to information and resources. *Access control* is used to restrict the actions or operations that an authorized user or entity can perform. An *auditing* system collects data about system and user activities that can be later analyzed to identify access violations or to establish accountability.

Various access control approaches have been proposed in the literature, including discretionary access control (DAC), mandatory access control (MAC), role-based access control (RBAC), and attribute-based access control (ABAC) (Joshi et al. 2001). In a DAC approach, a user is allowed to grant the privileges he already has to other users; whereas in a MAC approach, a classification scheme for subjects and objects is used, and a set of rules on different classes is defined in order to grant accesses. To avoid the unauthorized flow of sensitive information, the MAC model, also referred to as the multilevel model, can enforce *no read-up* (no write-up) and *no write-down* (no read-down) rules at a given level to capture *confidentiality* (integrity) requirements (Joshi et al. 2001). In an RBAC approach, users are assigned to roles; permissions related to activities associated with a role are made available to a user who is assigned to that role. Similarly, in an ABAC approach, access control rules are defined on subject or object attributes, which need to be satisfied before access is granted. In LBS, various access control approaches may need to be applied based on the specific requirements of an application domain. For instance, an LBS associated with military government (e.g., in mission critical mobile application) would need some form of multilevel security to provide information flow security based on location of the subjects and objects (Ray and Kumar 2006).

10.2.1.2   Information Privacy

With the growth of pervasive computing technologies, more information about people is now stored and shared in digital forms, sometimes without their knowledge. Hence, privacy issues are becoming a significant technological as well as social and legal challenge. Privacy may be defined as "the right of individuals, groups, or institutions to determine for themselves when, how, and to what extent information about them is communicated" (Joshi, Joshi, and Chandran 2006). Thus, privacy is a person-centric issue and deals with the control that an individual has on the use and dissemination of his personal information. A key issue of the privacy problem is the anonymity of the users that aims at protecting a user's identity. The identity of a person, in general, is a collection of various pieces of personal data that uniquely identifies him. In some context, a subset of the identity-related information may be used to represent a person. Such a "partial identity" is typically also attached to a person with a pseudonym (Berthold and Köhntopp 2001), and may or may not uniquely identify the individual. The notion of privacy and identity is often inherently complex and may even be contradictory; furthermore, each application domain may have a different perspective on them. Central to the privacy issue in LBS is the location information associated with the user; inadequate protection of a user's location information can help identify a user's identity (Poolsappasit and Ray 2009).

Various security technologies that are indispensable in general for any information systems, such as LBS, include firewalls, intrusion detection systems, encryption techniques, and public key infrastructure (PKI) technologies. As LBS involve location information and in general mobility, these technological solutions raise additional challenges. For instance, in a mobile LBS environment, encryption techniques should factor in the power constraints of mobile devices. Issues related to general mobile application environments are beyond the scope of this chapter.

10.2.2   Ecosystem of Location Information in **LBS**

To understand the security and privacy issues in LBS, it is important to understand the types of location information that flow across entities in the process of enabling LBS. Three main entities primarily involved in an LBS are an LBS user, an LBS provider, and a location provider. The basic mechanism of LBS is that an LBS user provides his location information to an LBS provider on which the latter offers contextually useful services. Route guidance, emergency assistance, location-based advertising, LBSN, and location-based games are a few examples of LBS.

An LBS can be viewed as a query, $Q(L_u, S)$, that an LBS user u submits to an LBS provider P, where L_u is the location of u and S is the service for which u requests. In response, P provides the service S. For instance u's query may be as follows:

$$(L_u = <x_u, y_u>, S = \text{where is the nearest restaurant?})$$

P's response may be $R = (<x_1, y_1>, <x_2, y_2>)$, where $<x_1, y_1>$ and $<x_2, y_2>$ refer to the coordinates or the addresses of two restaurants closest to L_u.

An LBS user may provide his current location or have his mobile device (e.g., smartphone) automatically transmit location information to an LBS provider. Sometimes, computing a user's location may involve revealing his location information to third parties, namely *location providers*. For example, the cell phone provider may be a location provider, or a geoenabled browser like Google Chrome may play the role of a location provider. There are different positioning technologies, such as GPS, WiFi, and cellular radio, that can be used to determine a user's location (e.g., *longitude* and *latitude*) in location-enabled mobile devices. A *location provider* may also translate coarse location information to a geolocation or an exact physical address.

Depending on the application domain of an LBS, location information can have different meanings and values. Location information can be of different granularity, ranging from very precise (e.g., coordinates of a user) to very coarse-grained (e.g., spatial region where the user is located) information. Location information may also provide logical context, that is, symbolic names associated with physical locations, such as street address, floor in a building, and a football stadium. A user may also refer to a location in a virtual term such as home, work, and gym. Location information can also be deduced from a user's activities within an application. For example, a user's check-in information in an LBSN reveals his location information to others at a particular time.

10.3   SECURITY AND PRIVACY ISSUES IN LBS

10.3.1   Protection Requirements in **LBS**

Protection requirements of LBSs include those of a general mobile information system and those introduced by the special role of location information in enabling and delivery of the LBS. Key protection requirements in LBS include the following:

- *Disclosure of precise location information* where the key concern is protecting a person's location information. LBS users may be concerned about privacy when they are known to an LBS provider as it may collect their precise location information. Misuse of such information by an adversary may lead to physical threats (e.g., stalking, burglary) or may simply violate the privacy expectations of a user (e.g., not being tracked). Various location *obfuscation* techniques have been proposed to control such location information disclosure issues.

*   *Disclosure of association of users and location information* where the key concern is protecting the association of an identity with its location information. Location information can be very revealing about the identity of users, even if the identities are decoupled from the location information (i.e., acting as quasi-identifier). For instance, when a user usually uses a service in a private place such as his house, an easy lookup using other publicly available information, such as property owners' directory or phonebook, may reveal the user's identity, and put the privacy of his location and other service-related attributes (such as contents of the queries) at jeopardy. Various *anonymization* and *pseudonymization* techniques are typically used to protect inference of such associations.
*   *Location-based access control* where the key concern is ensuring authorized access to resources based on location information associated with users and resources. Several access control models address the need to protect access to sensitive information or to enable controlled dissemination of sensitive information, including location information. For example, in an LBSN, a user may want to set an access policy to let only a specific subset of users to know about his check-in location.
*   *Location authentication* where the key concern is ensuring that location information is authentic in the first place. Faked location information may enable delivery of a sensitive service to a wrong person for wrong purpose.

Duckham and Kulik (2006) define location privacy as a special type of information privacy which is concerned with the claim of individuals to determine for themselves when, how, and to what extent location information about them is communicated to others. Thus, achieving location privacy can be considered as the ability to prevent unauthorized parties from learning one's current or past location (Beresford and Stajano 2003). Whereas the current location helps an adversary to physically locate a person, past locations can help the adversary to build a profile of that person. From an attack perspective, the former attack is known as *snapshot attack* and the latter attack is known as *historical attack*. If an adversary cannot associate an identity with location information, disclosed sensitive information may not breach anyone's privacy.

As discussed earlier, within an LBS, location information is shared with the LBS provider and location provider. Both the LBS provider and location provider may be untrustworthy and the user may endanger her security and privacy by sharing her current location with others. An attacker may also impersonate an LBS provider and make an LBS user believe that the attacker is a genuine LBS provider in order to track the LBS user's location. An LBS user may feel discomfort in releasing information of having used a particular service. In this case, the user's identity should not be revealed. Information on a user's current location can lead to computer-aided crimes like harassment and theft. Since an LBS is rendered through a two-way communication between an LBS user and an LBS provider, an adversary may hijack the location transmission channel, and therefore, the protected communication channel is an important aspect of LBS security. On the other hand, an adversary may build a profile of a user based on correlation of LBS requests made at different times in the past by that user. For example, by tracking LBS queries for the closest restaurant, an attacker can know about the food habits of users and places they travel. LBS providers or location providers may observe the behaviors of LBS users and sell that information of consumer habits to third parties and spammers. As a result, an LBS user may become a victim of discrimination, denial of service, or unsolicited ad.

An LBS user can be identified by means of the locations contained in the queries he sends to service providers. Location information can be used to identify anonymous users. *Restricted space identification* (RSI) and *observation identification* (OI) are two of the location-based identity disclosure attacks (Gruteser and Grunwald 2003a). An RSI attack can be described as follows: if attacker A knows that location L exclusively belongs to a user u, then from the location information of LBS service request Q, A knows that u has sent the request. For example, consider the owner of a suburban house sending an LBS request from his garage or driveway. The coordinates can be correlated with a database of geocoded postal addresses, such as those provided by geocoding, to identify the residence. An address lookup in phone or property listings then can easily reveal the owner and the likely originator of the message. An OI attack can be defined in the following way: by observing the current location L of user u and service request Q from L, an attacker A learns that u has sent the request.

Even though a key goal of many security and privacy approaches is to protect users' location information against unwanted or accidental disclosure, the security and privacy requirements may be more subtle than that. Smith, Consolvo, et al. (2005) and Consolvo et al. (2005) conclude that people reveal their location based on who is requesting, why they want to know their location (i.e., purpose), when and where they are, and how they feel about the requester at the time of request. In particular, a user may want to disclose his location information, but he does not want to be monitored or profiled for commercial purposes. A user may not want his location information to be used for certain purposes, for example, law enforcement or dispute resolution. As a result, sometimes it may not be clear what to protect against. Sometimes, a user may want to disclose his past location information but not the present.

10.3.2   CLASSIFICATION OF LBS BASED ON PROTECTION NEEDS

Liu (2007) classifies LBS in to the following three types based on the perceived need of privacy in LBS: anonymous LBS, identity-driven LBS, and pseudonym-driven LBS. In an *anonymous LBS* a user's identity information is not required for service delivery. For instance, alerts related to a query: "alert me when I am near an Italian restaurant" represents a service that only needs a user's location and not his identity. In an *identity-driven LBS*, the services cannot typically be provided without proper user's identity information, in addition to his location information. For instance, consider the query: "turn on the projector when Alice enters his classroom and allow all the students in the classroom to download the slides" where a service needs to validate the identities of Alice and the students, as well as their corresponding location information. Some LBSs may need identity information of an LBS user but not his exact identity. Examples include navigation and road tolling services. Such LBSs are considered *pseudonym-driven*.

10.4 SECURITY AND PRIVACY APPROACHES FOR LBS

In this section, we present an overview of the existing approaches that address the various issues mentioned in Section 10.3. We first present various location privacy solutions based on obfuscation and anonymization techniques, including the anonymization of emerging LBSNs. The goal of various anonymization and pseudonymization approaches is essentially to make it harder for the LBS provider to know the actual identity of an LBS user to whom the location information in a location query belongs. The idea behind anonymization is to create ambiguity about a user's identity by grouping that user with other users using spatial or temporal cloaking. We then discuss various location access control approaches that aim at providing authorization rules centered on location information related to users and resources.

10.4.1 Location Obfuscation Techniques

Obfuscation of location information helps in protecting users' location information. Obfuscation of location essentially degrades the quality of location information without altering a user's identity. Spatial obfuscation can be achieved by means of three different kinds of uncertainties: inaccuracy, imprecision, and vagueness (Duckham and Kulik 2006). *Inaccuracy* refers to a lack of truth in the information; *imprecision* refers to the lack of specificity in the information; and *vagueness* is lack of determinacy in information. Obfuscation is formalized by Duckham and Kulik (2005, 2006) based on a graph $G = (V, E)$, where the vertices V are the locations and edges E represent adjacency between locations. The position of a user is represented as a vertex $l \in V$, and an obfuscation of his position is a set $O \subseteq V$, with $l \in O$. They also propose a negotiation algorithm for more accurate answers to proximity queries. The negotiation allows the LBS to ask for more precise location from a user in case he needs a more accurate result. In the case the user prefers not to release more precise location, the algorithm decides to return one of the potential results based on some criteria, for example, closest to one of the points in the obfuscated set, along with a *confidence value*. Such a protocol is advantageous where the LBS is not willing to disclose all the potential results.

To provide a concrete example of an obfuscation technique, we briefly discuss a technique based on a region graph proposed by Jafarian et al. (2009). A region represents a spatial area characterized by a symbolic name, such as "22nd Street," as illustrated in Figure 10.1. The set of regions creates a directed acyclic graph (DAG) in which nodes represent regions and edges represent containment relationships among them. To obfuscate a user's location information, his measured location is mapped to a region and the region is obfuscated based on the privacy preferences specified by the user for that region. When a region, A, is specified to be obfuscated

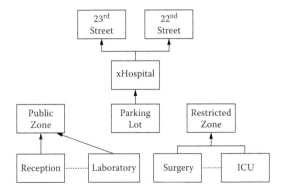

FIGURE 10.1 Examples of regions in a region graph. (From Jafarian, Jafar Haadi, Ali Noorollahi Ravari, Morteza Amini, and Rasool Jalili. "Protecting location privacy through a graph-based location representation and a robust obfuscation technique," In *Information Security and Cryptology*, 116–133, Berlin: Springer-Verlag, 2009.With permission.)

to a region type, *C*, represented as *obf*(*A*) = *C*, by a user, *u*, it means that if *u* is in *A* and her current location is queried, the nearest ancestor of *A* (e.g., Surgery) with type *C* (e.g., itself or restricted zone) is returned. If no such ancestor is found, *A* will be returned. If two qualified ancestors with the same distance to *A*, for example, *xHospital*'s two ancestors are *22ⁿᵈ Street* and *23ʳᵈ Street*, are found, then one of them will be randomly selected as the result.

Ardagna et al. (2007) formalize the concept of location obfuscation for regions of circular shape, allowing users to define a privacy requirement of minimum radius, and the services define a radius as *accuracy* requirement. They consider three different obfuscation operators: *enlarging the radius*, *shifting the center*, and *reducing the radius*. These operators can be employed individually or in combination to provide higher level of protection. Ardagna et al. also analyze the robustness to deobfuscation attacks and experimentally show that their approach provides better protection than a simple enlargement strategy that is most commonly used.

The *PROBE* framework proposed by Damiani, Bertino, and Silvestri (2010) provides *personalized obfuscation* by considering sensitivity and statistical distribution of locations. It models semantic locations in terms of different types of location features (e.g., hospital, residential building). In a privacy profile, a user defines sensitive feature types and an associated sensitivity threshold for each that can be tolerated for any reported location. PROBE generates an obfuscated map for each user in which obfuscated regions satisfy the user's privacy profile, that is, the sensitivity of a region with respect to different feature types is limited by the predefined threshold value in the users' profile for the respective types. The obfuscated map can be uploaded to the mobile device and used offline, that is, there is no need to rely on a trusted third party. The PROBE approach incorporates three different greedy algorithms for actual obfuscation, namely:

1. The *region-based* method progressively expands the regions of sensitive features by a column or row until the required sensitivity threshold is achieved for each feature type.

2. The *division-based* method subdivides each feature in four quadrants and follows the region-based method to obfuscate each quadrant.
3. The *Hilbert-based* method obfuscates by adding one cell at a time following a Hilbert space-filling curve.

Based on the experimental results of Damiani, Bertino, and Silvestri (2010), the accuracy of each method has been measured based on the average size of obfuscated area. Among the three methods, the region-based method has the worst accuracy. The Hilbert-based method assures much better accuracy than the other two in the case of strong privacy constraint (low sensitivity threshold).

10.4.2 PSEUDONYMIZATION TECHNIQUES

Pseudonymity provides a form of traceable anonymity so that the association of location information and user's actual identity can only be established under specified and controlled circumstances. The pseudonym, in lieu of the user's actual identity, allows communication between him and the LBS provider. However, a nontrusted LBS provider with some background information such as an LBS user's home and workplace can compare these locations with the pseudonymized data in order to identify the LBS user. In other words, pseudonymity cannot withstand location-based identification.

Beresford and Stajano (2003) propose the *mix-zone* model of privacy protection that aims to avoid a user being identified by the locations he visits. The model assumes a trusted middleware system, positioned between the underlying location system(s) and untrusted third-party applications. Applications register interest in a geographic space (e.g., hospital, coffee shop, university buildings), namely, *application zone*, with the middleware. Users register interest in a particular set of LBS applications and the middleware limits the location information received by applications inside the application zone. Each user has one or more unregistered geographical regions, called *mix zones*, where no application can trace user movements. Once a user enters a mix zone, the user's identity is mixed with all other users in the mix zone. The pseudonym of any given user changes whenever the user enters a mix zone, as shown in Figure 10.2. Formally, let us consider mix zone Z with a set, A, of users. For every user, u, getting out of mix zone, it is equally probable for u to be a particular user, $j \in A$, having the probability of $1/|A|$. The aim of the mix zone model is to prevent tracking of long-term user movements.

10.4.3 ANONYMIZATION TECHNIQUES

To address the issue of anonymity within the context of LBS, researchers have adopted the anonymization principles from the database community that have focused on tackling the crucial need to anonymize data sets before releasing them. The k-anonymity principle, which is widely used for such a purpose, requires that each record is indistinguishable from at least $(k - 1)$ other records to protect the data set from *linking* attacks. Subsequently the proposed anonymity principles provide

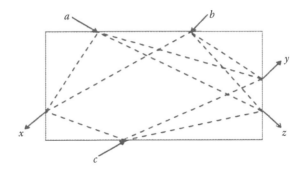

FIGURE 10.2 Example of motion of three people (a, b, c) through a mix zone. (From Beresford, Alastair R., and Frank Stajano, "Location privacy in pervasive computing," *Pervasive Computing (IEEE)* 2, no. 1 (2003): 46–55. With permission.)

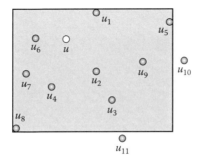

FIGURE 10.3 An example of an anonymizing spatial region (ASR) computed based on u's anonymity requirement of $k = 10$. (From Tan, Kar Way, Yimin Lin, and Kyriakos Mouratidis, "Spatial cloaking revisited: Distinguishing information leakage from anonymity." *11th International Symposium on Advances in Spatial and Temporal Databases,* Springer-Verlag, 2009, 117–134. With permission.)

more advanced protection; for instance, l-diversity ensures the diversity of sensitive values in equivalency classes.

In spatial anonymization techniques, a user's exact location is cloaked into a spatial region such that the cloaked region satisfies the (user-specified) anonymity requirements, typically k-anonymity. This cloaked spatial region, called *anonymizing spatial region* (ASR), is the minimum bounding rectangle area that contains at least k users including the user, u, that submits a query. Figure 10.3 shows a cloaking example where user u, shown as a white circle, requests for k-anonymity with $k = 10$. The figure shows the computed *ASR*, assuming that the anonymizing set of u additionally contains users $u_1, ..., u_9$.

10.4.3.1 Trusted Third-Party-Based Anonymization Approaches

Several anonymization mechanisms for queries involve using a trusted third party (TTP), the *anonymization server*, as a proxy to ensure k-anonymity. Here, a user submits his query along with his precise location to the anonymizer, which in turn sends the query with an ASR to the server, and redirects the response back to the user. The potential noise introduced to the response due to the spatial cloaking needs to be vetted either by the anonymizer or the user client itself. Next, we briefly overview the key solutions based on such an approach.

An approach in this category is the *adaptive interval cloaking* (AIC) algorithm proposed in Gruteser and Grunwald (2003b). The AIC algorithm recursively subdivides the area around a requester subject into four quadrants, until the number of subjects in the area that includes the requester fall below threshold k. The previous quadrant, which still meets the threshold, is selected as the cloaked area. The authors also propose a *temporal cloaking* approach, orthogonal to *location cloaking*, which delays a request until k subjects visit an area, and then reports that area with a random cloaking interval that includes the request time. *New Casper*, proposed by Mokbel, Chow, and Aref (2006), allows users to individually define their preferences of k and minimum cloaking area A_{min}. The anonymization scheme is based on a pyramid data structure comprising of a grid of location cells with ascending resolution toward higher levels, that is, each grid cell at a level corresponds to four cells in the next level. The users are registered in the highest level cells. The cloaking algorithm traverses the levels from the highest level all the way to a cell that contains at least k number of users and has an area larger than A_{min}.

PrivacyGrid, proposed in Bamba et al. (2008), is another grid-based location anonymization framework that supports user-specified parameters for location k-anonymity, l-diversity, and maximum spatial/temporal cloaking. Although it is similar to New Casper, it has a better cloaking performance since it allows flexible cloaking strategies (i.e., bottom-up, top-down, and hybrid), and dynamic expansion of grid cells (instead of static quad-tree scheme in New Casper). *CliqueCloak*, proposed by Gedik and Liu (2005, 2008), groups at least k queries together and submits them all together to the LBS using the same cloaked location. However, as the algorithm relies on finding a clique on a constraint graph, it faces severe performance issues. Moreover, it needs to delay the queries in order to submit them in groups and does not minimize the size of the cloaked areas.

10.4.3.2 TTP-Free Collaborative Approaches

Instead of employing the TTP-based architecture, a collaborative approach can be taken to generate ASRs. In the mechanism proposed by Solanas and Martínez-Ballesté (2008), a user finds $k–1$ other companions in a certain distance, and together they compute a *centroid* value of their bounding area and send it as their location to an LBS. The technique supports masking individual user locations using Gaussian noise, avoiding location disclosure for a nonmoving companion by employing probabilistic homomorphic cryptography, and a chain-based protocol to avoid the threat of collusion between a malicious user and the LBS. The *Prive* approach proposed by Ghinita, Kalnis, and Skiadopoulus (2007) uses distributed anonymization architecture with an anonymization algorithm based on the Hilbert filling curve. The authors define a reciprocity requirement for cloaked locations, based on if a user happens to be in a cloaked location of another user's query; the latter user would also be located in queries issued by the former user. In a complementary work, Kalnis et al. (2007) propose an alternative to the *Hilbert cloak* approach, called the *nearest neighbor*

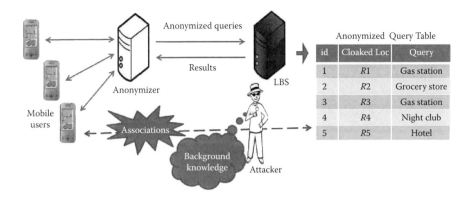

FIGURE 10.4 An example of a TTP-based anonymization approach for LBS. Background knowledge may need to be assumed to thwart attacks aimed at location privacy. (From Masoumzadeh, Amirreza, and James Joshi, "An alternative approach to k-anonymity for location-based services," *The 8th International Conference on Mobile Web Information Systems (MobiWIS)*, 2011a, 522–530. With permission.)

cloak, which is not vulnerable to an adversary's heuristic that the user that is closer to the center of a submitted cloaked region (ASR) is most probably the query issuer.

Takabi et al. (2009) propose a TTP-free approach where users are registered with different location providers, and location providers collaborate using a cryptographic protocol based on homomorphic cryptography to ensure that there exists enough number of users in a user's query region to satisfy *k*-anonymity. If the location is provided by users' cellular networks, the protocol can be implemented very inexpensively in terms of architectural components. The main strength of the proposed protocol is that neither users nor location providers need to fully trust each other. They consider different threats and attack scenarios including malicious users, location providers, and directory server, and propose countermeasures to prevent or detect misbehavior.

Another approach is not sending an exact location to the LBS provider at the first place. Based on the theory of private information retrieval, Ghinita et al. (2008) propose an approach where a user sends the generalized query to the LBS provider; the LBS replies with a set of results from which the requester retrieves the actual result. Since the LBS provider is unaware of the exact location of the user, location privacy is protected.

Shankar et al. (2009) propose a decentralized anonymization approach called *SybilQuery*. The key idea in SybilQuery is that for a given query, the client generates *k*–1 sybil location queries that are statistically similar. It may be easy to identify the real location from *k* such queries if an attacker can estimate the path or the other *k*–1 location looks unrealistic. Kido et al. (2005) propose a similar scheme where dummy locations are sent along with the true location. The LBS provider processes all requests and sends all answers back to the requester.

10.4.3.3 Anonymization of Continuous Queries

Some LBS support location-based continuous queries where the user needs to continuously provide his location to have up-to-date results. However, anonymization in

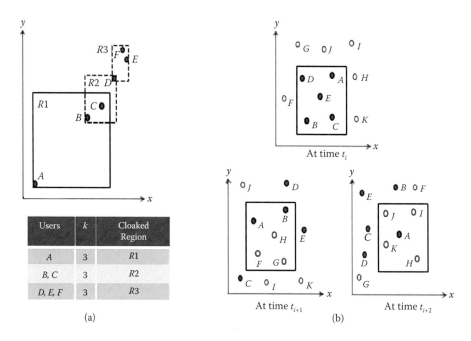

FIGURE 10.5 Privacy attacks on users with (a) publicly known location and (b) continuous queries. (From Chow, Chi-Yin, and Mohamed F. Mokbel, "Enabling private continuous queries for revealed user locations," *Advances in Spatial and Temporal Databases.* Springer, 2007, 258–275. With permission.)

the case of continuous queries, which is also referred to as trajectory anonymization, is more complex than that of snapshot queries. Chow and Mokbel (2007) show that keeping track of ASRs that a user provides for such continuous queries can help an adversary to violate k-anonymity requirement. This is because the set of common users in a series of ASRs for a continuous query can potentially decrease by submitting each new ASR.

Chow and Mokbel (2007) also show that for the users whose location is publicly available (e.g., due to the nature of the service), most of the state-of-the-art LBS anonymization approaches fail to provide privacy for user's query. They propose an algorithm that is safe in providing k-anonymity in both cases of continuous queries and publicly known locations. Figure 10.5a shows that, for user A with a publicly known location, since he is the only user that would use ASR R1, his query can be identified by an adversary that has access to the anonymized queries. Figure 10.5b shows the sequence of ASRs submitted by user A in three consecutive time points for a continuous query, and how the attacker can narrow to the target user by considering only common users in the sequence. A network-aware trajectory anonymity model is proposed by Gkoulalas-Divanis, Verykios, and Bozanis (2009). The framework provides two types of k-anonymity for a user trajectory, which are chosen depending on the current location of the user as well as her subsequent locations until the completion of the service provisioning. In k-present trajectory anonymity (the weaker property), k–1 other subjects close to the user that could have issued the request are selected. In k-frequent trajectory anonymity (the stronger property), an added requirement is that the current route of the user should be frequent for the k–1 other subjects. The framework builds a network topology of user movements and mines patterns of user trajectories based on user movement history.

10.4.3.4 An Alternative Approach to LBS k-Anonymity

Masoumzadeh and Joshi (2011a) show that the widely adopted interpretation of k-anonymity for LBS in the existing solutions, such as in Bamba et al. (2008), Ghinita, Kalnis, and Skiadopoulus (2007), Kalnis et al. (2007), and Mokbel, Chow, and Aref (2006), is not consistent with the original definitions of the k-anonymity principle (Samarati 2001; Sweeney 2002), which in turn results in not delivering the expected anonymity to the LBS users, as explained next.

Let the relations *AQ(location, query)* and *UL(user, location)* represent, respectively, the submitted anonymized queries to the LBS and the exact locations of the LBS users. As the LBS provider is not considered trusted, the relation *AQ* is considered known to the adversary. The predominant interpretation of k-anonymity for LBS is as follows.

Definition 10.1: LBS k-anonymity

Relation *AQ* is LBS k-anonymous iff for every query in *AQ* there exist at least k users in *UL* whose locations match the query's location. ∎

However, as the exact population of individuals that are represented in an external data set (attacker's background knowledge as shown in Figure 10.4) is not known to the data anonymizer, a safer approach has been followed to assure k-anonymity (Samarati 2001). Assuming that an individual is only associated with one tuple in a privacy-sensitive relation that needs to be anonymized, the following definition ensures that, for each tuple in a k-anonymous relation, there are at least k individuals that would be matched based on the quasi-identifier (Samarati 2001; Sweeney 2002). Note the following as the original definition of k-anonymity.

Definition 10.2: k-anonymity

Let P be a relation and QI be the quasi-identifier associated with it. P is said to satisfy k-anonymity iff each sequence of values in $P[QI]$ occurs at least k times in $P[QI]$. ∎

By comparing LBS k-anonymity (Definition 10.1) to the above definition, Masoumzadeh and Joshi (2011a) show that *AQ* is the privacy-sensitive relation with the quasi-identifier location, which can be linked to the location in *UL*. It can be

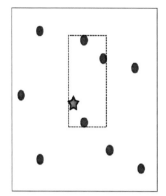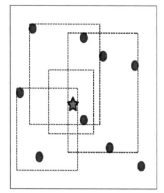

FIGURE 10.6   (a) Location cloaking in LBS k-anonymity and (b) LBS (k,T)-anonymity. (From Masoumzadeh, Amirreza, and James Joshi, "An alternative approach to k-anonymity for location-based services," *The 8th International Conference on Mobile Web Information Systems (MobiWIS)*, 2011, 522–530. With permission.)

observed that LBS k-anonymity captures the k-anonymity requirement by matching at least k user locations in *UL* for every query's location in *AQ*. However, it fails to follow the safeguard implied in Definition 10.2. Note that Definition 10.2 requires at least k occurrences of each sequence of quasi-identifier in order to rule out any assumptions regarding the population in the linkable external information. In the context of LBS, this means that there should be at least k queries with the same cloaked location for every existing location in the *AQ* relation. Definition 10.1 clearly does not ensure this property. Masoumzadeh and Joshi (2011a) propose an alternative formulation of k-anonymity for LBS. Although LBS k-anonymity ensures k users for each tuple in *AQ*, the proposed approach ensures k queries in *AQ* for each issuing user. An example is shown in Figure 10.6, where the star-shaped point is the location of a user that issues a query and $k = 4$. LBS k-anonymity ensures that there are four users in the user's cloaked location (Figure 10.6a), while the new approach requires four queries to cover the user's location (Figure 10.6b). By considering query issuance time in a time period as the background knowledge, a relaxed version of such an approach is defined as follows (Masoumzadeh and Joshi 2011a):

Definition 10.3: LBS (k,T)-anonymity

Relation *AQ* is LBS (k,T)-anonymous iff, for any submitted query q at time t issued by user u, there exist at least k–1 other queries in any time window of size at least T that includes t.   ∎

Based on Definition 10.3, Masoumzadeh and Joshi (2011a) formulate a simplified version of the LBS (k,T)-anonymization problem and propose a greedy algorithm to solve it.

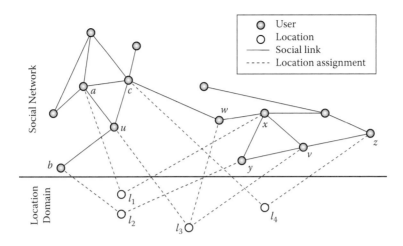

FIGURE 10.7 A model for geo-social networks. (From Masoumzadeh, Amirreza, and James Joshi, "Anonymizing geo-social network datasets," *4th ACM SIGSPATIAL International Workshop on Security and Privacy in GIS and LBS*, 2011, 25–32. With permission.)

10.4.3.5 Anonymization of LBSNs

The location rich social network data collected in LBSNs, which are also known as geosocial networks (GSNs), such as Foursquare, could be of research interests for various purposes. However, such data sets are at the risk of user reidentification and consequently privacy violation of the involved users if they are not adequately anonymized before sharing. Masoumzadeh and Joshi (2011b) study anonymization of geosocial network data sets based on the adversarial knowledge on location information of their users. To the best of our knowledge, this is the first anonymization approach proposed in the literature that is specific to the GSNs. The approach is based on the following definition of the GSN, which is also illustrated in Figure 10.7.

Definition 10.4: GSN

A geosocial network is a 4-tuple $GSN = \langle V, E, L, l \rangle$, where V is a set of users, $E \subseteq V \times V$ is a set of relationships between users, L is a domain for location information, and $l: V \rightarrow L$ is a function that assigns location information to users. ∎

The location domain is abstract in the definition of GSN. It may refer to geographic coordinates, street addresses, logical locations, or even more complex types, such as a number of top locations associated with a user. Figure 10.7 illustrates a small GSN data set. Masoumzadeh and Joshi (2011b) define two levels of location equivalency. If two users have the same location information they are called l-equivalent. Two users are called l^2-equivalent if, in addition to themselves, their adjacent users in the social network are also l-equivalent. Based on these definitions, a GSN is called l_k-anonymized (l_k^2-anonymized) if there are at least $k–1$ other users who are l-equivalent (l^2-equivalent) to each user.

As a suitable location model for GSNs such as Foursquare, Masoumzadeh and Joshi (2011b) introduce TL_m, a model that captures top m locations of users, which can act as a signature for reidentifying them. This location model is motivated by previous studies that show mobile users are uniquely reidentifiable based on their top two or three locations. In TL_m, each location value is a rectangular region, and the distance between two values TL_m is defined as follows. A clustering-based l_k-anonymization algorithm based on TL_m is proposed by Masoumzadeh and Joshi (2011b). The distance metric for clustering is defined as follows.

Definition 10.5

The distance between TL_m values t and s is calculated by the following formula:

$$D(t,s) = \min_{C \in P(\langle 1,\ldots,m \rangle)} \sum_{\langle i,j \rangle \in \langle 1,\ldots,m \rangle \times C} MBRA(t.r_i, s.r_j)$$

where MBRA calculates the area of minimum bounding rectangle of two regions. The proposed algorithms in Masoumzadeh and Joshi (2011b) achieve l_k-anonymization by a clustering strategy based on TL_m values and perform edge perturbation in the social network to achieve l_k^2-anonymization. ∎

10.4.3.6 Location l-diversity and Road Segment s-diversity

It is possible that a cloaked location that supports k-anonymity may have just one geographical address, such as a health/religious center, revealing sensitive location that all the k users are associated with a single location of interest (e.g., health/religious center). Bamba et al. (2008) propose location l-diversity to ensure the diversity of location values in such equivalency classes. A location, L, is said to have l-diversity if there are at least $l > 1$ different geographical addresses that are associated with L. Similarly, when mobility is associated with roads and walking paths, it is possible that a cloaked region providing k-anonymity is associated with a specific road. This allows inferring sensitive information that all the k entities are in the same road segment. Wang and Liu (2009) propose road segment s-diversity that focuses on ensuring that each cloaked region includes at least $s > 1$ different road segments.

In all the anonymity approaches such as k-anonymity, l-diversity, and road segment s-diversity, a key issue is determining the appropriate values of k, l, and s.

10.4.4 LOCATION-BASED ACCESS CONTROL (LBAC) APPROACHES

In some services, such as location-based friend finding or LBSNs, it is critical to determine who should have access to what location information and under what conditions. Additionally, based on the location of a user, his access to certain resources or information, including the location information of other users or resources, may need to be controlled. This process is commonly referred to as access control. Several researchers have proposed access control models where access is conditioned on the location of the requesting users or the objects they want to access (Bertino et al. 2005; Chandran and Joshi 2005). In general, the key location-based conditions that are useful for the LBAC accesses include (Ardagna et al. 2006): (a) position-based conditions on location of users and resources; (b) movement-based conditions on the mobility of entities; and (c) location specific interaction- or relation-based conditions; for instance, conditions on relative positioning/distance of entities, colocation of entities, or the density of users in a given area. Next, we overview the key research efforts focused on LBAC approaches.

10.4.4.1 GEO-RBAC Model

The GEO-RBAC model, proposed by Bertino et al. (2005), supports location constraints to address location-based access control needs. GEO-RBAC is based on the notion of spatial role that is a geographically bounded organizational function. The boundary of a role is defined as a geographical feature, such as a road, a city, or a hospital. This spatial boundary specifies the spatial extent in which the role is valid. It is important to note that a spatial role is enabled when the location condition associated with it is fulfilled; but the user has to actually activate the enabled roles that he has been assigned to within a session to acquire the permissions associated with these activated roles. The location constraints and the authorization are integrated through such enabled roles. For example, *<ChildrenPass,LowesTheater>* is a role spatial constraint. The role *ChildrenPass* will be enabled if the user is in the location *LowesTheater*. Toward this end, the central idea of the GEO-RBAC model is the role schema and role instance, as follows (Bertino et al. 2005):

- *Role schema*—A role schema is a tuple *<r, ext, loc, mloc>*, where *r* is the name of the role, *ext* is the feature type of the role extent, *loc* is the feature type of the logical positions, and *mloc* is the mapping function that maps a real position into a logical position of type *loc*.
- *Role instance*—The role instances are generated from role schema. Given a role schema, r_s, an instance r_i of r_s is a pair *<r, e>*, where $r = r_s.r$ and $e \in F$, such that $FT\_Type(e) = r_s.ext$, where *F* represents features and *FT_Type* represents feature types.

Figure 10.8 summarizes the GEO-RBAC model. Here, R_i and R_s represent the sets of role instances and role schemas, respectively; RPOS is the set of real positions; SPAS and SPAI are *permission-to-spatial role schema assignment* and *permission-to-spatial role instance assignment* relations, respectively. Sets *U, SES, PRMS, OPS,* and *OBJ* are the sets representing *users, sessions, permissions, operations,* and *objects*, respectively.

10.4.4.2 Location- and Time-Based RBAC (LoT-RBAC)

Chandran and Joshi (2005) propose the *location- and time-based RBAC* (LoT-RBAC) model to specify access control based on location and time. LoT-RBAC consists of the location context model and allows physical, logical, and relative location as well

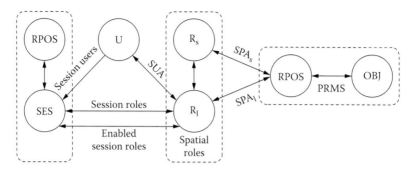

FIGURE 10.8 Core GEO-RBAC. (From Bertino, Elisa, Barbara Catania, Maria Luisa Damiani, and Paolo Perlasca, "GEORBAC: A spatially aware RBAC," *Proceedings of the 10th ACM Symposium on Access Control Models and Technologies,* New York, NY: ACM, 2005, 29–37. With permission.)

as motion (e.g., a mobile user requesting mobile resources as in a location based friend search) to be incorporated in the access control specification. It allows defining location hierarchy among physical and logical locations to define role hierarchies; it allows both role-activation (*A*-hierarchy) and permission inheritance (*I*-hierarchy) hierarchies based on the hybrid hierarchy types proposed in Joshi, Bertino, et al. (2005). LoT-RBAC allows specifying location context of users, roles, and permissions (i.e., location associated objects). The key features provided by the LoT-RBAC model include the following:

- *Location and temporal constraints on role enabling and disabling*—These allow specification of the location where a role is enabled as well as the intervals and durations in which a role is enabled.
- *Location and temporal constraints on user-role and role-permission assignments*—These allow specifying the location of the users and the location of the permissions (or the associated object) as the precondition for the assignment of roles and the intervals/durations in which a user or permission is assigned to a role.
- *Location-based activation constraints*—These constraints allow specification of location and duration/cardinality restrictions on the activation of a role. The location context specifies the location of the role in the constraint.
- *Run-time events*—These allow an administrator and users to dynamically initiate various role events, or enable the duration or activation constraints at a specified location.
- *Constraint enabling events*—These enable or disable location and duration constraints and location activation constraints mentioned earlier.
- *Triggers*—The extended trigger allows specification of the location context of the events in the trigger in addition to temporal conditions.

A key issue in LoT-RBAC is the support for flexible event-based policies and dependency constraints that are adopted from the earlier proposed generalized temporal RBAC model (Joshi, Bertino, et al. 2005). Further, LoT-RBAC emphasizes that LBAC approaches should consider the location information related to both the users/subjects and objects/resources. For instance, in a mobile location-based collaborative environment, users may need to allow each other to access resources stored in their mobile devices based on the location of their resources (characterized by the device location), in addition to considering the location of the users who access the resources.

10.4.4.3   Other LBAC Approaches

Several other research efforts have focused on developing LBAC approaches. Poolsappasit and Ray (2009) present an authorization system for an LBS that allows specification and enforcement of location-based privacy preferences. They emphasize various factors, namely, identity or role (i.e., who is requesting the location information), usage (i.e., purpose of this request), time (i.e., time of the request), and location (i.e., location of the requested object). Their approach requires a user to express his location privacy preference in terms of these factors. These factors form the query context of location query l, which is specified by the tuple $< I_l, U_l, T_l, L_l >$, where I_l represents the identity or role of the requester, U_l denotes the usage requirement of the requester, T_l specifies the time when the query is placed, and L_l is the location of the requested object. The individual entities in the context corresponding to identity, usage, time, and location are referred to as context elements. The query context determines the location information that can be revealed in response to the query. The policy owner specifies the privacy preference as a set of tuples of the form $<c, loc_c>$, where c is the context of the query and loc_c is the location that is revealed in response to the query. Poolsappasit and Ray also provide some implementation and experimental results.

Ardagna et al. (2006) propose reference architecture for deploying a LBAC system, which is shown in Figure 10.9. Its main components include the (a) *user* whose access request needs to be authorized; (b) *business application,* which offers the services to authorized people; (c) *access control engine* that is responsible for evaluating access requests as per the LBAC policies; and (d) *location providers* that

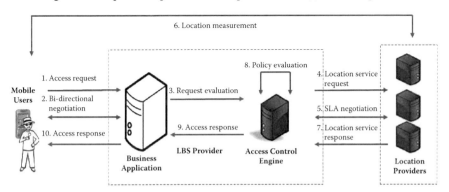

FIGURE 10.9 An LBAC architecture. (From Ardagna, Claudio A., Marco Cremonini, Ernesto Damiani, Sabrina De Capitani di Vimercati, and Pierangela Samarati, "Supporting location-based conditions in access control policies," *Proceedings of the 2006 ACM Symposium on Information, Computer and Communications Security.* New York, NY: ACM, 2006, 212–222. With permission.)

provide the location context information. Figure 10.9 depicts the access authorization process through the numbered steps shown. Once the access response is generated, the user can now proceed to access the authorized resources. This can be extended to support in general other existing LBAC approaches with extension related to authentication and other functions.

Geospatial eXtensible Access Control Markup Language (GeoXACML) has been introduced by the Open Geospatial Consortium (OGC) as an extension of the XACML Policy Language (Open Geospatial Consortium 2008) to support logical conditions based on geographic information. Various spatial features such as typological (e.g., containment, overlap), and geometric (e.g., intersection, boundary, distance or area) are supported and constraints can be specified based on these. Zhang and Parashar (2003) propose an approach that extends RBAC with location-awareness to support grid-based distributed applications. A cryptographic approach to geospatial access control has been proposed by Atallah, Blanton, and Frikken (2007); here, the issue of key management and derivation to facilitate access control is the primary focus. In this approach, a geographic space is considered as a grid of $m \times n$ cells, and policies are defined to indicate which rectangular areas, made of cells, are authorized for which users. Keys are associated with cells and are derived based on the spatial areas authorized for a user. Atluri, Shin, and Vaidya (2008) address the issue of efficient security policy enforcement for mobile environments and propose a model based on entities in motion with spatiotemporal attributes. They consider authorizations based on *moving subjects and static objects*, *static subjects and moving objects*, and *moving subjects and moving objects* (Atluri, Shin and Vaidya, 2008). They use a special index structure called S^{PPF} to capture past, present, and future locations of moving entities. They also define *locate* and *track* rights.

One key issue with LBAC approaches is the specification and verification of the LBAC policies. GeoXACML leverages the existing XACML standard. Although several abstract models have been proposed, little work exist that concretely address policy verification issue. Toahchoodee and Ray (2008) propose to use Alloy, a formal specification language supporting first-order logic, to formally specify spatiotemporal access control policies, and automatically analyze them. In their work, they support spatiotemporal constraints, hybrid hierarchy, and various separation of duty (SoD) constraints, providing a flexible framework. In this work, they leverage their earlier work on spatiotemporal access control models (Ray and Kumar 2006; Ray, Kumar, and Yu 2006; Ray and Toahchoodee 2007).

10.4.4.4 Access Control for LBSNs

LBSNs are becoming increasingly popular and users' security and privacy concerns related to exposure of their location check-ins including their sensitive, private location information is a growing concern. Jin et al. (2012) analyze the support for access control that the various LBSNs provide to safeguard users' location information. In particular, they focus on four popular LBSNSs that provide a location check-in feature: Foursquare, Facebook Place, Google Latitude, and Yelp. They first generalize a check-in model based on their analysis of these features in LBSNSs and then they comparatively analyze the access control mechanisms provided.

TABLE 10.1

Check-in Features in Facebook Place, Google Latitude, Foursquare, and Yelp

| | A User's Check-in(c) | Resources in a Check-in c | | | | A User's Check-in History (C_u) | A User's Future Check-ins (F-C_u) |
		I	*t*	*OU*	*M*		
Facebook Place	√	√	√	√	√	√	
Google Latitude	√	√	√			*	
Foursquare	√	√	√		√	#	√
Yelp	√	√	√		√	*	

Notes: √ means the feature is available; * means the last record in C_u; # means latest records in C_u

Source: Jin et al. 2012.

TABLE 10.2

Access Control Policy for a User Check-in c in Facebook Place, Google Latitude, Foursquare, and Yelp

	Access Control Policy Components				
	S	*U*	*a*	*d*	*r*
Facebook Place	Facebook	Public; Private; Specific Friends	Create; Delete; Modify; Read	NA	Allow; Deny
Google Latitude	Google Latitude	Public; Friends; Private	Create; Read	NA	Allow
Foursquare	Foursquare; Facebook; Twitter	Public; Friends; Private	Create; Read	NA	Allow
Yelp	Yelp; Facebook; Twitter	Friends; Private	Create; Read	NA	Allow

Source: Jin et al. 2012.

In their abstract model a user's check-in, c, is modeled as a 5-tuple $<u, l, t, OU, M>$, where u represents the creator and the owner of c; l represents the location information; t represents the timestamp indicating when c occurs; OU represents other users that u is with when he creates c; and M represents messages (comments or tips) that are associated with c. A user u can post a tip as $m \in M$ when u creates c. Other users, such as u's friends, can also post messages as comments for m. These comments are also included in M. Based on this, a policy, p, for a check-in, c, is modeled as a 6-tuple $<c, S, U, a, d, r>$, where S is a system set (including LBSNSs and other social network systems) where c is created and posted (e.g., a message could be created in Foursquare and posted in Twitter); U is a user set for whom p applies in the specified S (it is to be noted that U cannot contain the users who are not in any system in S); a represents an action for c such as create, modify, delete, and read; d represents the available duration of c for users in U in S. After the duration d, c will no longer be available for them; and r represents the access decision made by u for c. The decision can be allow or deny.

Table 10.1 and Table 10.2 summarize the result of Jin et al.'s (2012) comparison of access control support in the four LBSNs. In particular, there is a lack of fine-grained access control support for protecting information such as l, t, OU, and M separately in a check-in, c, in these four LBSNs exposing significant privacy risks.

The LBAC approaches discussed earlier are yet to be fully employed in various LBSs, including in the LBSNs. With growing private information being shared through these LBSNs, a challenge is to provide an easy-to-use framework that allows fine-grained protection of information and resources based on user-specific location information.

10.4.5 LOCATION AUTHENTICATION

It is important to realize that authenticity of location information provided by users is a basic requirement before other security and privacy concerns can be addressed. One example threat to location authentication in LBS is a *location cheating* attack (He, Liu, and Ren 2011). In this attack an LBS user (attacker) deceives the LBS provider about his location information in order to receive services illegally from the LBS provider or carry out other malicious activities. In an LBSN, for instance, location cheating can be a serious problem (Jin et al. 2012). To attract more users, LBSN services may provide real-world rewards to the user, when a user checks in at a certain venue or location. For example, Foursquare offers rewards such as points, badges, mayorship, or incentives such as a free cup of coffee at Starbucks. Such offers may encourage even general users to cheat on their locations. He, Liu, and Ren (2011) show how an LBSN user can falsify his true location. For example, a malicious user can easily tamper with the Foursquare's client application running on his smartphone for collecting his location from the embedded GPS receiver. Gonzalez-Tablas Ferreres, Álvarez, and Garna (2008) define location authentication as follows: "location authentication is the process whereby one party is assured (through acquisition of collaborative evidence) of a second party's location in a protocol, and the second party must have participated in the protocol (that is, was active when or immediately before the evidence was acquired)."

Bertino and Kirkpatrick (2009) discuss several approaches to address location authentication issues. An approach to establishing location information is to use GPS, in which case the device determines the coordinates and sends them to the LBS. A key disadvantage is that the user's device needs to be trusted. A malicious user can supply a fake location to wrongfully achieve access to services. Another approach could be to employ a location certifier that a device could connect and get a token from it verifying the location of the device. With appropriate cryptographic techniques, location forgeries could be avoided (Bertino and Kirkpatrick 2009). Yet another approach is to create a proxy workstation that is secure to connect to communicate with the LBS providers. In such a case, a mobile user would then need to bind the authentication process with this secure workstation. Such binding may be possible through the use of the Trusted Platform Module (TPM) (Trusted Computer Group 2011) that stores a key that is provided to trusted applications. A hardware-based approach that is an alternative to TPM is the use of *physically unclonable functions* (PUF) (Suh and Devadas 2007; Atallah, Bryant, et al. 2008). The PUF technique is based on using the inherent limitation of chip manufacturing processes because of which no two chips can be produced perfectly identical (Suh and Devadas 2007). Because of such variations in physical characteristics of devices a physical measurement based on a device's unique physical characteristics can be used as a unique signature of the device; this can then be used to generate a unique key for device identification process.

Location authentication schemes can be infrastructure based or terminal based (Gonzalez-Tablas Ferreres, Álvarez, and Garna 2008). For an *infrastructure-based scheme*, the goal of an attacker is to manipulate the reference nodes to incorrectly verify a node's location information by affecting one of the three elements of a tuple (id, l, t), where id is the identity of the node, l is the location, and t is the time. Gonzalez-Tablas Ferreres, Álvarez, and Garna (2008) also identify several threats to location authentication, including those that have been mentioned earlier. One threat is the *impersonation attack*, where an attacker node, A, tries to impersonate an honest node, N, and claim that N is at a particular location, l, when actually A is not there. Another threat is the *distance fraud* where a node N compromised by an attacker could try to make the other party, P, believe that it is closer to P than it actually is. Challenge response approaches that employ time of arrival (ToA) schemes may be able to thwart such an attack. However, in such approaches if the clock could be manipulated or the attacker can send an advance response, then the protocol could be compromised. An attacker may also launch an *absolute location fraud*, where the attacker manipulates the absolute location information of its victim; thus the victim will appear to be in a different place than it actually is. A compromised node may also establish its current location to be one that actually was the location where it was in the recent past; it is called a *time fraud* (Gonzalez-Tablas Ferreres, Álvarez, and Garna 2008). Similarly, sybil attacks and device cloning present another set of possible threats. Sybil attacks are popular attacks against ad hoc networks that may be adopted to spread misinformation about location information by establishing fake nodes. Here, an attacker may compromise several reference nodes and make them collude to determine a target node's location incorrectly (Gonzalez-Tablas Ferreres, Álvarez, and Garna 2008). Device cloning is a growing concern such as in products that use radio-frequency identification (RFID) or other embedded devices (Atallah, Bryant, et al. 2008). By cloning a device one may easily fake the location of a cloned device. Approaches such as PUF have been recently used to address such cloning issues in RFID and small embedded devices (Atallah, Bryant, et al. 2008; Bertino and Kirkpatrick 2009). Attackers may also manipulate signals (e.g., GPS signals) to subvert location authentication protocols (Gonzalez-Tablas Ferreres, Álvarez, and Garna 2008). In general, in satellite-based schemes, ToA is very crucial and an attacker may manipulate the transmission signals and data to delay ToA associated with a targeted node, thus affecting the location information determined for that node.

Ren, Lou, and Kim (2006) propose a scheme that is resistant to eavesdropping, impersonation, and service spoofing attacks. The scheme uses a trusted third party to provide mutual authentication between users and services providers. The certificate obtained from the authenticator allows the user to obtain the service while remaining anonymous. Zhang, Li, and Trappe (2007) propose a power-modulated challenge response approach to verify the location claim of a node. Here, a set of transmitters is selected to issue challenges that a claimant node should be able to witness or not witness based on its claimed location. Based on the correct responses of the claimant node, the location information is verified. Abdelmajid et al. (2010) propose *N-Kerberos*, which essentially extends the existing Kerberos authentication protocol to include location information as a new authentication factor.

10.5 SUMMARY AND RESEARCH DIRECTIONS

The use of location information in applications, such as local search, tour guides, mobile games, advertising, mobile commerce, social networking, offers great added value. As a result, we are witnessing many innovative ways to use location information in new LBSs. However, location information has many privacy and security implications for LBS users. In this chapter, we have discussed key security and privacy concerns in LBSs and overviewed existing approaches that address these challenges. Location privacy and security are very complex topics. They have many facets from protecting location information to protecting identity to protected communication channel between LBS providers and LBS users. On the other hand, location cheating by an LBS user may harm the business of the LBS provider.

We believe the following research directions are the key to providing a secure and safe LBS.

- Although there are a good amount of research efforts related to anonymization/pseudomization techniques, one key issue is how much privacy protection is necessary for a given application. In particular, setting the right value of k, l, and s in various anonymity schemes is important. Further, while the number and use of LBSNs are growing at an accelerated speed, there exist countable few techniques for anonymizing LBSN structures. Anonymization techniques that have been developed for general social networks need to be revisited to address this issue. There is an inherent trade-off between the utility of LBSs that users wish to receive and the location privacy they can afford to risk. Even when the location information is exposed without the user being identified, the user's identity and individual information can often be inferred by analyzing the location information. Approaches that consider such trade-offs are essential. Further, a proper understanding of all types of threats in anonymization techniques is critical.
- Location-based access control is a crucial component of a secure LBS environment and the existing work has primarily focused on developing flexible but abstract models. Practical systems that employ these are yet to be developed. Another research direction is the development of user-friendly policy specification and administration approaches. Similarly, policy verification frameworks that can support automated and continuous policy conflict analysis efficiently are needed. Growth in the use of LBSNs and adoption of location information in all types of sensitive applications, such as healthcare and homeland security, make the need for effective privacy-aware LBAC approaches a critical and immediate need.
- Location authentication or location verification is a fundamental issue and there exist threats related to this at the device level, protocol level, and application level. Work in this domain is still in infancy.

Although this chapter has focused on the technological aspects, there are legal and social factors that are critical before technological solutions can be effectively deployed. Another crucial issue not discussed in this chapter is the usability and human factors issues, which are becoming more critical with the pervasive applications that LBSs represent. Use of a different set of smaller devices adds another layer of usability and human factors challenges to all the research challenges mentioned.

ACKNOWLEDGMENT

This research has been supported by the U.S. National Science Foundation award IIS-0545912 and DUE-0621274.

REFERENCES

Abdelmajid, N. T., M. A. Hossain, S. Shepherd, and K. Mahmoud. "Location-based Kerberos Authentication protocol." *IEEE Second International Conference on Social Computing (SocialCom)*. IEEE, 2010, 1099–1104.

Ardagna, Claudio A., Mariana Cremonini, Sabrina De Capitani di Vimercati, and Pierangela Samarati. "An Obfuscation-Based Approach for Protecting Location Privacy." *IEEE Transactions on Dependable and Secure Computing (IEEE)* 8, no. 1 (2011): 13–27.

Ardagna, Claudio A., Marco Cremonini, Ernesto Damiani, Sabrina De Capitani di Vimercati, and Pierangela Samarati. "Supporting location-based conditions in access control policies." *Proceedings of the 2006 ACM Symposium on Information, Computer and Communications Security.* New York, NY: ACM, 2006, 212–222.

Ardagna, Claudio A., Mariana Cremonini, Ernesto Damiani, Sabrina De Capitani di Vimercati, and Pierangela Samarati. "Location privacy protection through obfuscation-based techniques." *IFIP WG 11.3 Working Conference on Data and Applications Security.* Springer, 2007, 47–60.

Atallah, Mikhail J., Marina Blanton, and Keith B. Frikken. "Efficient techniques for realizing geo-spatial access control." *Proceedings of the 2nd ACM Symposium on Information, Computer and Communications Security.* New York, NY: ACM, 2007, 82–92.

Atallah, Mikhail J., Eric D. Bryant, John T. Korb, and John R. Rice. "Binding software to specific native hardware in a VM environment: The PUF challenge and opportunity." *Proceedings of the 1st ACM workshop on Virtual Machine Security.* New York, NY: ACM, 2008, 45–48.

Atluri, Vijayalakshmi, Heechang Shin, and Jaideep Vaidya. "Efficient security policy enforcement for the mobile environment." *Journal of Computer Security (IOS Press)* 16, no. 4 (2008): 439–475.

Bamba, Bhuvan, Ling Liu, Peter Pesti, and Ting Wang. "Supporting anonymous location queries in mobile environments with PrivacyGrid." *Proceedings of the 17th International Conference on World Wide Web.* New York: ACM, 2008, 237–246.

Beresford, Alastair R., and Frank Stajano. "Location privacy in pervasive computing." *Pervasive Computing (IEEE)* 2, no. 1 (2003): 46–55.

Berthold, Oliver, and Marit Köhntopp. "Identity management based on P3P." *International Workshop on Designing Privacy Enhancing Technologies: Design Issues in Anonymity and Unobservability.* New York: Springer-Verlag, 2001, 141–160.

Bertino, Elisa, Barbara Catania, Maria Luisa Damiani, and Paolo Perlasca. "GEORBAC: A spatially aware RBAC." *Proceedings of the 10th ACM Symposium on Access Control Models and Technologies.* New York, NY: ACM, 2005, 29–37.

Bertino, Elisa, and Michael Kirkpatrick. "Location-aware authentication and access control—Concepts and issues." *International Conference on Advanced Information Networking and Applications*. IEEE, 2009, 10–15.

Chandran, Suroop Mohan, and James B. D. Joshi. "LoT-RBAC: A location and time-based RBAC model." *Web Information Systems Engineering*. Springer, 2005, 361–375.

Chow, Chi-Yin, and Mohamed F. Mokbel. "Enabling private continuous queries for revealed user locations." *Advances in Spatial and Temporal Databases*. Springer, 2007, 258–275.

Consolvo, Sunny, Ian E. Smith, Tara Matthews, Anthony LaMarca, Jason Tabert, and Pauline Powledge. "Location disclosure to social relations: Why, when, & what people want to share." *Proceedings of the SIGCHI Conference on Human Factors in Computing Systems*. New York, NY: ACM, 2005, 81–90.

Damiani, Maria Luisa, Elisa Bertino, and Claudio Silvestri. "The PROBE framework for the personalized cloaking of private locations." *Transactions on Data Privacy (ACM)* 3, no. 2 (2010): 123–148.

Duckham, Matt, and Lars Kulik. "A formal model of obfuscation and negotiation for location privacy." *3rd International Conference on Pervasive Computing*. Springer, 2005, 152–170.

Duckham, Matt, and Lars Kulik. "Location privacy and location-aware computing." In *Dynamic & Mobile GIS: Investigating Change in Space and Time*, edited by J. Drummond et al. CRC Press, 2006, 34–51.

Gedik, Bugra, and Ling Liu. "Location privacy in mobile systems: A personalized anonymization model." Proceedings of the 25th IEEE International Conference on Distributed Computing Systems. IEEE, 2005, 620–629.

Gedik, Bugra, and Ling Liu. "Protecting location privacy with personalized k-anonymity: Architecture and algorithms." *IEEE Transactions on Mobile Computing* 7, no. 1 (2008): 1–18.

Ghinita, Gabriel, Panos Kalnis, Ali Khoshgozaran, Cyrus Shahabi, and Kian-Lee Tan. "Private queries in location based services: anonymizers are not necessary." *Proceedings of the 2008 ACM SIGMOD International Conference on Management of Data*. ACM, 2008, 121–132.

Ghinita, Gabriel, Panos Kalnis, and Spiros Skiadopoulus. "PRIVE: Anonymous location-based queries in distributed mobile systems." *Proceedings of the 16th International Conference on World Wide Web*. ACM, 2007, 371–380.

Gkoulalas-Divanis, Aris, Vassilios S. Verykios, and Panayiotis Bozanis. "A network aware privacy model for online requests in trajectory data." *Data & Knowledge Engineering* 68, no. 4 (2009): 431–452.

Gonzalez-Tablas Ferreres, Ana Isabel, Benjamín Ramos Álvarez, and Arturo Ribagorda Garna. "Guaranteeing the authenticity of location information." Pervasive Computing (IEEE) 7, no. 3 (2008): 72–80.

Gruteser, Marco, and Dirk Grunwald. "A methodological assessment of location privacy risks in wireless hotspot networks." In *Proceedings of the First International Conference on Security in Pervasive Computing*. Springer, 2003a, 10–24.

Gruteser, Marco, and Dirk Grunwald. "Anonymous usage of location-based services through spatial and temporal cloaking." *Proceedings of the 1st International Conference on Mobile Systems, Applications and Services*. New York: ACM, 2003b, 31–42.

He, Wenbo, Xue Liu, and Mai Ren. "Location cheating: A security challenge to location-based social network services." *Proceedings of the 31st International Conference on Distributed Computing Systems*. Washington, DC: IEEE, 2011, 740–749.

Jafarian, Jafar Haadi, Ali Noorollahi Ravari, Morteza Amini, and Rasool Jalili. "Protecting location privacy through a graph-based location representation and a robust obfuscation technique." In *Information Security and Cryptology*, 116–133. Berlin: Springer-Verlag, 2009.

Jin, Lei, Xuelian Long, James B. D. Joshi, and Mohd Anwar, "Analysis of Access control mechanisms for users' check-ins in location-based social network systems, 2012." *Workshop on Issues and Challenges Social Computing in Conjunction with IEEE IRI*, 2012, 712–717.

Joshi, James B. D., Elisa Bertino, Usman Latif, and Arif Ghafoor. "A generalized temporal role-based access control model." *IEEE Transactions on Knowledge and Data Engineering* 17, no. 1 (2005): 4–23.

Joshi, James B. D., S. R. Joshi, and S. M. Chandran. "Identity management and privacy issues." *Encyclopedia of Digital Government*, 2006.

Joshi, James B. D., Walid G. Aref, Arif Ghafoor, and Eugene H. Spafford. "Security models for web-based applications." *Communications of the ACM* 44, no. 2 (2001): 38–44.

Kalnis, Panos, Gabriel Ghinita, Kyriakos Mouratidis, and Dimitris Papadias. "Preventing location-based identity inference in anonymous spatial queries." *IEEE Transactions on Knowledge and Data Engineering* 19, no. 12 (2007): 1719–1733.

Kido, Hidetoshi, Yutaka Yanagisawa, and Tetsuji Satoh. "Protection of location privacy using dummies for location-based services." 21st International Conference on Data Engineering. IEEE, 2005.

Liu, Ling. "From data privacy to location privacy: Models and algorithms." 33rd *International Conference on Very Large Data Bases*, 2007, 1429–1430.

Masoumzadeh, Amirreza, and James Joshi. "An alternative approach to k-anonymity for location-based services." *The 8th International Conference on Mobile Web Information Systems (MobiWIS),* 2011a, 522–530.

Masoumzadeh, Amirreza, and James Joshi. "Anonymizing geo-social network datasets." *4th ACM SIGSPATIAL International Workshop on Security and Privacy in GIS and LBS*, 2011b, 25–32.

Mokbel, Mohamed F., Chi-Yin Chow, and Walid G. Aref. "The New Casper: Query processing for location services without compromising privacy." *32nd International Conference on Very Large Data Bases*, 2006, 763–774.

Open Geospatial Consortium. Geospatial eXtensible Access Control Markup Language (GeoXACML). Bloomington: Open Geospatial Consortium, 2008.

Poolsappasit, Nayot, and Indrakshi Ray. "Towards achieving personalized privacy for location-based services." *Transactions on Data Privacy* 2, no. 1 (2009): 77–99.

Ray, Indrakshi, and Mahendra Kumar. "Towards a location-based mandatory access control model." *Computers & Security* 25, no. 1 (2006): 36–44.

Ray, Indrakshi, Mahendra Kumar, and Lijun Yu. "LRBAC : A location-aware role-based access control." *International Conference on Information Systems Security*, 2006, 147–161.

Ray, Indrakshi, and Manachai Toahchoodee. "A Spatio-temporal Role-Based Access Control Model." Conference on Data and Applications Security. IFIP, 2007. 211–226.

Ren, Kui, Wenjing Lou, Kwangjo Kim, and Robert Deng. "A novel privacy preserving authentication and access control scheme for pervasive computing environments." *IEEE Transactions on Vehicular Technology* 55, no. 4 (2006): 1373–1384.

Samarati, Pierangela. "Protecting respondents' identities in microdata release." *IEEE Transactions on Knowledge and Data Engineering* 13, no. 6 (2001): 1010–1027.

Shankar, Pravin, Vinod Ganapathy, and Liviu Iftode. "Privately querying location-based services with sybilquery." *International Conference on Ubiquitous Computing*, 2009, 31–40.

Smith, Ian, Sunny Consolvo, Anthony LaMarca, Jeffrey Hightower, James Scott, Timothy Sohn, Jeff Hughes, Giovanni Iachello, and Gregory D. Abowd. "Social disclosure of place: From location technology to communication practice." *3rd International Pervasive Computing Conference*, 2005, 151–164.

Solanas, Agusti, and Antoni Martínez-Ballesté. "A TTP-free protocol for location privacy in location-based services." Computer Communications (Elsevier) 31, no. 6 (2008): 1181–1191.

Suh, G. Edward, and Srinivas Devadas. "Physical unclonable functions for device authentication and secret key generation." *Proceedings of the 44th Annual Design Automation Conference*, 2007, 9–14.

Sweeney, Latanya. "k-anonymity: A model for protecting privacy." *International Journal on Uncertainty, Fuzziness and Knowledge-Based Systems* 10, no. 5 (2002): 557–570.

Takabi, Hassan, James Joshi, and Hassan Karimi. "A collaborative k-anonymity approach for location privacy in location-based services." *5th International Conference on Collaborative Computing: Networking, Applications and Worksharing (CollaborateCom)*, 2009, 1–9.

Tan, Kar Way, Yimin Lin, and Kyriakos Mouratidis. "Spatial cloaking revisited: Distinguishing information leakage from anonymity." *11th International Symposium on Advances in Spatial and Temporal Databases.* Springer-Verlag, 2009, 117–134.

Toahchoodee, Manachai, and Indrakshi Ray. "On the formal analysis of a spatio-temporal role-based access control model." *Data and Applications Security*, 2008, 17–32.

Trusted Computer Group. "Trusted platform module main specification." 2011.

Wang, Ting, and Ling Liu. "Privacy-aware mobile services over road networks." *Proceedings of the VLDB Endowment* 2, no. 1 (2009): 1042–1053.

Zhang, Guangsen, and Manish Parashar. "Dynamic context-aware access control for grid applications." *Fourth International Workshop on Grid Computing.* IEEE, 2003, 101–108.

Zhang, Yu, Zang Li, and W. Trappe. "Power-modulated challenge-response schemes for verifying location claims." *Global Telecommunications Conference (GLOBECOM '07).* IEEE, 2007, 39–43.

11 How Using Location-Based Services Affects Our Understanding of Our Environment

Kai-Florian Richter

CONTENTS

ABSTRACT

Location-based services (LBS) are an exciting paradigm for how we interact with our devices, other people, and our environment. They can assist with a variety of tasks, everywhere and anytime. The integration of sensors and computing power in today's smartphones destines them to become an integral part of our daily life. However, this bears the danger that we tend to rely on their availability and, thus, become dependent on these LBS, and in the end will be unable to perform these services ourselves if they should become unavailable. This chapter will discuss this issue of users' growing dependency on LBS, specifically with respect to navigation assistance, which is one of the predominant LBS. The discussion will focus on the (lack of) spatial learning as a key cognitive effect of LBS usage, the keyhole problem as one reason for this lack of learning, and some avenues to mitigate the negative cognitive effects of using LBS.

11.1 INTRODUCTION

"Drivers don't always find way with GPS,"* "GPS routed bus under bridge, company says,"† and "Metro crash may exemplify paradox of human–machine interaction"‡ are some recent news headlines illustrating that all is not well when it comes to using location-based services (LBS). To add to this, since the widespread use of these services, especially of navigation services, different local governments feel compelled to put up additional signage to counter the services' guidance, particularly for truck drivers (see Figure 11.1 for an example).

In light of these seeming failures of LBS (and automation in general), this chapter will discuss the underlying cognitive mechanisms that explain some of these failures. As will become evident, the main reasons are a lack of attention, a lack of receiving an overview of what is going on, and the apparent safety of being assisted by the service. All this leads people to "switching off their brains," that is, focusing their cognitive resources on something else. When the services suddenly fail, their users are lost; in the case of navigation services both literally in the environment they are navigating in, and figuratively as of why the service failed and what to do next.

FIGURE 11.1 Warning sign to counter navigation systems' instructions that do not match with the actual road situation. (From Flickr, © by John Sargent [Flickr user jackharrybill], published under a Creative Commons license.)

---

* Maria Sciullo, "Drivers don't always find way with GPS," *Pittsburgh Post-Gazette*, March 28, 2012, http://www.post-gazette.com/stories/sectionfront/life/drivers-dont-always-find-way-with-gps-228069/.
† Jennifer Langston, "GPS routed bus under bridge, company says," Seattlepi.com, April 17, 2008, http://www.seattlepi.com/local/article/GPS-routed-bus-under-bridge-company-says-1270598.php.
‡ Shankar Vedantam, "Metro crash may exemplify paradox of human–machine interaction," *The Washington Post*, June 29, 2009, http://www.washingtonpost.com/wp-dyn/content/article/2009/06/28/AR2009062802481.html.

The next section will further delve into the theme of this chapter by explaining the effects of automation on human task performance. Section 11.3 will detail this further for the use of navigation services. In Section 11.4, the keyhole problem will be discussed as one of the reasons for our difficulties to acquire knowledge about our environment and some ways "out of the hole" will be highlighted. Finally, Section 11.5 will broaden the discussion to some general approaches of allowing both assistance and autonomy of users in using LBS.

11.2   THE IRONY OF AUTOMATION*

People acquire new knowledge and skills all the time. Some kind of learning is involved in most everything we do. This may be deliberate learning, as when studying for an exam, or incidental learning, for example, when picking up spatial relationships between states while browsing through an atlas. While learning already happens with first exposure to knowledge or a task, it usually is strongest with several repeats over time. Moreover, learning may happen almost accidentally, but still involves paying attention and focus on the matter at hand to achieve a learning effect (Hyde and Jenkins 1973; Newell and Rosenbloom 1981; Anderson 1999).

Now what happens if automation comes into play? In the context of this chapter, automation refers to "a device or system that accomplishes (partially or fully) a function that was previously [...] carried out (partially or fully) by a human operator" (Parasuraman, Sheridan, and Wickens 2000, p. 287). With automation, part of the learning is taken away. Instead of learning how to perform the task itself, we learn how to get the machine (be it a computer, an industrial production machine, or another form of automation) to perform the task for us. This is not necessarily problematic in itself. In many cases, the automated process is more efficient (e.g., faster, cheaper, or safer, or even all of these) and it opens up resources for the human to do something else. However, as pointed out by Bainbridge (1983), in industry processes, for example, human operators are usually expected to step in if the automatic process fails. But since the operators were not previously involved in the process, they have a hard time understanding what just happened, where (in terms of the sequence of processing steps) the process has stopped or went wrong, and what needs to be done to recover from failure. The operators of the machine are out of the loop, not having had to pay attention while everything was going according to plan. Now, attention needs to be shifted, which takes time and may cause stress, especially because it happens in an abnormal situation.

This effect, as described for industry processes (e.g., machine operation), is transferable to any form of automation. Taking over from a device that was previously performing a task is difficult or maybe even impossible, depending on the kind of task. Considering navigation services, one typical such recovery need is localizing oneself without the help of the Global Positioning System (GPS), for example. The next section will discuss this further.

---

* This section's heading is taken from Bainbridge, Lisanne, 1983, Ironies of automation, *Automatica* 19 (6):775–779.

11.3 LOCATION-BASED SERVICES AND SPATIAL LEARNING: THE CASE OF NAVIGATION SERVICES

This section will provide some background about how people acquire knowledge about their environment (Section 11.3.1) and about general principles of giving route directions (Section 11.3.2). Based on this background, it will then discuss the backseat driver effect of using navigation services (Section 11.3.3), that is, why we do not acquire much knowledge when using these services (Section 11.3.4).

11.3.1 Spatial Learning and Mental Processing of Spatial Information

People learn about a physical environment through interaction with either the environment directly or its representations thereof. For example, by traveling through your new neighborhood, you will gain knowledge about it, as you would by looking at a map or listening to others talking about it.

According to the spatial learning model of Siegel and White (1975), in a new environment the first kind of knowledge that people acquire is *landmark knowledge*. In the context of this model, landmarks are places in the environment that people recognize as having been there before. The next step in spatial learning is connecting these places, that is, learning how to get from one place to another. This constitutes *route knowledge*. Only in the final step people learn to integrate these different routes into an overall picture of the environment—their *survey knowledge*, which allows people to calculate shortcuts between places, among others. These knowledge acquisition steps do not necessarily happen exactly in this strict order (Montello 1998), but Siegel and White's model provides a useful understanding of how people acquire knowledge about their environment. In particular, the important role that landmarks play as anchor points in this knowledge has been confirmed in several studies (Sadalla, Burroughs, and Staplin 1980; Hirtle and Jonides 1985; Couclelis et al. 1987); people organize their configurational knowledge of an environment around these landmarks.

When a map is used to learn an environment, people tend to achieve survey knowledge from the beginning, since this is what the map depicts. However, using information learned from a map for interacting with an environment is not as straightforward as it may seem. Maps provide a top-down view; people experience moving in an environment from an egocentric perspective. Among others, this introduces orientation biases, that is, it is difficult for people to adopt an orientation that differs from that of the map (Lloyd and Cammack 1996; Richardson, Montello, and Hegarty 1999).

As already stated in Section 11.2, learning an environment requires repeated exposure and an actual mental processing of the perceived information, that is, it needs to be paid attention to. Without this processing, the perceived information remains shallow and fades out before it can be retained in memory.

11.3.2 In-Advance and Incremental Route Directions

Providing navigation instructions to users is one of the prime use cases of location-based services. On mobile devices, instruction presentation has to account for the limitations of these devices. This mostly affects graphical presentations due to the limited screen size but may also consider background noise in city environments when delivering spoken instructions (Kray et al. 2003).

In general, there are two approaches for presenting instructions on how to follow a route: either *in-advance* or *incrementally* (Habel 2003; Richter 2008). In-advance directions present all instructions before the trip starts. This is what route service Web sites, such as Google Maps or Microsoft Bing, do, which were originally designed to work primarily on desktop computers. With in-advance directions, users get an overview of what to expect along a route, typically in form of a map, and a complete set of detailed instructions on how to reach their destination presented all at once. Also, these directions can be calculated for any valid origin location, regardless of a user's current location; there is no need to establish a user's location for the instruction generation to work.

Incremental directions provide stepwise instructions on actions to be performed. Here, only the instruction that is relevant next is given, usually close to where it becomes relevant, that is, close to the next decision point (even though it is often possible to browse through the instructions by some forwarding and reversing operation). Mobile navigation services typically provide incremental instructions, both because they can exploit colocation of system and user (it is possible for the service to time instructions according to user progress), and because it better suits the limited screen size of mobile devices. Since these instructions assume a colocation of service and user, successful assistance requires a real-time localization of the user (e.g., through GPS).

There are essentially three options to incrementally present navigation guidance: (1) present local information (provide information only on the upcoming decision point); (2) perform automated adaptation to the currently appropriate level of detail (switch automatically between overview and detailed information); and (3) allow users to decide on level of detail (provide means for zooming and panning information). Mobile navigation services predominantly use the first option, though the other options may be present as well, at least to some extent (e.g., car navigation systems zooming in to complex intersections, or the zooming and panning operations in Google's Android navigation application).

11.3.3 THE BACKSEAT-DRIVER EFFECT IN NAVIGATION SERVICES

Being able to issue instructions when and where they are needed, it seems that navigation services offer a failsafe, easy-to-use navigation experience. And indeed, most people manage to reach their destination most of the time using these services. Still, there is a lot of anecdotal evidence that these services do fail on occasions, which may be due to several reasons. For example, the map data these services use may be outdated, generating instructions that do not match with the actual environment (Karimi, Roongpiboonsopit, and Kasemsuppakorn 2011). Sometimes the generated instructions are difficult to understand and are confusing, either because the real-world situation is much more complex than the instruction implies or because the way an instruction is phrased does not match with the user's understanding of the situation (Hirtle et al. 2010). For example, the car navigation system owned by the author provides change of directions by using turn right versus keep right versus bear right that does not always match with the author's concepts of these direction changes; these discrepancies have repeatedly led to some confusion in navigating with the device.

More important, however, using such navigation services, users hardly acquire any spatial knowledge. It is very difficult for users to retrace their route without the system, similar to the backseat driver effect we all experience when we are driven somewhere. This effect has been shown in several recent studies and there are clear cognitive reasons for it, as will be discussed next.

11.3.4 WHY WE DO NOT LEARN WHILE THE SERVICE NAVIGATES

Several studies have investigated the effects of using mobile navigation services on navigation performance and spatial learning. In these studies participants navigate a real or virtual environment that is (largely) unknown to them. Their wayfinding is assisted by a service, or some nonelectronic assistance, such as a map, in case of comparative studies. Studies then collect data on navigation performance of the participants. Navigation performance is measured by traveled distance, time taken, and the number of navigation errors made. Faster travel with a few deviations from the shortest path is considered to indicate successful assistance by the service.

Studies may also look at the spatial knowledge participants acquire while using the service. This may test for route knowledge (i.e., the linear knowledge subjects have about the way they have traveled) or survey knowledge (i.e., the metric knowledge they have acquired about the layout of the environment) (Montello 1998). Sorting images of intersections in their order of appearance along a route, pointing tasks, or drawing sketch maps can all be used to test for this knowledge. If participants score high on these tests it is taken as an indication of successful assistance from the service.

In Krüger et al. (2004), participants navigated through a zoo using either a PDA or a head mounted clip-on. Both devices incrementally displayed photographs of each intersection. The photos were either augmented by overlayed graphical lines indicating the direction to be taken or by verbal commands describing the action to be performed. Across both modalities, participants performed fairly well, and their directional knowledge was good as well. However, survey knowledge of participants was poor in all conditions.

In another study in the same setting, participants navigated either using a PDA or printed maps (Münzer et al. 2006). The PDA had three different modes:

1. Only visual information was presented. An animation showed the relation between the previous, current, and next intersection when approaching an intersection. A line drawn on the photo showing the intersection then indicated the road to take.
2. The same as the first mode, but together with showing the photo a verbal instruction was given.
3. Only the photo (without highlighting the road to take) and verbal instruction is given.

The printed maps only showed part of the environment at the same time. The study showed that map users acquired much better route and survey knowledge than the

PDA users. The presentation mode had no influence on the performance; animations did not help.

In another study, participants navigated a multilevel building in a virtual environment setting (Parush, Ahuvia, and Erev 2007). They either had continuous access to a map showing their position or could request to see it any time. In both conditions participants either had to solve location quizzes (e.g., indicate your current position on the map) or not. Sixteen runs were performed with assistance and a final transition run without any assistance. Each run asked the participants to navigate to a particular location in the environment, indicated by an object located there (e.g., a plant, or a cabinet). Participants with continuous position indication performed best with regard to excess distance. However, for those requesting a map, excess distance and number of requests decreased with increasing number of runs, indicating that learning took place. The quizzes had no immediate effect on performance, but again learning took place, as participants got better in responding to the quizzes with an increasing number of runs. For the transition run, those having had continuous position indication and no quizzes performed worst, while those requesting maps and having quizzes performed best.

Ishikawa et al. (2008) tested the influence of assistance medium on wayfinding performance using three groups: participants who had traveled a test route once before, participants using paper maps, and participants using a GPS-based handheld navigation system. The groups traveled six different routes; at the end of each route they had to point back to the origin. After all routes, participants drew a sketch map of the environment. Participants using the device traveled longer distances, were slower, and made more stops; their configurational and topological understanding of the environment was worse. Although the device allowed users to find their way, it was less effective than maps or direct experience as support for smooth navigation.

Taken together, these studies clearly demonstrate that using mobile navigation devices lead users to "turn off their brain." Information that is available in principle—both presented by the device and perceived in the environment—is not processed mentally to a sufficient level. Consequently, users have great difficulty acquiring route and survey knowledge as the navigation experience remains shallow from a cognitive perspective (even if it is successful from a task perspective, that is, users reach their destination). As discussed in Section 11.2, this is a common phenomenon in automation. Users focus solely on when the devices issue new instructions, which decouples the actions to be performed from their spatial embedding (Leshed et al. 2008). This decoupling dominates any benefits that may be expected from presenting information in multiple modalities (Mayer and Moreno 2003; Dickmann 2012). The next section will discuss one reason for this effect in further detail: the lack of sufficient overview information while being assisted by a navigation service.

11.4   THE KEYHOLE PROBLEM: WHAT IS GOING ON WHILE I AM DELVING IN?

A major problem with mobile devices is the small screen estate that can be used to present information. While recent smartphones have larger screens (including touchscreens) than previous phones, and with higher resolutions, there is still a clear limit to the amount of information that can reasonably be displayed on these screens. As a consequence, it is usually either possible to show an overview of the presented information that is so small that any details are lost or to zoom in to a level where required details become recognizable, but all overview is lost. This is known as the *keyhole problem* (Bartram et al. 1995; Woods 1995).

Most mobile navigation services use detailed, zoomed-in, information views, that is, they usually just show a small part of the environment the user is navigating in. This is because of their small screen space that hardly allows for presenting a larger area without losing readability of information at the same time. This causes users to lose their sense of their current location, which impedes their decision-making ability related to the surrounding environment. Users only see individual views of their immediate surroundings without any information on how these relate to the route they are currently following, let alone the overall environment, which makes it hard to develop an integrated mental representation of the presented information (see Section 11.3.1). This issue was confirmed by Dickmann (2012) who had participants navigate a route using either a city map or a navigation service and then tested for their survey knowledge. Clearly, those using the city map gained more (and more reliable) survey knowledge.

Over the years, different approaches have been suggested to overcome the keyhole problem, initially often with desktop-sized displays in mind. For example, one option is to include a small overview window into the larger detailed view (Hornbaek, Bederson, and Plaisant 2002). However, for mobile devices this takes away precious screen space, and the small overview window is unlikely to be readable on these small screens. Schmid et al. (2010) introduced *route aware maps*, maps that focus on a route for navigation but present enough spatial context to keep a user sufficiently oriented to find their way should they inadvertently deviate from the route (for an example, see Figure 11.2). These maps are not explicitly designed for mobile devices, and in their current stage would likely need additional zooming and panning operations to become usable on small screens.

Although interfaces for map/information manipulation (e.g., zooming and panning) allow users to be more involved in the information presentation and to gain more overview in principle, especially panning has been shown to have detrimental effects on task performance and user satisfaction (Gutwin and Fedak 2004). Such shortcomings call for other methods for displaying information beyond what is currently visible on the screen. Baudisch and Rosenholtz (2003) and Gustafson et al. (2008) developed Halo and Wedge, respectively. For points of interest (POIs) that are outside the current view, these methods indicate distance and direction to these points on the borders of the current view. Halo uses circles centered at a POI such that they just reach onto the display. The length of the arc, that is, the part of the circle that is visible on the display, indicates the distance to the POI and direction is indicated based on which border the arc enters the display. Wedge works the same way but uses cones instead of arcs as cones reduce overlap and generally take less space per element.

Although both Halo and Wedge provide an idea of where POIs are in relation to a user's current position, they do not provide any information on how to get there (because that information is outside of the display). Taking up principles of static

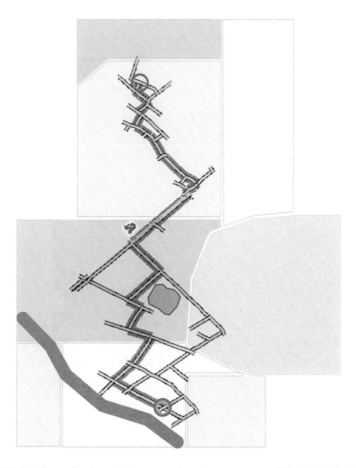

FIGURE 11.2 (See color insert.) An example route-aware map. Origin is at the top, desti-nation is at the bottom (denoted by the circles). The main route is shown in bold; alterna-tive routes help to recover from potential wayfinding errors, as do increased details around origin and destination, and the inclusion of landmark and region information. (Modified from Schmid, Falko, Denise Peters, and Kai-Florian Richter, 2008, You are not lost—You are somewhere here, in *You-Are-Here-Maps: Creating a Sense of Place through Map-Like Representations*, edited by A. Klippel and S. Hirtle, Workshop at International Conference Spatial Cognition 2008.)

you-are-here (YAH) maps, Schmid et al. (2010) designed YAH^x maps. These maps present information to various degrees of detail, divided into three zones: the imme-diate surroundings of a user shown in detail (thus, providing survey information), the zone further out containing route information to selected important landmarks, and the outer zone pointing to landmarks beyond the display horizon (similar to Halo and Wedge). An empirical evaluation demonstrated that users of YAH^x maps are much better equipped to locate their current position with respect to the overall environment. An accompanying zoom function, which changes the display accord-ing to the three zones, limits the amount of interaction required for this localization. Figure 11.3 shows examples of Halo, Wedge, and YAH^x maps.

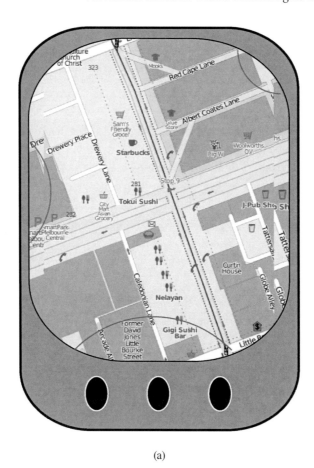

(a)

FIGURE 11.3 **(See color insert.)** Three different approaches to overcome the keyhole problem: (a) Halo, (b) Wedge, and (c) YAH[x] maps. (Data from OpenStreetMap.)

11.5 THE WAY FORWARD: EDUCATIONAL ASSISTIVE LOCATION-BASED SERVICES

The results of the study by Parush et al. (2007) clearly demonstrate that users who are actively involved in the navigation process learn more of their environment. This is confirmed by work of Richter et al. (2010), who compared navigation performance of people using the iPhone navigation application versus people using the Android navigation application. The iPhone app zooms and pans to the next decision point while navigating, whereas the Android app only shows the next instruction but leaves zooming and panning to the user. Study participants were tasked to navigate an unknown route with the help of either app, and afterward to retrace the same route without further assistance. Although participants using the iPhone app made fewer navigation errors, participants using the Android app were better in retracing the route. The participants using the Android app also showed better survey knowledge in distance estimation tests. Again, this demonstrates that those users who are more

(b)

FIGURE 11.3 (continued)   **(See color insert.)**

engaged in the assistance process (i.e., Android users) learn more about their environment—even if in this case assistance by the device seems somewhat less helpful.

Another potential path for designing applications that increase user engagement emerges from the realm of user-generated content or UGC (Krumm, Davies, and Narayanaswami 2008), also termed volunteered geographic information (VGI) in a geographic context (Goodchild 2007). In general, approaches using UGC mechanisms exploit data submitted by users to improve the services they can provide. For example, OpenStreetMap[*] provides a topographic data set of the world, which is entirely based on data users add to the OpenStreetMap database (Haklay and Weber 2008). Other services, such as FixMyStreet,[†] collect and forward requests to responsible authorities to ease communication and reduce time to act—in this case informing local councils about issues, such as potholes or broken street lights. Richter and Winter (2011a, 2011b) present OpenLandmarks, a mobile application that, based

---

[*] http://www.openstreetmap.org.
[†] http://www.fixmystreet.com/.

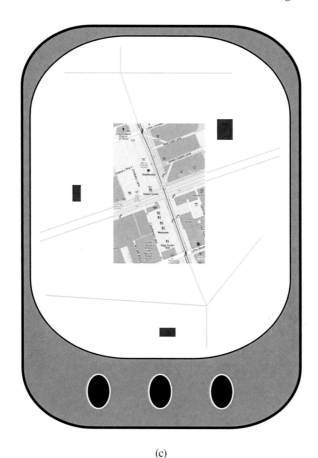

(c)

FIGURE 11.3 (continued)   (See color insert.)

on OpenStreetMap data, allows users to mark geographic features (buildings) as landmarks. Overall, such approaches tap into users' knowledge and experiences, which may be exploited to improve the services (Bilandzic, Foth, and De Luca 2008; Richter and Winter 2011a).

Approaches using UGC mechanisms could also be combined with a gamification of the services. Gamification refers to the use of (video)game elements in nongaming applications to improve user experience and engagement (Deterding et al. 2011). These game elements ensure that users stay engaged in the given task, while the results of the user interaction with the service may be used for other, more serious purposes beyond the "game" (Ahn and Dabbish 2008).

Bell et al. (2009), for example, use a location-based game to collect photos that depict an area in an easily recognizable way. While the players of the game can earn points, the photos are used to augment a navigation service, similar to the OpenLandmarks application. Both approaches share the idea that services providing information on how to navigate an environment should better link their instructions to the environment itself to become easier to follow and more engaging for the user (Dale, Geldof, and Prost 2005; Leshed et al. 2008). Since the required data are hard to collect with traditional, authoritative methods, UGC mechanisms are ideally suited toward this end (Richter, forthcoming). Also because such applications ask users to provide information about their neighborhood (or other areas), use of these applications raises users' awareness of their surroundings and, thus, will have a spatial learning effect (see Section 11.3.1). But providing data to improve a service is a secondary task compared to navigating an environment; thus, approaches combining UGC aspects and assistance provision would need to be carefully balanced in order not to impede the original purpose of the service to an extent where users stop using it.

In conclusion, to avoid any of the problems and dangers highlighted in Sections 11.1 and 11.2, LBS design should aim for ways of presenting information that is useful in the given situation and also fosters active mental processing of that information (Section 11.3). This will increase users' confidence of "doing the right thing" and decrease their dependency on the device (Willis et al. 2009). LBS designed with such principles in mind will need to be able to adapt their assistance and their information presentation to the current situation as well as adapt to the current context. The context is primarily defined through the task at hand, but likewise capabilities and preferences of the user, the (type of) device, and the environment the task is performed in play a role. Figure 11.4a illustrates a process-oriented context model reflecting this need for adaptation. This model (cf. Richter, Dara-Abrams, and Raubal 2010) focuses on the tasks from the users' perspective, looking at their level of involvement, the effects of automation on understanding the situation, and at personalization. In the model, the task determines which adaptations in interaction with and presentation of the device are sensible and are performed either by the user or the device. The model allows for identifying the level of user involvement in different tasks performed with an LBS, which is based on the activity theory for cartography (Dransch 2001). Each task is divided into activities, goals, and subgoals, with different (sequences of) activities fulfilling different subgoals that eventually lead to the overall goal. Such a detailed analysis of tasks reveals those activities that are crucial to keep users globally oriented and involved. Thus, when implementing LBS, special attention may be paid to these activities in order to avoid the issues identified in this chapter. Overall, such an approach will lead to a modular architecture for LBS (see Figure 11.4b) that specifically focuses on the processes required to perform a task. With this architecture, it becomes possible for each activity, performed by each task, to set the balance between user involvement and automation, depending on cognitive considerations, technical requirements, and user preferences.

11.6   SUMMARY

This chapter discussed cognitive effects of using navigation services, or more generally the interplay between automation and autonomy. Clearly, location-based services are useful; they fulfill user needs and deliver desired services—to many they are truly a boon. They are not going to go away, and they should not. But at the same time, there are clear detrimental effects for spatial learning and being able to cope with errors in and loss of service. A lack of spatial learning may arguably be taken

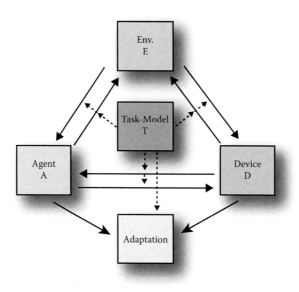

(a)

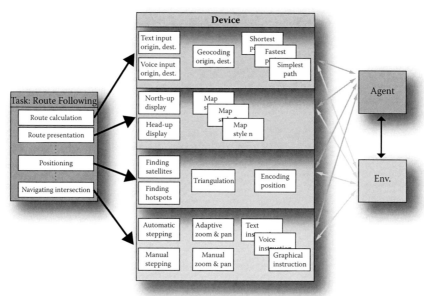

(b)

FIGURE 11.4 **(See color insert.)** (a) Task-based context model. (b) Modular LBS architecture. (Modified from Richter, Kai-Florian, Drew Dara-Abrams, and Martin Raubal, 2010, Navigating and learning with location based services: A user-centric design, in the *7th International Symposium on LBS and Telecartography*, edited by G. Gartner and Y. Li, 261–276.)

as a logical consequence of the (technological) changes of our times, similar to what happened to our abilities of (mental) calculations with the mass introduction of electronic calculators in the 1970s. But the lack of involvement in the navigation process may lead to situations that are dangerous for users as illustrated in the introduction, making them a bane in these situations. This is a problem that needs to be addressed.

Lack of involvement and of overview information have been identified as key challenges that would need to be overcome to remove these issues. This would increase safety of use, and likely also user satisfaction, which would lift the bane and make location-based services a true boon of our current and future times.

REFERENCES

Ahn, Luis von, and Laura Dabbish. 2008. Designing games with a purpose. *Communications of the ACM* 51 (8):58–67.

Anderson, John R. 1999. *Cognitive Psychology and Its Implications*. New York: Worth Publishers.

Bainbridge, Lisanne. 1983. Ironies of automation. *Automatica* 19 (6):775–779.

Bartram, Lyn, Albert Ho, John Dill, and Frank Henigman. 1995. The continuous zoom: A constrained fisheye technique for viewing and navigating large information spaces. In *Proceedings of the 8th Annual ACM Symposium on User Interface and Software Technology*. Pittsburgh, PA: ACM.

Baudisch, Patrick, and Ruth Rosenholtz. 2003. Halo: A technique for visualizing off-screen objects. In *CHI'03: Proceedings of the SIGCHI Conference on Human Factors in Computing Systems*. New York, NY: ACM.

Bell, Marek, Stuart Reeves, Barry Brown, Scott Sherwood, Donny MacMillan, John Ferguson, and Matthew Chalmers. 2009. EyeSpy: Supporting navigation through play. In *Proceedings of the 27th International Conference on Human Factors in Computing Systems*. New York, NY: ACM.

Bilandzic, Mark, Marcus Foth, and Alexander De Luca. 2008. CityFlocks: Designing social navigation for urban mobile information systems. In *DIS '08: Proceedings of the 7th ACM Conference on Designing Interactive Systems*. New York, NY, USA: ACM.

Couclelis, H., R. G. Golledge, N. Gale, and W. Tobler. 1987. Exploring the anchor-point hypothesis of spatial cognition. *Journal of Environmental Psychology* 7:99–122.

Dale, Robert, Sabine Geldof, and Jean-Philippe Prost. 2005. Using natural language generation in automatic route description. *Journal of Research and Practice in Information Technology* 37 (1):89–105.

Deterding, Sebastian, Miguel Sicart, Lennart Nacke, Kenton O'Hara, and Dan Dixon. 2011. Gamification: Using game-design elements in non-gaming contexts. In *Extended Abstracts Proceedings of the 2011 Annual Conference on Human Factors in Computing Systems*. Vancouver, British Columbia, Canada: ACM.

Dickmann, Frank. 2012. City maps versus map-based navigation systems: An empirical approach to building mental representations. *The Cartographic Journal* 49 (1):62–69.

Dransch, Doris. 2001. User-centred human-computer interaction in cartographic information processing. In *20th International Cartographic Conference*. Beijing, China.

Goodchild, Michael. 2007. Citizens as sensors: The world of volunteered geography. *GeoJournal* 69 (4):211–221.

Gustafson, Sean, Patrick Baudisch, Carl Gutwin, and Pourang Irani. 2008. Wedge: Clutter-free visualization of off-screen locations. In *CHI'08: Proceeding of the 26th Annual SIGCHI Conference on Human Factors in Computing Systems*. New York, NY: ACM.

Gutwin, Carl, and Chris Fedak. 2004. Interacting with big interfaces on small screens: A comparison of fisheye, zoom, and panning techniques. In *GI '04: Proceedings of Graphics Interface 2004*. Waterloo, Ontario, Canada: Canadian Human-Computer Communications Society.

Habel, Christopher. 2003. Incremental generation of multimodal route instructions. In *Natural Language Generation in Spoken and Written Dialogue*. Palo Alto, California.

Haklay, Mordechay, and Patrick Weber. 2008. OpenStreetMap: User-generated street maps. *Pervasive Computing* 7 (4):12–18.

Hirtle, Stephen C., and John Jonides. 1985. Evidence of hierarchies in cognitive maps. *Memory & Cognition* 13 (3):208–217.

Hirtle, Stephen C., Kai-Florian Richter, Samvith Srivinas, and Robert Firth. 2010. This is the tricky part: When directions become difficult. *Journal of Spatial Information Science* (1):53–73.

Hornbaek, Kasper, Benjamin B. Bederson, and Catherine Plaisant. 2002. Navigation patterns and usability of zoomable user interfaces with and without an overview. *ACM Transactions on Computer-Human Interaction* 9 (4):362–389.

Hyde, Thomas S., and James J. Jenkins. 1973. Recall for words as a function of semantic, graphic, and syntactic orienting tasks. *Journal of Verbal Learning and Verbal Behavior* 12 (5):471–480.

Ishikawa, Toru, Hiromichi Fujiwara, Osamu Imai, and Atsuyuki Okabe. 2008. Wayfinding with a GPS-based mobile navigation system: A comparison with maps and direct experience. *Journal of Environmental Psychology* 28:74–82.

Karimi, Hassan A., Duangduen Roongpiboonsopit, and Piyawan Kasemsuppakorn. 2011. Uncertainty in personal navigation services. *The Journal of Navigation* 64 (2):341–356.

Kray, Christian, Katri Laakso, Christian Elting, and Volker Coors. 2003. Presenting route instructions on mobile devices. In *International Conference on Intelligent User Interfaces (IUI'03)*. New York: ACM Press.

Krüger, Antonio, Illhan Aslan, and Hubert Zimmer. 2004. The effects of mobile pedestrian navigation systems on the concurrent acquisition of route and survey knowledge. In *Mobile Computer Interaction—MobileHCI 2004*, edited by S. Brewster and M. Dunlop. Berlin: Springer.

Krumm, John, Nigel Davies, and Chandra Narayanaswami. 2008. User-generated content. *Pervasive Computing* 7 (4):10–11.

Leshed, Gilly, Theresa Velden, Oya Rieger, Blazej Kot, and Phoebe Sengers. 2008. In-car GPS navigation: Engagement with and disengagement from the environment. In *CHI '08: Proceeding of the 26th Annual SIGCHI Conference on Human Factors in Computing Systems*. New York, NY: ACM.

Lloyd, Robert, and Rex Cammack. 1996. Constructing cognitive maps with orientation biases. In *The Construction of Cognitive Maps*, edited by J. Portugali. Dordrecht, Netherlands: Kluwer Academic Publishers.

Mayer, Richard E., and Roxana Moreno. 2003. Nine ways to reduce cognitive load in multimedia learning. *Educational Psychologist* 38 (1):43–52.

Montello, Daniel R. 1998. A new framework for understanding the acquistion of spatial knowledge in large-scale environments. In *Spatial and Temporal Reasoning in Geographic Information Systems*, edited by M. J. Egenhofer and R. G. Golledge. New York: Oxford University Press.

Münzer, Stefan, Hubert D. Zimmer, Maximilian Schwalm, Jörg Baus, and Ilhan Aslan. 2006. Computer-assisted navigation and the acquisition of route and survey knowledge. *Journal of Environmental Psychology* 26:300–308.

Newell, Allan, and Paul S. Rosenbloom. 1981. Mechanisms of skill aquisition and the law of practice. In *Cognitive Skills and their Acquisition*, edited by J. R. Anderson. Hillsdale, NJ: Erlbaum.

Parasuraman, Raja, Thomas B. Sheridan, and Christopher D. Wickens. 2000. A model for types and levels of human interaction with automation. *IEEE Transactions on Systems, Man and Cybernetics, Part A* 30 (3):286–297.

Parush, Avi, Shir Ahuvia, and Ido Erev. 2007. Degradation in spatial knowledge acquisition when using automatic navigation systems. In *Spatial Information Theory*, edited by S. Winter, M. Duckham, L. Kulik, and B. Kuipers. Berlin: Springer.

Richardson, Anthony E., Daniel R. Montello, and Mary Hegarty. 1999. Spatial knowledge acquistion from maps and from navigation in real and virtual environments. *Memory & Cognition* 27 (4):741–750.

Richter, Kai-Florian. 2008. *Context-Specific Route Directions: Generation of Cognitively Motivated Wayfinding Instructions*. Amsterdam, The Netherlands: IOS Press.

Richter, Kai-Florian. Forthcoming. Prospects and challenges of landmarks in navigation systems. In *Cognitive and Linguistic Aspects of Geographic Space— New Perspectives on Geographic Information Research*, edited by M. Raubal, A. U. Frank and D. Mark. Berlin: Springer.

Richter, Kai-Florian, Drew Dara-Abrams, and Martin Raubal. 2010. Navigating and learning with location based services: A user-centric design. In *7th International Symposium on LBS and Telecartography*, edited by G. Gartner and Y. Li, 261–276.

Richter, Kai-Florian, and Stephan Winter. 2011a. Citizens as database: Conscious ubiquity in data collection. In *Advances in Spatial and Temporal Databases*, edited by D. Pfoser, Y. Tao, K. Mouratidis, M. Nascimento, M. Mokbel, S. Shekhar, and Y. Huang. Berlin: Springer.

Richter, Kai-Florian, and Stepan Winter. 2011b. Harvesting user-generated content for semantic spatial information: The case of landmarks in OpenStreetMap. In *Proceedings of the Surveying & Spatial Sciences Conference*. Wellington, New Zealand.

Sadalla, Edward K., Jeffrey Burroughs, and Lorin J. Staplin. 1980. Reference points in spatial cognition. *Journal of Experimental Psychology: Human Learning and Memory* 6 (5):516–528.

Schmid, Falko, Colin Kuntzsch, Stephan Winter, Aisan Kazerani, and Benjamin Preisig. 2010. Situated local and global orientation in mobile you-are-here maps. In *12th International Conference on Human Computer Interaction with Mobile Devices and Services*. New York, NY: ACM.

Schmid, Falko, Denise Peters, and Kai-Florian Richter. 2008. You are not lost—You are somewhere here. In *You-Are-Here-Maps: Creating a Sense of Place through Map-Like Representations*, edited by A. Klippel and S. Hirtle. Workshop at International Conference Spatial Cognition 2008.

Schmid, Falko, Kai-Florian Richter, and Denise Peters. 2010. Route aware maps: Multigranular wayfinding assistance. *Spatial Cognition & Computation* 10 (2–3):184–206.

Siegel, Alexander W., and Sheldon H. White. 1975. The development of spatial representations of large-scale environments. In *Advances in Child Development and Behaviour*, edited by H. W. Reese. New York: Academic Press.

Willis, Katharine S., Christoph Hölscher, Gregor Wilbertz, and Chao Li. 2009. A comparison of spatial knowledge acquisition with maps and mobile maps. *Computers, Environment and Urban Systems* 33 (2):100–110.

Woods, D. 1995. Toward a theoretical base for representation design in the computer medium: Ecological perception and aiding human cognition. In *Global Perspectives on the Ecology of Human–Machine Systems*, edited by J. Flach, P. Hancock, J. Caird, and K. Vicente. Hillsdale, NJ: Erlbaum.

12 The Open Geospatial Consortium and Location Service Standards

Carl Reed

CONTENTS

ABSTRACT

This chapter describes the location services standards work of the Open Geospatial Consortium (OGC). The OGC began location-based standards work in 1999. This early work resulted in the development and approval of the OGC Open Location Services Core Interface standards. This standard has been widely, if not quietly, implemented. Since then, the OGC has enhanced the OLS standard as well as defined a range of standards that have been and can be used in location services (LS) applications. These include Open GeoSMS, the Geography Markup Language, the OGC sensor standards, and most recently standards work for augmented reality applications. Specifically, this chapter will discuss: the evolution of the LS standards work in the OGC; key OGC standards that have been deployed in LS applications; where OGC standards "fit" in the LS stack; how OGC standards are being used in LS applications; how OGC standards have been integrated into LS standards developed by other standards organizations, such as the Internet Engineering Task Force (IETF) and the Open Mobile Alliance (OMA); how OGC LS standards are evolving for use in mobile smart devices; and recent work for augmented reality (AR) and location-enabled short message service (SMS).

We begin with an overview of the key OGC standards that have been deployed and used in location service applications as well as incorporated into other standards developed by other standards organizations. These short descriptions provide a reference for the information provided in subsequent sections of this chapter.

12.1 KEY OPEN GEOSPATIAL CONSORTIUM (OGC) STANDARDS RELEVANT TO LOCATION SERVICES

The following are short descriptions of key Open Geospatial Consortium (OGC) standards that have been developed and used in the locations services (LS) industry. Many of these standards, such as OpenLS and GeoSMS, are described in much greater detail later in this chapter.

12.1.1 GEOGRAPHY MARKUP LANGUAGE (GML)

Geography Markup Language (GML) is an XML grammar for expressing geographical features. GML serves as a modeling language for geographic systems as well as an open interchange format for geographic transactions on the Internet. As with most XML-based grammars, there are two parts to the grammar: the schema that describes the document and the instance document that contains the actual data. Elements of GML have been incorporated into dozens of other standards with the goal of having common and consistent mechanisms for expressing and communicating location.

A GML document is described using a GML schema. This allows users and developers to describe generic geographic data sets that contain points, lines, and polygons. However, the developers of GML envision communities working to define community-specific application schemas based on an agreed information or content model. Using application schemas and related content models, users can refer to roads, highways, and bridges instead of points, lines, and polygons. If everyone in a community agrees to use the same schemas they can exchange data easily. GML is also an ISO Standard.

12.2 OGC KML

KML (formerly Keyhole Markup Language) is an XML language focused on geographic visualization, including annotation of maps and images. Geographic visualization includes not only the presentation of graphical data on the globe but also the control of the user's navigation in the sense of where to go and where to look. In 2006, Google submitted KML to the OGC for consideration as a standard. KML was the first instance of a de facto standard being submitted into the OGC standards process and as such the OGC modified its standards approval process to accommodate standards that have been developed externally from the OGC and then submitted into the OGC process. KML was approved as an OGC standard in 2008.

12.2.1 WEB MAP SERVICE (WMS)

Web Map Service (WMS) provides a simple HTTP interface for requesting geo-registered map images from one or more distributed geospatial databases. A WMS request defines the geographic layer(s) and area of interest to be processed. The response to the request is one or more georegistered map images (returned as JPEG, PNG, etc.) that can be displayed in a browser application. The interface also supports the ability to specify whether the returned images should be transparent so that layers from multiple servers can be combined or not. The WMS Interface standard is widely implemented and available as an app for both the Android and iPhone operating systems. WMS is also an ISO Standard.

12.2.2 SENSOR OBSERVATION SERVICE (SOS)

Access to sensors for real-time data is an increasing requirement in location services. Since 2000, the OGC has worked on defining a suite of standards for describing, tasking, and accessing sensors that are accessible via some network. The OGC Sensor Observation Service (SOS) standard defines an interface and operations that when implemented enable access to observations from sensors and sensor systems in a standard way that is consistent for all sensor systems including remote, in situ, fixed, and mobile sensors. SOS leverages the Observation and Measurements (O&M) specification for modeling sensor observations and the SensorML specification for modeling sensors and sensor systems.

12.2.3   OGC Open Location Service (OpenLS) Core

The OGC Open Location Services (OpenLS): Core Services, Parts 1–5, also known as the GeoMobility Server (GMS), defines an open platform for location-based services and specifies five core service interfaces. The standard also outlines the scope and relationship of OpenLS with respect to other specifications and standardization activities. The primary objective of OpenLS is to define access to the core services and abstract data types (ADTs) that comprise the GeoMobility Server, an open location services platform. The development of the OpenLS Core Service interface standard is now described in greater detail.

12.3   THE OGC AND LOCATION SERVICES: OPENLS (1999–2003)

The first OGC location services standards activity was titled OpenLS. This section describes the evolution of the OpenLS activity in the OGC and then provides a description of the reference architecture, information model, and the core interface specifications. The first phase of the OpenLS initiative was completed in 2003 with the adoption by the OGC members of the core OpenLS interfaces. The OpenLS standards have been widely implemented and are used in a number of deployed applications.

In 1999 and early 2000, the OGC Planning Committee (the Management Committee back then) held a number of special Market Architecture Sub-Committee (MASC) meetings to discuss new areas for OGC standards development activities. Market forces dictated that the OGC needed to remain responsive to interoperability needs beyond the traditional GIS marketplace.

In early 2000, the MASC recommended the following OGC actions related to location services:

1. That the location-based services (LBS) market becomes a major standards focus for the balance of 2000. This includes identifying the interfaces necessary to meet the requirements of this market. The MASC agreed that the interfaces and technology being defined and built for the IP test beds provide the foundation for the location services market. In addition, there were a number of recommendations and actions related to LBS.
2. Create a new OGC Location Services Domain Working Group. The function of this working group was to discuss and document requirements and use cases for the OpenLS standard specification.

3. Brand the name OpenLS (Open Location Services).
4. Action and agreement with a proposal that the OGC cannot wait to define key interfaces (functional) for LBS. Therefore, a set of straw man interfaces was to be defined in the near term. OGC staff definition of an initial set of interfaces was seminal to this work.
5. Ensure that the existing architecture frameworks can support the requirements of LBS.
6. That in 2001 a location services interoperability test bed be started.

Based on these recommendations and agreement by the full OGC Planning Committee and Technical Committee, the OGC embarked on standards work related to open location services. The primary goal of the OpenLS Initiative was to allow the successful provision and integration of geospatial data and geoprocessing resources into the location services and telecommunications infrastructure. The next step in the process was to successfully complete the OpenLS interoperability test bed activity.

12.3.1   What Is an OGC Test Bed?

From the OGC Web site, "test beds are fast-paced, multi-vendor collaborative efforts to define, design, develop, and test candidate interface and encoding standards. These draft standards are then reviewed, revised, and, potentially, approved in the OGC Standards Program." Functionally, a test bed begins when a number of OGC members define a set of interoperability requirements and use cases. These sponsoring organizations also provide some level of funding to offset management costs, participant travel costs, and so forth. Based on the defined requirements and use cases, a request for proposals is generated and publicly released. Any organization can respond to one or more requirements in the Request for Proposal (RFP). Selected organizations, known as participants, then work collaboratively with the sponsoring organizations to drive to a successful completion of the test bed. A major OGC test bed typically is completed within one year of concept development. Since 1998, the OGC has managed dozens of successful test beds and related interoperability pilots.

12.3.2   The OpenLS Test Bed

The OpenLS test bed was announced in July 2001. Sponsors of the test bed included Hutchison 3G UK, ESRI with SignalSoft, Oracle with Webraska, Sun Microsystems, and In-Q-Tel.* The participating OGC member organizations were Autodesk, Cquay Inc., ESRI, Hutchison 3G UK Limited, Intergraph IntelliWhere, Ionic Software, MapInfo, Navigation Technologies, Oracle, Sun Microsystems, and Webraska. The objectives of the test bed were to:

---

* Started in 1999, In-Q-Tel IQT was created to bridge the gap between the technology needs of the intelligence community (IC) and new advances in commercial technology.

- specify interoperable location service interfaces between application service, network, and device;
- test and integrate end-to-end solutions; and
- submit interfaces to OGC Standards Development Process and/or other specification and standards bodies.

The initial OLS test bed completed in early 2002 and a series of interoperability program reports were submitted to the OGC for member review. These reports defined a reference architecture, an information model, and a set of draft interface specifications for geocoding, reverse geocoding, directory services, a gateway service, presentation services, and routing.

12.3.3 THE OPENLS REFERENCE ARCHITECTURE

The OpenLS effort defined an overarching architecture titled "The GeoMobility Server" (Figure 12.1). The GeoMobility server is an element offering basic functions on which location-based applications are built (the OpenLS core services). This server uses open interfaces to access network location capacity (provided through a GMLC, for instance) and provides a set of interfaces allowing applications hosted on this server, or on another server, to access the OpenLS core services.

The GeoMobility server also provides content such as maps, routes, addresses, points of interest, and traffic. It can also access other local content databases via the Internet. In summary, the GeoMobility server contains:

- The core services and their OpenLS interfaces.
- The OpenLS information model, consisting of ADTs.
- Possibly, a set of local applications built upon the core services and accessing them through OpenLS interfaces.
- Content such as map data, points of interest, and routes used by the core services. This content can also be hosted on other servers and accessed through the Internet.
- Possibly other supporting functions for personalization, context management, billing, logging, and so on.

Figure 12.2 depicts the role of the GeoMobility server.

12.3.4 THE OPENLS INFORMATION MODEL

The OpenLS Core Services exchange content in the form of well-known OpenLS ADTs. The basic information construct is used by the GeoMobility server and associated core services. ADTs are well-known data types and structures for location information. ADTs are defined as application schemas that are encoded in XML for Location Services (XLS). Collectively these ADTs comprise the OpenLS information model (Figure 12.3).

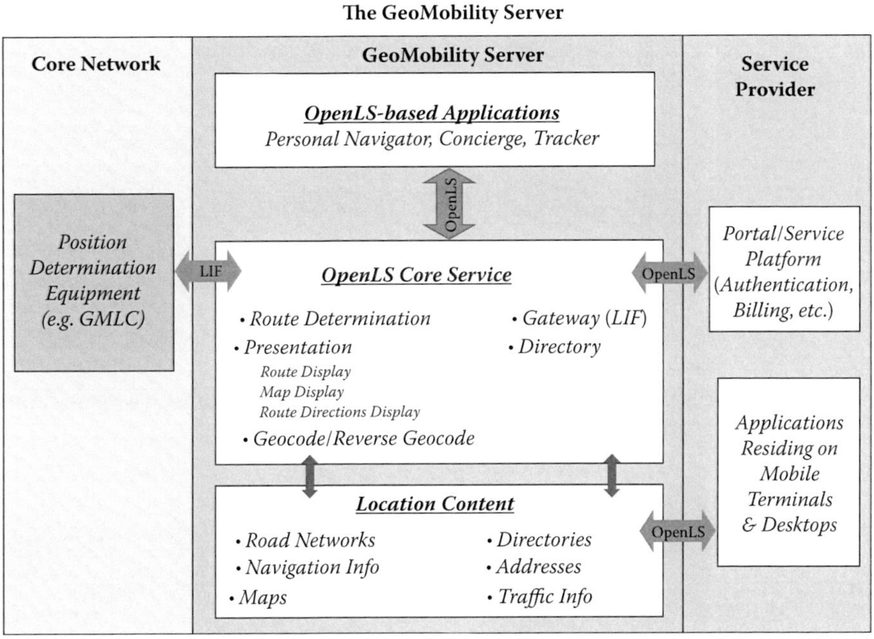

The GeoMobility Server
Core Network
GeoMobility Server
Service Provider
OpenLS-based Applications
Personal Navigator, Concierge, Tracker
OpenLS
Position Determination Equipment (e.g. GMLC)
LIF
OpenLS Core Service
• Route Determination
• Presentation
Route Display
Map Display
Route Directions Display
• Geocode/Reverse Geocode
• Gateway (LIF)
• Directory
OpenLS
Portal/Service Platform (Authentication, Billing, etc.)
Location Content
• Road Networks
• Navigation Info
• Maps
• Directories
• Addresses
• Traffic Info
OpenLS
Applications Residing on Mobile Terminals & Desktops

FIGURE 12.1 The GeoMobility server.

Role of OpenLS GeoMobility Server
Provide Subscribers with Location-Based Application Services/Content
Service Platform (Portal)
Internet
GeoMobility Server
• Navigation
• Discovery
• Presentation
• Maps
• Routes
• Directories
• Points of Interest
• Addresses
On the go
At work or home
GMLC/MPC
Location Server (PDE)

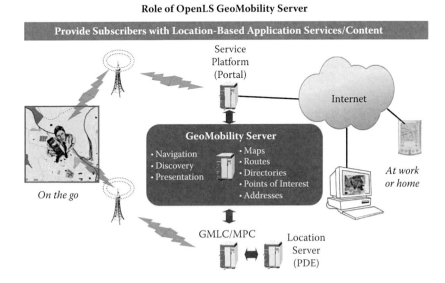

FIGURE 12.2 The role of the GeoMobility server

The OpenLS Information Model	
Position ADT	Point location in well-known coordinate system
Address ADT	Street address or intersection
Point of Interest (POI) ADT	The location where someone can find place, product, or service
Area of Interest (AOI) ADT	A polygon, bounding box, or circle used as a search template
Location ADT	A location (Position, Address, or POI)
Map ADT	The portrayal of maps and feature overlays (routes & POI)
Route Summary ADT	Metadata pertaining to a route
Route Geometry ADT	Geometry data for a route
Route Maneuvers ADT	Navigation maneuver data for a route
Route Directions ADT	Turn-by-turn navigation instruction for route

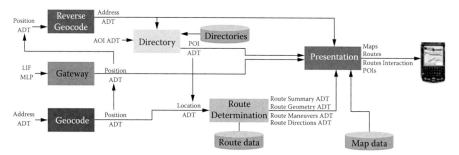

FIGURE 12.3   The OpenLS information model.

12.3.5   THE OPENLS CORE SERVICES

The Core Services are location-based application services that form the services framework for the GeoMobility server. This section describes the core services as defined in the OpenLS initiative.

12.3.5.1   The Directory Service

The Directory Service provides subscribers with access to an online directory to find the nearest, or a specific, place, product, or service. Through a suitably-equipped OpenLS application, the subscriber starts to formulate the search parameters in the service request, identifying the place, product, or service that they seek by entering the name, type, category, keyword, phone number, or some other "user-friendly" identifier. A position must also be employed in the request when the subscriber is seeking the nearest place, product, or service, or if they desire a place, product, or service at a specific location or within a specific area. The position may be the current mobile terminal position, as determined through the Gateway Service (this service translates MLP into an OpenLS "Position"), or a remote position determined in some other manner. The directory type may also be specified (e.g., yellow pages, restaurant guide, etc.). Given the formulated request, the Directory Service searches the appropriate online directory to fulfill the request, finding the nearest or specific place, product, or service, depending on the search criteria. The service returns one or more responses to the query (with locations and complete descriptions of the place, product, or service, depending on directory content), where the responses are in ranked order, based on the search criteria.

12.3.5.2   Gateway Service

The Gateway Service is the interface between the GeoMobility server and the location server that resides in the Gateway Mobile Location Center (GMLC) or Mobile Phone Center (MPC), through which OpenLS services obtain position data for mobile terminals. This interface is modeled after the MLP specified in LIF 3.0.0 for Standard Location Immediate Service.

12.3.5.3   Location Utility Service (Geocoder/Reverse Geocoder)

The Location Utility Service performs as a geocoder by determining a geographic position, given a place name, street address, or postal code. The service also returns a complete, normalized description of the place (which is useful, say, when only partial information is known). The service also performs as a reverse geocoder by determining a complete, normalized place name/street address/postal code, given a geographic position. Both the geocoder and reverse geocoder may return zero, one, or more responses to a service request, depending on subscriber request information, the algorithm being employed, and the match criteria.

12.3.5.4   Presentation Service

The Presentation Service renders geographic information for display on a mobile terminal. Any OpenLS application may call upon this service to obtain a map of a desired area, with or without map overlays that depict one or more OpenLS abstract data types (see later), such as route geometry, point of interest, area of interest, location, position, and/or address. The service may also be employed to render route directions from Route Maneuver List ADT and/or Route Instructions List ADT.

12.3.5.5   Route Determination Service

The Route Determination Service determines a route for a subscriber. The subscriber must use a navigation application service to set up the use of the service. They must indicate the start point (usually the position acquired through the Gateway Service, but this could be a planned trip from a specified location, say, from their home), and they must enter the endpoint (any location, like a place for which they only have the phone number or an address, or a place acquired through a search to a Directory Service). The subscriber may optionally specify waypoints, in some manner, the route preference (fastest, shortest, least traffic, most scenic, etc.), and the preferred mode of transport. The subscriber may optionally store a route for as long as needed, thus requiring the means to also fetch a stored route.

12.3.6   Test Bed Completion and Adopted Standards

The OpenLS test bed was successfully completed in 2002. The team submitted a number of reports to the OGC membership. These reports were a suite of candidate standards documents known as OpenLS core services (see earlier). The OGC members continued to work on these documents and in 2003 asked the OGC membership for a formal adoption vote to have the OpenLS Core Service Interfaces become official OGC standards. The documents were formally approved as OGC standards in late 2003. By this time, numerous OGC Members, such as Oracle, MapInfo, Esri, Intergraph, and TelContar,* and non-OGC members, such as Verizon and Sprint, had implemented the OGC OpenLS standards.

12.3.7 THE OPENLS INITIATIVE AND COORDINATION WITH THE LOCATION INTEROPERABILITY FORUM (LIF)

The Location Interoperability Forum (LIF) was established in 1999.† The goal for the LIF activity was to provide LBS providers with a standard set of interfaces that hide the different implementations of the location server. LIF approached location interoperability from the wireless network viewpoint, which at that time had no shortage of unresolved issues. The result of LIF's effort was the Mobile Location Protocol (MLP). From the standard, "The Mobile Location Protocol (MLP) is an application-level protocol for getting the position of mobile stations (mobile phones, wireless personal digital assistants, etc.) independent of the underlying network technology, that is, independent of location derivation technology and bearer. The MLP serves as the interface between a Location Server and a Location Services (LCS) Client. This specification defines the core set of operations that a Location Server should be able to perform." The first version of this standard was approved in 2002 by the LIF membership.

Recognizing that the OGC OpenLS initiative could not exist in isolation and understanding that there existed a standards stack for location services (Figure 12.4), the OGC determined that it needed to actively participate in the LIF activity. As can be seen in Figure 12.4, the OGC's focus is on the application, data, and presentation layers of the standards stack. Both OGC staff and a number of OGC member organizations participated in the definition and approval of the MLP API. The primary reasons for OGC participation were (1) to ensure that OGC OpenLS services operated in the standards stack and specifically with the MLP, and (2) to ensure that the modeling and encoding of the location elements in the MLP were consistent with existing OGC and ISO standards. In 2000, there were at least 12 common member organizations participating in both LIF and OGC activities.

As expected, there was considerable overlap between LIF and OGC concerning wireless location interoperability.‡ This overlap provided for expanded awareness and adoption of the combined standards. For example, LIF incorporated the location requirements of 3GPP, thus furthering a more uniform set of standards across multiple organizations and "adjacent" industries. Even between OGC and LIF, there was mutual support for a single set of location interoperability standards. OpenLS, for example, will include the MLP by reference (e.g., providing a better understanding on how to access LIF standards). This is, in effect, an endorsement by OGC of LIF's standard, resulting in a more focused effort by the industry behind location interoperability. At the same time, the LIF community determined that various existing

---

* Now DeCarta.
† LIF is now part of the Open Mobile Alliance.
‡ Ubiquitous Wireless Location Interoperability, Directions Magazine, 2002.

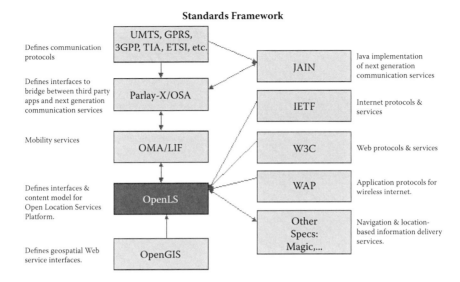

FIGURE 12.4 OGC and the LS Standards framework.

OGC standards, such as the GML and standard mechanisms for encoding coordinate reference information, were well suited to modeling and encoding location payloads as part of the MLP API.

Version 3.2 of the MLP Specification was approved by the OMA membership in July 2011. The current version uses OGC reference for all coordinate reference system elements and GML for encoding CRS references and GML for encoding complex geometries (informative).

12.3.8 How OpenLS Works in an Operational Environment

Figure 12.5 provides a typical request–response use case. A subscriber wants to access a list of restaurants nearby its location. He or she does not have any positioning capabilities on his or her device. The position of the subscriber is provided by network elements (GMLC/MPC).

The following sequence of events occurs:

- A subscriber accesses the telecom operator portal via the wireless network.
- The portal identifies the service requested by the subscriber and directs the request to a "Restaurant Guide" application running on the GeoMobility server; this application can also be hosted on a third-party server.
- The requested service sends a request to a GMLC to get the position of the subscriber's mobile terminal.
- The GMLC provides an estimation of the mobile terminal position.
- Knowing the subscriber position, the "Restaurant Guide" application sends a request, including the returned mobile terminal position in the form of a position ADT, to a Directory Service that returns a list of restaurants within a certain radius around the estimated subscriber position.

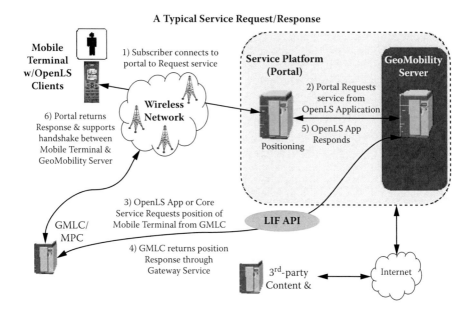

FIGURE 12.5   A typical service request–response.

- The list of restaurants is sent back to the subscriber.
- An application running on the mobile terminal allows the subscriber to view and interact with the list. [The next actions the subscriber is likely to take are to select a restaurant from the list and then ask for the directions and route map to assist them in driving from where they are to the chosen restaurant.]

The MLP defined by OMA (formerly LIF) is used to access GMLC/MPC positioning information. GMS resources are accessed through OpenLS APIs.

12.3.9   CURRENT STATUS OF OPENLS (2004 TO PRESENT)

Since approval in 2003, OpenLS has undergone two revision cycles to reflect implementation experience and changing market requirements. Version 1.2 is the current released version of the OpenLS Core Service Interface specifications.[*]

12.4   THE OGC COLLABORATION WITH OTHER STANDARDS ORGANIZATION AND LOCATION SERVICES

The OGC has a very strong belief in the value of collaborating with other standards organizations, especially when location elements are a common requirement. No standard exists in isolation. Harmonization of how location content is modeled and encoded through the LS standards stack is critical to ensuring interoperability, ease of implementation, and so forth.

---

[*] http://portal.opengeospatial.org/files/?artifact_id=22122

The OGC has collaborated with other standards organizations since 1994 with steadily increasing activity since 1999. We now collaborate with over 20 different standards setting organizations. Through these collaborations, OGC standards are strengthened by other groups providing requirements and use cases. Standards developed in other standards organizations benefit from OGC expertise in location services, GIS, geospatial services, and sensor systems. Further, many standards developed in other standards organizations are using elements from OGC standards, particularly the Geography Markup Language.

In addition to the use of GML by OMA for the MLP, the following are examples of the use of GML by other standards organizations.

12.4.1 THE INTERNET ENGINEERING TASK FORCE (IETF)

Since 2004, OGC staff has been involved in the Internet Engineering Task Force (IETF) GEOPRIV Working Group. From the GEOPRIV Charter:[*] "The GEOPRIV working group is chartered to continue to develop and refine representations of location in Internet protocols, and to analyze the authorization, integrity, and privacy requirements that must be met when these representations of location are created, stored, and used."

As a result of this collaboration, a GML application schema was developed for expressing a limited set of geometry types in IETF RFCs (requests for comments). The IETF developed a concept known as "location object." There are two types: civic and geodetic. The geodetic location object type is encoded using the GML application schema.[†] Since the initial definition for use with *Presence Information Data Format* (PIDF), the geodetic location (GML) object is now incorporated into a number of Internet standards including *HTTP Enabled Location Delivery* (HELD), *LoST—A Location-to-Service Translation Protocol, Filtering Location Notifications in the Session Initiation Protocol* (SIP), and others. Further, there is now a location extension for *Dynamic Host Configuration Protocol (DHCP) Options for Coordinate-Based Location Configuration Information* that uses the same information model as GML. The DHCP binary location encoding easily maps to a full GML encoding.

12.4.2 NATIONAL EMERGENCY NUMBER ASSOCIATION (NENA): NEXT GENERATION 911

"NENA serves the public safety community as the only professional organization solely focused on 9-1-1 policy, technology, operations, and education issues."[‡] Next Generation 9-1-1 (NG 911) is a system comprised of Emergency Services IP networks (ESInets), IP-based software services and applications, databases, and data management processes that are interconnected to public safety answering point (PSAP) premise equipment. The system provides location-based routing to the appropriate

---

[*] https://datatracker.ietf.org/wg/geopriv/charter/
[†] portal.opengeospatial.org/files/?artifact_id=21630 and http://tools.ietf.org/html/rfc4119
[‡] http://www.nena.org/

emergency entity. The system uses additionally available data elements and business policies to augment PSAP routing. The system delivers geodetic or civic location information and the call back number.

Due to the work by the IETF related to location-enabled standards that support a range of applications, including emergency services, the NENA NG 911 community has determined that GML is a preferred mechanism for sharing GIS data between PSAPs and counties. For example, the *NENA Standard Data Formats For ALI Data Exchange & GIS Mapping*[*] are encoded as GML (I2 schema repository). We should note that a number of other OGC standards have been identified as of potential high importance to the NG 911 activity, including CityGML, Geosynchronization, and the OGC Web Feature Service.

12.4.3 ORGANIZATION FOR THE ADVANCEMENT OF STRUCTURED INFORMATION STANDARDS (OASIS)

Since 2005, OGC staff has been participating in the Organization for the Advancement of Structured Information Standards (OASIS) Emergency Management (EM) Technical Committee (TC). OGC staff chairs the GIS subcommittee. In mid-2011, the OASIS EM TC approved a profile of GML (version 3.2.1) for use in OASIS EM standards. This profile is already used in EDXL.[†] Hospital Availability Exchange (HAVE) already uses GML. The next version of CAP (Common Alert Protocol) will use the OASIS GML Simple Features (gsf) profile.

12.5 W3C POINT OF INTEREST (POI) COLLABORATION

Since 2010, OGC staff has been a very active participant in the W3C POI Working Group. The mission of the POI Working Group is "to develop technical specifications for the representation of "Points of Interest" information on the Web." The September 2011 POI face-to-face meeting was hosted by the OGC at an OGC Technical Committee meeting held in Boulder, Colorado. The current public draft utilizes the OGC/ISO geometry model and states that GML is the preferred encoding language for POI location elements.[‡]

12.6 RECENT OGC LOCATION SERVICES STANDARDS WORK: OPEN GEOSMS (2009 TO PRESENT)

OpenLS provides a standard set of service interface that typically are implemented as part of the back end servers used to implement location services framework. As such, the end user never sees nor is even aware that the standards are being used. Further, the OpenLS interfaces, while not complex, are not necessarily lightweight.

---

[*] http://www.nena.org/resource/collection/6366E817-C855-4776-AF3A-F9F715D1AF12/NENA_02-010-v-8.2_Data_Formats_for_ALI_MSAG_GIS.pdf

[†] http://www.oasis-open.org/apps/org/workgroup/emergency-gis/download.php/42737/edxl-gsf-v1.0-wd07.odt

[‡] http://www.w3.org/2010/POI/documents/Core/core-20111216.html

Therefore, in 2009 the OGC members began considering much more lightweight, consumer-facing location service standards. The Open GeoSMS activity began in late 2009 when the Taiwan Industrial Technology Research Institute (ITRI) contributed Open GeoSMS as a candidate standard to the OGC.

The OGC members formed an official Standards Working Group (SWG) to edit and move the candidate standard through the OGC standards track. After a required public comment period, review by the OGC Architecture Board, and comment and approval by the OGC members, the first version of Open GeoSMS standard was approved in September 2011 and released in January 2012. There is already extensive implementation of this standard. This is one of the reasons that the GeoSMS standards process required almost 2 years—quite long. The SWG wanted to make sure that there was significant implementation and testing of the standard prior to formal approval.

12.6.1 What is GeoSMS?

The OGC Open GeoSMS Standard provides developers with an extended SMS encoding and interface to facilitate communication of location content between different LBS devices or applications. SMS is the open communication service standard most commonly used in phone, Web, and mobile communication systems for the exchange of short text messages between fixed line or mobile phone devices. The lightweight and easy-to-implement Open GeoSMS Standard facilitates location interoperability between mobile applications and the rapidly expanding world of geospatial applications and services that implement OGC standard interfaces, encodings, and best practices.

12.6.2 Example Implementations of GeoSMS

Open GeoSMS is already in use in a number of deployed commercial applications in Taiwan as well as in several disaster response applications. ITRI donated software (an Android application) based on Open GeoSMS to the open source Sahana Disaster Management System[*] for use in disaster warning, reporting, and relief, and ITRI has offered the software for free download on the Android market. The Sahana application establishes a synchronization mechanism between Sahana and one or more mobile devices. It provides communication between rescue teams and members, and provides location updates to Sahana for relief and rescue coordination. There is also an Open GeoSMS-enabled Ushahidi[†] Android app.

12.6.3 Example of a GeoSMS-Enabled SMS

Sam had a flat tire near an unknown village. Sam cannot describe where he is located to the towing service. Luckily, he installed an Open GeoSMS service application on his phone and that application sent out his location directly to a call center. The

---

[*] http://sahanafoundation.org/
[†] http://ushahidi.com/

corresponding Open GeoSMS could be http://maps.geosms.cc/showmap?geo=
23.9572,120.6860&GeoSMS&I NEED TOWING SERVICE NOW.

12.7   CURRENT OGC ACTIVITIES RELATED TO LOCATION SERVICES

This section describes current OGC standards activities that are germane to the
Location Services domain.

12.7.1   OGC AND AUGMENTED REALITY STANDARDS WORK

Augmented reality (AR) allows the user to see the real world with virtual objects
superimposed upon or composited with the real world. The potential application
domains for AR are vast, including medical, manufacturing, visualization, path
planning, entertainment, and military (Azuma, 1997).

Recently AR applications have emerged as popular on mobile devices equipped
with sensors for orientation and location (iPhones, Android devices, etc.). "In fact,
location based services (LBS)—used with smart phones and other types of mobile
technology—are a major driving force behind AR's entering the mainstream" (IEEE
Computer, 2009).

From widely available standards and the numerous potential applications, clearly
for the AR industry to grow, there must be further research to agree on standards,
profiles suited to AR, and for there to be discussion among AR experts on many different levels of development. Although extending existing standards will be highly
beneficial to achieve the ultimate objectives of the community, there also must be
room for the inclusion of new ideas and evolving technologies. Therefore, standardization meetings for finding the best interconnection and synergies have emerged
(i.e., International AR Standards Meetings).

In 2010, the OGC became an active participant in an International Augmented
Reality Standards Coordination group.* This group has a discussion list as well
as three to four face-to-face meetings per year. The goal of this group is to avoid
duplication of standards effort as well as to enable dialogue among and between the
various standards organizations that represent standards requirements for different
portions of the AR standards track. These standards coordination meetings allow
all players to not only discuss requirements, use cases, and issues but also the establishment of focused working groups that address specific AR standards issues. This
approach fosters and enhances the process of standardization of AR in specialized
fields in order that the community develops seamless and stable working products
in the market.

More specifically, the OGC has a focused AR standards development activity
known as "ARML, Augmented Reality Markup Language." Originally, ARML 1.0
was submitted to the OGC by Wikitude. ARML 1.0 is a descriptive, XML-based
data format, specifically targeted for mobile AR applications. ARML focuses on
mapping georeferenced POIs and their metadata, as well as mapping data for the

---

* http://www.perey.com/ARStandards/

POI content providers publishing the POIs to the AR application. In 2011, the OGC members approved the formation of an ARML 2.0 standards activity.

ARML 2.0 will focus on the following topics:

- KML, ARML 1.0, and other comparable data formats are purely descriptive formats. ARML 2.0 will also allow dynamic parts to modify the properties defined in the descriptive part.
- Define a set of events a developer can react to and execute custom functionality on occurrence of such events.
- Extend the rather basic POI presentation options (visual representation) to more sophisticated visualizations like 3D objects, lines, and polygons. Most likely, the geometry model of KML will be reused to represent POI geometries.
- Will provide connecting ports to other widely used AR tracking methods (mainly visual tracking, but also audio tracking, etc.). However, standardize tracking is not part of ARML 2.0 as such.

The ultimate goal of ARML 2.0 is to provide an extensible standard and framework for AR applications to serve the AR use cases currently used or developed. With AR, many different standards and computational areas developed in different working groups come together. ARML 2.0 needs to be flexible enough to tie into other standards without actually having to adopt them, thus creating an AR-specific standard with connecting points to other widely used and AR-relevant standards. The plan is for ARML 2.0 to be approved as an OGC standard by mid 2013.

12.7.2 OGC INDOORGML

A very hot topic of discussion and application development is for indoor location services, such as indoor navigation. Beyond the issue of locating a mobile device in an indoor environment, the indoor environment provides a complex set of standards requirements, such as how navigation directions are supplied (take the elevator to the third floor and turn right), semantics (my first floor is not your first floor), special zones (heating, security, WiFi, etc.) as well as the lack of a standard for modeling and encoding floor plans. Finally, outdoor navigation tends to be 2D, whereas indoor navigation needs to be 3D!

Understanding these complexities, in January 2012, the OGC members started an IndoorGML standards activity. The purpose of this IndoorGML Standard Working Group is to develop an application schema of OGC GML and progress the document to the state of an adopted OGC standard.[*] The goal of this candidate standard is to establish a common model and schema framework for indoor navigation applications. This SWG is starting with the OGC discussion paper "Requirements and Space-Event Modeling for Indoor Navigation." This document summarizes the requirements and basic model for a standard for indoor navigation. The objective for IndoorGML is to represent and exchange the geoinformation that is required to build and operate indoor navigation systems. IndoorGML will provide the essential

---

[*] https://portal.opengeospatial.org/files/?artifact_id=47562

model and data for important applications like building evacuation, disaster management, personal indoor navigation, indoor robot navigation, indoor spatial awareness, indoor location-based services, and the support for tracking of people and goods. IndoorGML provides a framework for the flexible integration of different localization technologies and allows the ad hoc selection of the appropriate navigation data according to the capabilities of the mobile device and the offered localization technologies of a building.

12.7.3 INTERNET OF THINGS

A very recent activity in the OGC is to consider how (or if) existing OGC standards contribute to the development and evolution of the Internet of Things. Connecting our world with accessible networks is scaling to trillions of everyday objects. The Internet of Things and Pervasive Computing are research names for this development. Planetary Skin, Smarter Planet, and CeNSE are several corporate names. The Internet will be augmented with mobile machine-to-machine communications and ad hoc local network technologies. At the network nodes, information about objects will come from barcodes, radio-frequency identifiers (RFIDs), and sensors. The location of all objects will be known and the objects will interact extensively with fixed and mobile clients.

Recognizing the growing importance of the Internet of Things (IoT), the OGC hosted a special meeting on the OGC Sensor Web Enablement (SWE) and IoT standardization during the OGC Technical Committee Meeting, March 2012, in Austin, Texas. This is the latest in a series of special meetings the OGC has convened in order to discuss geospatial and location-based standards requirements for the IoT.*

Based on these meetings and discussions, the OGC is planning to "Initiate an OGC Web of Things (WoT) standardization activity suitable for consumer IoT devices/sensors, such as implemented using RESTful interfaces." The plan is to also involve existing IoT companies to get broad adoption. This activity is being implemented through initiation of an OGC Standards Working Group (SWG) for "Sensor Web for the IoT/WoT." The first meeting of this group was June 2012. The scope of work for this SWG is to develop an OGC candidate standard for tasking and accessing sensors in an IoT/WoT environment.

12.8 SUMMARY

The OGC has a long history in developing standards that enable interoperable access to geospatial services, sharing of location-enabled content, and the integration of geospatial and location services, and content into enterprise architectures. A specific focus on standards for location services began in 1999. This initial focus was very successful and has allowed the OGC to progress to working on a variety of other standards activities targeted at meeting the requirements of the new generation of location service platforms and technologies.

---

* http://www.ogcnetwork.net/IoT

With changing technology infrastructures, requirements for mass market (consumer) applications, and requirements for the integration of additional (often real time) content, such as from sensors, as well as new media types, the OGC has embarked on a variety of new standards initiatives, such as Augmented Reality and the Sensor Web for the Internet of Things activities. These newer standards will tend to be more lightweight and defined based on REST* principles. However, these standards activities are grounded in existing OGC, ISO, and ITU standards. The goal is to ensure consistent and harmonized expression and transfer of location content throughout workflows and the standards stack.

REFERENCES

Location Interoperability Forum. 2011. "Mobile Location Protocol, LIF TS 101 Specification Version 3.0.0." http://www.openmobilealliance.org/lifdownload.html, accessed June 6, 2012.

NENA Standard Data Formats for ALI Data Exchange & GIS Mapping. 2011. http://ebookbrowse. com/gdoc.php?id=323690922&url=ae3a7c2e89a04f83a340ddc4277f36be, accessed July 6, 2012.

OASIS Common Alert Protocol (CAP), version 1.2. 2010. http://docs.oasis-open.org/emergency/ cap/v1.2/CAP-v1.2-os.html, accessed July 6, 2012.

OASIS Emergency Data Exchange Language Hospital Availability Exchange 1.0. 2008. EDXL-HAVEv1.0. Accessed July 6, 2012.

OASIS Emergency Data Exchange Language Resource Messaging 1.0. 2008. (EDXL-RMv1.0) incorporating Draft Errata. Accessed July 6, 2012.

OGC Geography Markup Language (GML) Encoding Standard, version 3.2.1. 2007. (Also ISO 19139). http://www.opengeospatial.org/standards/gml, accessed August 6, 2012.

OGC Geography Markup Language (GML) Encoding Standard, version 3.3. 2012. http:// www.opengeospatial.org/standards/gml, accessed August 6, 2012.

OGC KML, version 2.2. 2007. http://www.opengeospatial.org/standards/kml, accessed August 6, 2012.

OGC Location Services (OpenLS): Part 6—Navigation Service. OGC 08-027r7. 2008. http:// www.opengeospatial.org/standards/ols, accessed August July 6, 2012.

OGC Location Services (OpenLS) Interface Standards. OGC 07-074. 2007. http://www.open-geospatial.org/standards/ols, accessed August 6, 2012.

OGC Open GeoSMS Encoding Standard—Core, version 1.0. 2012. http://www.opengeo-spatial.org/standards/opengeosms, accessed August 6, 2012

OGC OpenLS Working Group. 2002. "Open GIS Consortium's OpenLS Initiative: Event, Architecture and Spec Overview." Presented at the OGC Technical Committee Meeting, September 10, 2002.

OGC Sensor Observation Service (SOS) Interface Standard, version 2.0. OGC 12-006. 2012. http://www.opengeospatial.org/standards/sos, accessed August July 6, 2012.

OGC Web Map Service (WMS) Interface Standard version 1.3. OGC 06-142. 2006. (Also ISO 19128). http://www.opengeospatial.org/standards/wms, accessed August 6, 2012.

---

* http://en.wikipedia.org/wiki/Representational_state_transfer

Index

Printed and bound by CPI Group (UK) Ltd, Croydon, CR0 4YY
18/10/2024